NEARSHORE SEDIMENT TRANSPORT

NEARSHORE SEDIMENT TRANSPORT

Edited by
Richard J. Seymour

Scripps Institution of Oceanography
University of California at San Diego
La Jolla, California

PLENUM PRESS • NEW YORK AND LONDON

Library of Congress Cataloging in Publication Data

Nearshore sediment transport / edited by Richard J. Seymour.
   p.   cm.
  Includes bibliographies and indexes.
  ISBN 0-306-43157-2
  1. Coast changes. 2. Marine sediments. I. Seymour, Richard J., 1929–
GB450.2.N43   1989   89-30746
551.3′6–dc19   CIP

PREFACE

This book represents the efforts of over a hundred individuals who planned and executed the NSTS field experiments, analyzed the billions of data points, and distilled their findings and insights into the summaries found here. Because these experiments were of a scope that will seldom, if ever, be duplicated, and because the program brought together many of the foremost field experimentalists in this country, we all felt from the beginning that it was important to preserve the outcome. This was done in two ways. First, the raw data were made available to any interested investigator within 18 months of the completion of each experiment. Secondly, both the methodology of the experiments and the findings from them were codified in the form of a monograph. This book is that result.

I have had the occasion recently (Sediments '87 Proceedings, Vol. 1, pp. 642-651) to assess the NSTS performance. I found that we made giant strides in our understanding of the surf zone hydrodynamics -- far more than our fondest expectations at the beginning. We were able to do less than we had hoped about the response of the sediment, largely because of a limited ability to measure it at a point. As I reported in the Sediments '87 assessment, we established a new state of the art in measurement techniques and we demonstrated the effectiveness of large, multi-investigator, instrument-intensive experiments for studying nearshore processes.

There are also a number of others who deserve credit for this volume: certainly the reviewers who labored diligently and anonymously to insure the quality of each chapter, David Duane of the Office of Sea Grant who performed the referee function on those sections to which I contributed, and finally -- the only truly indispensable contributor -- Martha Rognon. Ms. Rognon maintained our communications with the publisher and with the large number of contributors with characteristic efficiency and good humour. She designed the book, selected the typefaces, arranged the graphics and set the type (electronically) for the entire volume. It is clear to all of us who were gently guided and cajoled into their final output that we would never have produced this book without her.

Richard Seymour
La Jolla, California

CONTENTS

INTRODUCTION

David B. Duane

National Sea Grant Program
National Oceanic and Atmospheric Administration
Washington, D.C.

Richard J. Seymour

Scripps Institution of Oceanography
La Jolla, California

Arthur G. Alexiou

National Sea Grant Program
National Oceanic and Atmospheric Administration
Washington, D.C.

Models for predicting the transport of sediment along straight coastlines in general use in the mid-1970's were derived empirically from sparse measurement of both the forcing function (waves and currents) and the response function (sediment motions). In addition to the unsatisfactory nature of the basic measurements upon which they were based, the models were deficient because they failed to employ such potentially significant factors as wind stress, sediment size, bottom slope and spatial variations in waves and currents, including the effects of rip currents.

The economic impact of sediment transport in the nearshore regime is enormous. The costs for coastal dredging and shoreline protection can be measured in billions of dollars on a world scale. The need for improved predictive tools appeared to be universally accepted. The historical approach of research in the coastal zone, where one or two investigators working in the field obtained a few single point measurements over a limited time span, had proven inadequate for the development of satisfactory transport equations. A different approach was needed. It became apparent that several researchers working together in a coordinated series of field experiments over a span of several years using large arrays of instruments were required to make a substantial contribution to solving these problems.

Several events and factors converged in the mid-1970's to make that alternative possible. The National Sea Grant College Program, which by 1975 had achieved a good measure of success in building an organized network of Sea Grant institutions, desired to utilize the multi-institutional network to

1

address a research objective with a major national significance and impact. The U.S. Congress amended the Sea Grant Act in 1976 to include national projects. The first of 15 candidate research areas for funding as national projects was:

> *the development and the experimental verification of hydrodynamic laws governing the transport of marine sediments in the flow fields occurring in coastal waters.*

Advances in materials and electronics made possible, or likely, improved instruments for measuring forcing and response functions in the nearshore. An *ad hoc* group, convened by Richard J. Seymour of the Scripps Institution of Oceanography, met at the Fifteenth Coastal Engineering Conference (Honolulu, 1976) to plan a large scale and coordinated set of experiments leading to improved understanding of sediment transport in the nearshore. With considerable effort, a plan was drawn and submitted to the Office of Sea Grant (OSG). In 1977 OSG reviewed, accepted, and began funding that proposal which became the Nearshore Sediment Transport Study (NSTS).

The major objective of the NSTS program was to produce improved engineering models for predicting the motion of sediment along straight coastlines under the action of waves and currents in the nearshore zone and which could be simply employed (without recourse to large computers) and which would depend upon a few measurements or observations readily obtainable at reasonable costs.

Management of a technically complex program and the effective coordination of a large diverse group of investigators required the creation of a management concept substantially different from existing Sea Grant programs. The program employed a two-tier management structure (Figure 1). The National Sea Grant College Program Director appointed an advisory group referred to as the NSTS Review Group (Table 1A). This group, all experienced in field investigation of nearshore processes, reviewed proposals and made funding recommendations to the Director of Sea Grant, formulated overall program direction and reviewed the progress of the investigators. A second tier group, known as the NSTS Steering Committee, was formed of the senior investigators (Table 1B). The Chairman of this group functioned as the Project Manager and was an *ex officio* member of the Review Group. The Steering Committee formulated details of the project program, planned and executed cooperative field programs and conducted workshops to promulgate the findings of the study to the coastal engineering community.

To accomplish the NSTS objectives, it was agreed from the earliest planning sessions that the program must contain characteristics which set it apart from prior efforts. The most important of these attributes were that:

> (1) it was to be a broad-based program with many investigators from a large number of institutions. It was an attempt to bring together many experienced practitioners to plan and execute a

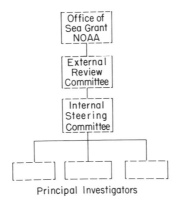

Figure 1. NSTS Program Management.

series of major field experiments which would vastly exceed the capabilities of a single investigator.

(2) it was to be a field-oriented program. Laboratory or numerical modeling were not foreseen as significant elements of the project.

(3) field experiments must attempt to measure, simultaneously, the details of both the velocity field within the surf zone and the sediment response to those velocities. This was recognition of the fact that there are significant spatial variations in sediment transport within the nearshore regime and that the ability to understand and predict these variations is a prerequisite to developing a universally applicable engineering model.

(4) measurements were to encompass a large number of parameters that were potentially significant to a predictive model.

(5) the final predictive model was to be two-dimensional. That is, it would predict cross-shore movement as well as longshore movement of sediment.

As originally envisioned, NSTS was to be a four-year program with three, possibly four, major field experiments conducted at the rate of one per year. The first was to be a sort of proof of concept; the second and third a Pacific and Atlantic coast experiment, and a fourth in the Great Lakes or Gulf Coast. In actuality, because of the unforeseen effects of both severe fiscal inflation on a fixed budget and the magnitude of the effort required to analyze the enormous data base produced, only two major field experiments (Torrey Pines and Santa Barbara in California), and a limited scope validation experiment (Rudee Inlet, Virginia) were completed. The latter experiment and the final two years of analyses and report preparation were accomplished with the fiscal support of the U.S. Army Corps of Engineers, the U.S. Geological Survey, the Office of Naval

Table 1. Participants in the Two-Tier Management
Structure for the Nearshore Sediment Transport Study

A. Review Group

	TERM
Mr. Arthur G. Alexiou, Office of Sea Grant	1977-82
Dr. James R. Allen, National Park Service	1981-82
Dr. David B. Duane, Office of Sea Grant, Chairman	1977-82
Dr. William R. James, U.S. Geological Survey	1977-79
Dr. Stephen P. Leatherman, National Park Service	1980-81
Dr. Bernard J. LeMehaute, Tetra Tech and University of Miami	1977-82
Dr. John C. Ludwick, Jr., Old Dominion University	1977-82
Dr. Alexander Malahoff, National Ocean Survey, NOAA	1977-82
Dr. Ned A. Ostenso, National Sea Grant Program	1977-82
Dr. Asbury H. Sallenger, Jr., U.S. Geological Survey	1977-82
Mr. Rudolph P. Savage, Coastal Engineering Research Center	1981-82
Mr. Thorndike Saville, Jr., Coastal Engineering Research Center	1977-81
Dr. James J. Saylor, Great Lakes Environmental Lab, NOAA	1977-82
Dr. Richard J. Seymour, Ex Officio	1977-82

B. Steering Committee

Dr. Richard J. Seymour Chairman/NSTS Project Manager	Scripps Institution of Oceanography
Professor Robert G. Dean	University of Delaware
Professor Douglas L. Inman & Director, Center for Coastal Studies	Scripps Institution of Oceanography
Professor Edward B. Thornton	Naval Postgraduate School

Research, and the U.S. National Park Service in addition to the Sea Grant funding. A substantial portion of the costs of editing and setting type for this volume were provided by a grant from the Foundation for Ocean Research, San Diego, California.

Despite the reduction from the planned level of effort, almost all of the basic program objectives were met. To understand the scope of NSTS, consider that the research was carried out at one time or another by ten principal investigators from five different universities or institutes and two government agencies; and, three significant field experiments were performed on two coasts. At peak activity, as many as 60 individuals and more than 100 instruments were involved simultaneously in field experiments. To apprise the community of progress in the program, four workshops were held. Available to the community, at cost of reproduction, for use in continued research are reports of the experiments and all of the data recorded on magnetic tape. (Inquiries should be directed to National Oceanographic Data Center, User Services Branch, NOAA/NESDIS E/OC21, Washington, D.C. 20235, telephone: (202) 673-5549). NSTS was funded from appropriations over a period of five fiscal years at a total of approximately 4 million dollars; a portion of a sixth year was required for the preparation of this final report.

This book is the culminating effort of NSTS. Its purpose is to bring together under one cover, the major aspects of the coordinated research, and by reference to previously prepared papers, to be the principle source document for NSTS. Of perhaps equal importance, this compilation provides a comprehensive review of the state of the art in measuring, analyzing and predicting surf zone dynamics and the resulting transport of sediment and may therefore serve as a valuable reference for students, researchers and coastal engineers.

Chapter 1

THE NSTS FIELD EXPERIMENT SITES

A. Torrey Pines Experiment

D. L. Inman, S. S. Pawka, and M. J. Shaw
Center for Coastal Studies
Scripps Institution of Oceanography

Torrey Pines Beach is part of the Torrey Pines State Preserve located along the coast between the communities of La Jolla and Del Mar, San Diego County, in the southern part of California. The coast is relatively straight with nearly north-south trending beaches backed by 90 m high wave-cut sea cliffs. The 2 km long section of beach selected for the experimental site is 6 km north of Point La Jolla headland and 3 km north of the head of Scripps branch of La Jolla Submarine Canyon (Figure 1A-1).

The Torrey Pines Beach site was selected for its straight coastline and gently sloping offshore bathymetry, and because of the previous studies of beach profile changes (e.g., Nordstrom and Inman, 1975; Winant *et al.*, 1975) and their relation to wave climate (e.g., Pawka *et al.*, 1976). The shelf here is approximately 3 to 4 km wide and slopes at about 1 in 50 to 1 in 100 out to the shelf break at a depth of about 100 m. In water depths of 20 m and less, the wave-cut shelf in this area is overlain by 1 to 5 m of sandy sediment (Moore, 1960; Inman and Bagnold, 1963).

The beach is near the southern end of the Oceanside Littoral Cell, a sedimentation compartment containing sources, transportation paths and sediment sinks, which extends for 84 km from the rocky headland at Dana Point (Figure 1A-2) to the heads of Scripps Submarine Canyon. The principal sources of sediment for the cell are the rivers, which periodically supplied large quantities of sandy material to the coast. The sand is transported along the coast by waves and currents until it is intercepted by Scripps Submarine Canyon, which diverts and channels the flow of sand into the adjacent submarine basin (Chamberlain, 1964; Inman, 1980; Inman, 1982).

The sediments on the beach and shelf are predominantly fine quartz sand with minor amounts of feldspars and heavy minerals. Light minerals, those with a specific gravity of less than about 2.85, usually comprise 90% of the beach sample, while heavy minerals total about 10%. Of the light minerals

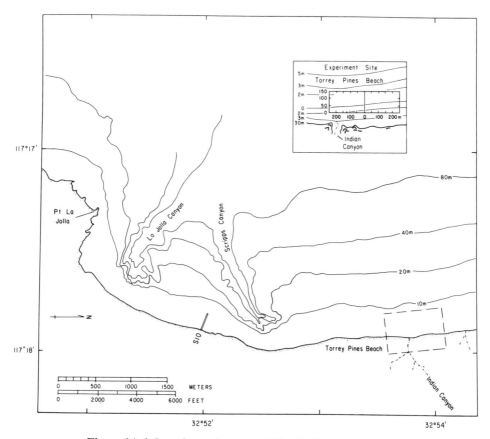

Figure 1A-1. Location and survey grid for the Torrey Pines Beach site.

approximately 88% is quartz, 10% feldspars, and 2% shell fragments and miscellaneous material. Of the total heavy minerals, hornblende comprises an average of 60%. There is a pronounced seasonal variation in the amount of heavy minerals in beach and nearshore samples. During the winter when the beaches are cut back, the heavy mineral content of the beach foreshore samples increases appreciably compared to summer conditions. At the same time, the heavy mineral content of the sands outside of the surf zone is reduced. An explanation for this apparent seasonal migration of heavy minerals is given by the transportation of light minerals from the beach foreshore to deeper water during the winter and back again during the summer (Inman, 1953; Nordstrom and Inman, 1975).

During some winter storms the beach sand is transported offshore and waves actively erode the sea cliffs. However, during the NSTS experiments there was always a cover of sand over the beach, and country rock was not exposed. The

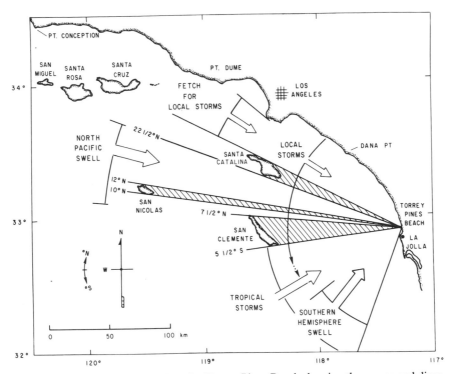

Figure 1A-2. Wave exposure chart for Torrey Pines Beach showing the source and direction.

seasonal cross-shore motion of sediment in the Torrey Pines and La Jolla Shores area has been studied in detail (e.g., Shepard and La Fond, 1940; Inman, 1953; Inman and Rusnak, 1956; Nordstrom and Inman, 1975; Winant *et al.*, 1975; Winant and Aubrey, 1976; Aubrey *et al.*, 1980). These studies show that during summer months when waves are relatively small, sand migrates back to the foreshore, increasing the height and width of the beach berm as well as the beach face slope. Nordstrom and Inman (1975) found that the seasonal migration of beach sand between the beach berm and the offshore portions of the beach profile amounted to 92 cubic meters per meter length of beach.

1A.1 Wave Climate

The Channel Islands, which border the Southern California coast, exert a dominant influence on the wave climate at Torrey Pines Beach. The islands significantly shelter the site from north swell as there is only a narrow (13 degree) window to the Northern Hemispheric generation regions (Figure 1A-2). The wave energy at the site is typically an order of magnitude lower than the deep ocean (unsheltered) level (Pawka, 1982). Refraction over the extensive shoals in the island region does spread some energy from sheltered sources to the site. This process is highly frequency dependent and is significant only for waves with frequencies less than 0.10 Hz. Torrey Pines Beach is relatively open to the Southern Hemisphere, although Point La Jolla does offer some sheltering from southern swell. In spite of the island sheltering, the northern wave components are climatologically more energetic than the southern waves because of the overwhelming dominance of the northern hemispheric sources in the offshore (unsheltered) wave climate. Thus the net wave-induced sediment transport at Torrey Pines Beach is usually from north to south.

The wave climate was sampled at Torrey Pines Beach from February 1973 to May 1974 with a linear array of four pressure sensors (Pawka *et al.*, 1976). The quantitative results were grouped in terms of four seasons. The frequency spectra of the summer months (June, July, August) display a consistent bimodal form. The narrow ($\approx$0.4 Hz) low frequency spectrum has peak periods predominantly in the range 14-17 seconds and approaches the site from southerly directions. These waves are generated by Southern Hemispheric cyclones and Eastern Pacific tropical disturbances. Refraction of the 17-second waves around Point La Jolla directs deep ocean energy from 180-230 degrees true into the narrow sector 238-245 degrees true at Torrey Pines Beach. This is a divergent refractive transformation. This process, coupled with the long distances to the storms, usually results in a narrow directional spectra (Figure 1A-3) for the southern swell. The southern swell peak has a mean spectral energy density of about $100 \, \mathrm{cm}^2$. The higher frequency peak is broader ($\approx$0.10 Hz width) with peak periods varying from 6-10 seconds. These higher frequency waves are generated by the relatively strong and persistent winds, from 300-330 degrees true, offshore of the Channel Islands. This energy reaches the site primarily through the northern quadrant window and has mode directions of 285-290 degrees true (Figure 1A-3). For the summer months, the high frequency waves are somewhat more energetic than the southern swell with a mean spectral energy density of $170 \, \mathrm{cm}^2$.

The wave climate during the fall months is very similar to the summer with the addition of occasional northern swell with periods of 12-16 seconds. These waves are generated by North Pacific cyclones and they reach the site through the northern quadrant window and by refraction over the Cortez and Tanner Banks (bearing 250-255 degrees true) resulting in a strongly bimodal directional spectrum. The southern directional mode is often comparable in energy to the northern mode at the lowest wave frequencies. The north swell spectral peak has a mean spectral energy density of about $250 \, \mathrm{cm}^2$ in this season.

North swell becomes the dominant source of low frequency energy in the winter months (December, January, February). The mean energy density of the north swell spectral peak is about $500 \, \mathrm{cm}^2$ during this season. Occasional southern swell, although a minor energetic component, does persist during the

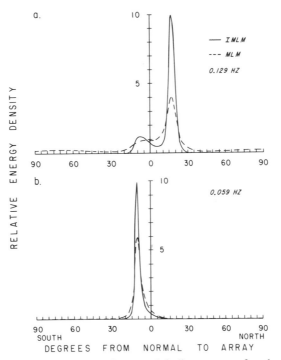

Figure 1A-3. Directional spectrum from a 1-2 linear array for the average of 8 consecutive 17.1 minute data segments on 10 June 1977. The estimates are for a mean depth of 9.6 m.

winter and spring months. Often the northern and southern swell are coincident, producing a trimodal directional spectrum at Torrey Pines Beach (two of the modes are due to island refraction of the northern swell). Relatively energetic waves are produced with the passage of storm fronts through the local region. Although infrequent, these wave conditions (typical energy density of 1000-2000 cm^2) are an important factor in the yearly energy budget. The local storm waves have a broad (0.10-0.15 Hz width) spectral peak with peak periods in the range of 7-10 seconds. The directional spectra are often trimodal with modes at 250-260 degrees true, 285-290 degrees true, and 300-310 degrees true (see inter-island fetch, Figure 1A-2).

Spring is a season of transition from the energetic winter conditions to the mild climate of the summer. The relatively frequent occurrence of local storm activity makes this season as energetic as the winter months. The transition from north to south swell dominance usually occurs late in the season (May).

In comparison with other ocean beaches, the wave intensity at Torrey Pines Beach is moderate, resulting in an average significant breaker height of about 65 cm and an average onshore flux of wave energy of about 750 watts per meter length of beach. This is equivalent to an average spectral energy density of about 300 cm^2. However, there are periods of intense wave activity at this coast. Wave energy densities up to 5000 cm^2 have been recorded at the site in the 1970's. Even more energetic conditions have been observed in this century. A tropical storm moved north into this region in September 1939, producing severe wave conditions. Hindcast calculations of the waves from this storm (Horrer, 1950) yield significant wave heights of about 9 m at this coast. These conditions would provide an onshore flux of wave energy of 450 kw per meter of beach length.

In summary, the Torrey Pines Beach site commonly has a complicated wave field with multiple modes in both the frequency and directional spectra. These mixed wave conditions are due to the persistence of northern and southern sources of wave energy and wave scattering by the island shoals. Strongly bimodal directional spectra, with the modes roughly equally balanced around normal incidence to the beach, are typically found with low frequency north swell. These types of wave conditions lead to a complicated forcing of surf zone dynamics.

1A.2 References

Aubrey, D. G., D. L. Inman and C. D. Winant, 1980, The statistical prediction of beach changes in Southern California, *Journal of Geophysical Research*, 85(C6): 3264-76.

Chamberlain, T. K., 1964, Mass transport of sediment in the leads of Scripps Canyon, California, R. L. Miller, ed., *Papers in Marine Geology*, Macmillan Company: 42-64.

Horrer, P. L., 1950, Southern hemisphere swell and waves from a tropical storm at Long Beach, California, Beach Erosion Board, *U. S. Army Bulletin*, 4(3): 1-18.

Inman, D. L., 1953, Areal and seasonal variations in beach and nearshore sediments at La Jolla, California, Beach Erosion Board, U.S. Army Corps of Engineers, Technical Memo 39, 134 pp.

_____. 1980, *Man's Impact on the California Coastal Zone*, Department of Boating and Waterways, Sacramento, California, 150 pp.

_____. 1982, Application of coastal dynamics to the reconstruction of paleocoastlines in the vicinity of La Jolla, California, In: P. M. Masters and M. C. Flemming (eds.), *Quaternary Coastlines and Marine Archaeology: Toward the Prehistory of Landbridges and Continental Shelves*, Academic Press.

Inman, D. L. and R. A. Bagnold, 1963, Littoral processes, In: M. H. Hill (ed) *The Sea*, Volume 3, *The Earth Beneath the Sea History*, John Wiley and Sons, New York: 529-553.

Inman, D. L. and G. A. Rusnak, 1956, Changes in sand level on the beach and shelf at La Jolla, California. Beach Erosion Board, U.S. Army Corps of Engineers, Technical Memo 82, 64 pp.

Moore, D. G., 1960, Acoustic-reflection studies of the continental shelf and slope off Southern California, *Geological Society of America Bulletin*, 71(8): 1121-36.

Nordstrom, C. E. and D. L. Inman, 1975, Sand level changes on Torrey Pines Beach, California, Coastal Engineering Research Center, U.S. Army Corps of Engineers, Miscellaneous Paper No. 11-75, 166 pp.

Pawka, S. S., 1982, Wave directional characteristic on a partially sheltered coast, Ph.D. Dissertation, Scripps Institution of Oceanography, University of California, San Diego, 246 pp.

Pawka, S. S., D. L. Inman, R. L. Lowe and L. Holmes, 1976, Wave climate at Torrey Pines Beach, California, Coastal Engineering Research Center, U.S. Army Corps of Engineers, Technical Paper 76-5, 372 pp.

Shepard, F. P. and E. C. LaFond, 1940, Sand movements along the Scripps Institution of Oceanography pier, *American Journal of Science*, 238: 272-285.

Winant, C. D. and D. G. Aubrey, 1976, Stability and impulse response of empirical eigenfunctions, *Proceedings*, Fifteenth Coastal Engineering Conference, July 11-17, 1976, Honolulu, Hawaii, American Society of Civil Engineers, New York: 1312-25.

Winant, C. D., D. L. Inman and C. E. Nordstrom, 1975, Description of seasonal beach changes using empirical eigenfunctions, *Journal of Geophysical Research*, 80(15): 1979-86.

Chapter 1

THE NSTS FIELD EXPERIMENT SITES

B. Santa Barbara Experiment

Christopher G. Gable
Scripps Institution of Oceanography

Santa Barbara is located on a sandy lowland on the coast of southern California, 153 km northwest of Los Angeles and 563 km southwest of San Francisco. It borders the Santa Barbara Channel which is bounded on the north and east by the mainland shoreline of Santa Barbara and Ventura Counties, on the south by the Channel Islands (San Miguel, Santa Rosa, Santa Cruz, and Anacapa) and on the west by the open waters of the Pacific Ocean (Figure 1B-1).

The coast in the vicinity of Santa Barbara trends in an east-west direction and is generally rugged. It is characterized by projecting headlands of rock and boulders with intervening coves having cobble covered shores or sandy pocket beaches backed by high bluffs. There are no large rivers, but there are numerous steep streams with torrential flows during rainy seasons that run through arroyos and empty into the ocean.

The climate in Santa Barbara is Mediterranean and is controlled primarily by the position and intensity of the semi-permanent Pacific high pressure system over the ocean to the west. During the summer this high pressure area covers the eastern North Pacific Ocean and deflects eastwardly moving storms to the north. During the winter months, this Pacific high migrates southward and weakens, allowing occasional frontal systems that originate in the Aleutians to move through southern California. The most intense extratropical storms, however, are those that develop between Hawaii and the California coast. These storms, because of their southerly position and intensity, often produce large westerly ocean swells and waves which move into the Santa Barbara Channel between Point Conception and San Miguel Island. A series of storms of this type hit the Santa Barbara area during the latter part of the NSTS experiment causing significant flooding due to intense rainfall, high winds, and extreme beach erosion from high waves coupled with extreme high tides.

The month-long intensive NSTS field experiment was located on Leadbetter Beach, which is adjacent to and west of Santa Barbara Harbor. The fifteen-

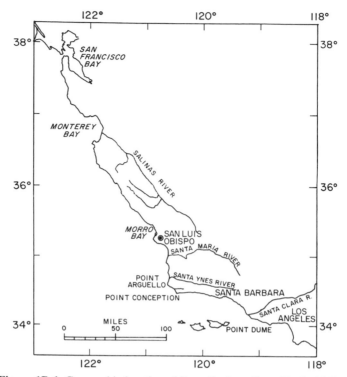

Figure 1B-1. Geographic location of Santa Barbara (from Trask, 1952).

month sediment trapping experiment measured sediment accumulations extending from the western end of Leadbetter Beach eastward to the tip of the sand spit in Santa Barbara Harbor. The harbor configuration at Santa Barbara offered an effective trap for longshore transport and an opportunity to acquire a very high quality data set on longshore transport.

Leadbetter Beach feeds the sandspit formed in the shadow of the breakwater. It provides a distinctive setting for a transport experiment in which the beach slope is steep, the surf zone narrow, the shore is protected by the islands from direct wave approach so that the wave incidence angle is always very high, the longshore current is nearly unidirectional and of high magnitude, and the site is adjacent to a nearly total trap for longshore transport of sediment. The depth contours in the vicinity of the experiment site are shown in Figure 1B-2. The sand is predominantly quartz, well sorted, and fine to medium in size.

Before the construction of the harbor breakwater, the flow of sand was uninterrupted and was transported naturally to the beaches to the south within the Santa Barbara littoral cell. In early 1930 when the harbor breakwater was completed, sand began accumulating west of the shore arm of the breakwater creating what is now Leadbetter Beach. Eventually, sand migrated along the breakwater and deposited in the lee (or shadow of the structure) forming a sand spit in the channel. As a result, the sand spit created a navigation problem as well as impounding the sand that previously was naturally supplied to the beaches to the south. This resulted in significant erosion to these downcoast beaches. Therefore, a dredging program was initiated in 1935 to remove material from the harbor and place it on the starved downcoast beaches to prevent further erosion. This dredging program is still in progress.

The waves at Santa Barbara are generated either between the Channel Islands and the coast (local wind waves from the south) or are generated in the ocean seaward of the islands. Wiegel (1959) reports that local storms from the southeast (125-145 degrees true) have a fetch of 145 km and generate waves toward the Santa Barbara coast with significant wave heights ranging from 2.4 to 4.9 m. The predominant waves are from the southwest and west (240-270 degrees true) that enter the Santa Barbara Channel between San Miguel Island and Point Conception. The waves from the southwest and west are residual elements of waves that originated under conditions of unlimited fetch. Leadbetter Beach is protected from swells generated by distant storms from all other active sectors by the Channel Islands and Point Conception. Wiegel (1959) reports that waves from the southwest and west range from 0.3 to 5.0 m but average about 0.9 m. The average wave period is 12 seconds, but ranges between 8 and 16 seconds. Based on fifteen months of wave data (October 1979 to December 1980) collected through the Coastal Data Information Program (CDIP), the nearshore significant wave height ranges from 0.1 to 2.0 m and the wave period ranges from 5 to 22 seconds with a modal period of 7 seconds. Seventy percent of the wave heights range between 0.3 and 0.6 m during this measurement period. Seasonally, there is an 18% and 7% probability that the significant wave height will exceed 1 and 1.5 meters, respectively, during the months of January, February, and March. Additionally, significant wave heights greater than 2.0 m occurred less than 12 hours throughout the data collection period. The wave gage was in 7.6 meters depth of water relative to mean sea level (MSL).

Santa Barbara Harbor and Leadbetter Beach are located approximately in the middle of the Santa Barbara littoral cell which extends from Point Conception to Point Mugu, California. The shoreline between Point Conception and Santa Barbara trends east-west and is composed of sedimentary rocks and shale bluffs fronted by sand and cobble beaches.

The major sources of sand for the beaches in Santa Barbara are streams and rivers entering the ocean. The Santa Maria and Santa Ynez Rivers are the main sources of sand for the beaches north of Point Conception. Bowen and Inman (1966) estimate that the net transport rate around Point Conception into the

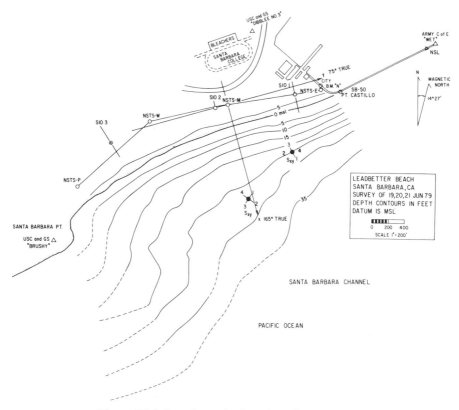

Figure 1B-2. Location and orientation of wave arrays.

Santa Barbara littoral cell is approximately 76,500 cubic meters per year. They also report that the net transport rate between Point Conception and Santa Barbara Harbor is 137,500 cubic meters per year and that the source of the additional material is that introduced into the littoral zone by the numerous small steep streams that flow down arroyos on the seaward slopes of the local Santa Ynez Mountains. Trask (1952) also reports that offshore sand deposits originated during flood years serve as reservoirs to replenish sand supplies to the beach during relatively dry years. Johnson (1953) reports that Santa Barbara Harbor has an average measured rate of sand entrapment of 214,000 cubic meters per year based on data between 1932 and 1951. Dean *et al.* (1982) report that the measured transport rate at Santa Barbara is 252,000 cubic meters per year based on the field observations associated with the NSTS Santa Barbara trap experiment.

The normal longshore current direction in the Santa Barbara area and at Leadbetter Beach is west to east. The predominant wind and wave direction from the west and southwest, coupled with local refraction effects, cause the waves to break at a high degree of incidence creating an easterly longshore current of high magnitude. However, the occasional southeasterly storms do create reversals in which the longshore current is to the west.

1B. References

Bowen, A. J. and D. L. Inman, 1966, Budget of littoral sands in the vicinity of Point Arguello, California. U. S. Army Coastal Engineering Research Center, Technical Memorandum No. 19, 41 pp.

Dean, R. G., E. P. Berek, C. G. Gable and R. J. Seymour, 1982, Longshore transport determined by an efficient trap, *Proceedings*, Eighteenth Coastal Engineering Conference, November 14-19, 1982, Cape Town, Republic of South Africa, American Society of Civil Engineers, New York, 2: 954-968.

Johnson, J. W., 1953, Sand transport by littoral currents. *Proceedings*, Fifth Hydraulics Conference, State University of Iowa Studies in Engineering, Bulletin 34: 89-109.

Trask, P. D., 1952, Source of beach sand at Santa Barbara, California, as Indicated by Mineral Grain Studies. U. S. Army Beach Erosion Board, Technical Memorandum No. 28, 24 pp.

Wiegel, R. L., 1959, Sand bypassing at Santa Barbara, California, *Journal Waterways Harbors Division*, Proceedings, American Society of Civil Engineers, 85(WW2), June, 1959: 1-30.

Chapter 1

THE NSTS FIELD EXPERIMENT SITES

C. Rudee Inlet Experiment

Robert G. Dean
University of Florida
(Formerly University of Delaware)

1C.1 Introduction

Rudee Inlet and the beaches to the north and south were selected as the site of the East Coast NSTS Total Trap experiment due to:

(1) the assessment that the Rudee Inlet and the associated deposition basin represented a near-complete trap,

(2) the longshore sediment transport was believed to be nearly unidirectional,

(3) the reasonable size of the system, and

(4) the favorable logistics of the area as provided by the cooperation of the Virginia Beach Erosion Committee.

1C.2 Characteristics of the Study Area

1C.2.1 Physical Setting

Rudee Inlet is located approximately 13 km south of Cape Henry which is at the entrance to Chesapeake Bay (Figure 1C-1). This inlet forms the southern boundary of the City of Virginia Beach, Virginia, and provides access to the Atlantic Ocean for shallow water vessels from marinas and residential and commercial docks fronting the connected inland waters. The inlet was first stabilized/protected by a short (60 m) pair of jetties in 1953. This jetty system was found to be ineffective in maintaining the desired channel depth and the north jetty was later (1967) extended an additional 240 m. In the same phase, a new southern weir jetty was constructed outside (south) of the original southern jetty. The weir section is oriented nearly perpendicular to the shoreline, and connects to an offshore rock section 85 m in length which is oriented approximately NE by SW. The weir section is constructed of wooden sheet piling and slopes from an elevation of 1.8 m MSL (mean sea level) to MSL over a distance of approximately 20 m, and is at MSL for the remaining 120 m length

21

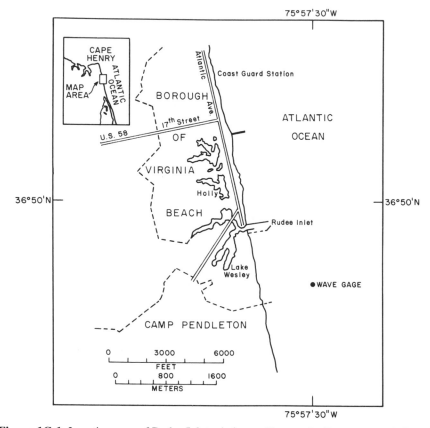

Figure 1C-1. Location map of Rudee Inlet relative to Chesapeake Bay entrance (adapted from U.S. Army Corps of Engineers, 1982).

of the weir. The purpose of the weir jetty system is to allow sediment to enter and deposit in the designated area between the two south jetties from which it is transferred to beaches to the north (Figure 1C-2).

A number of studies have been made of beach processes in this area. See, for example, Castel and Seymour (1982), U. S. Army Corps of Engineers (1982), Goldsmith *et al.* (1977) and Needham and Johnson (1972).

1C.2.2 Tides and Currents

The astronomical tide is semi-diurnal with neap and spring ranges of 1.0 m and 1.25 m, respectively.

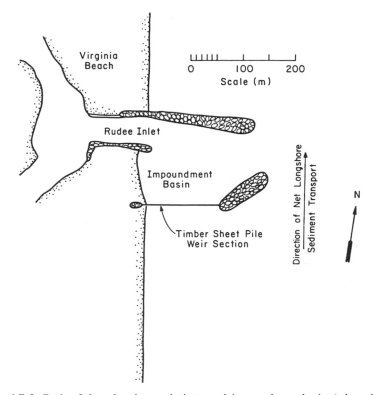

Figure 1C-2. Rudee Inlet, showing weir jetty and impoundment basin (adapted from Needham and Johnson, 1972).

The alongshore currents are predominantly wave driven within the surf zone; farther offshore, the currents are related to the tidal flows into and out of Chesapeake Bay.

The deepwater wave climate along the northeast coast of the United States is characterized by Northeasters during the winters and, with the exception of hurricanes, mild waves from the southeast during the summer months. In the vicinity of Rudee Inlet, this wave climate is modified both by the entrance to Chesapeake Bay and the hydrodynamic effects of the weir jetty.

1C.2.3 Wave Climate and Resulting Longshore Sediment Transport

The entrance to Chesapeake Bay tends to shelter the Rudee Inlet area from northeast waves by causing these waves to refract into the relatively deep entrance to Chesapeake Bay. The result, which is common to the localized downdrift areas of many bays, is that since waves originating from the updrift have a directional spread, during periods of northeaster occurrence, the nearshore waves south of the entrance may be directed toward the north. Extensive channel marginal shoals located off the updrift shore or ebb tidal shoals may also cause sheltering of the downdrift region. Regardless, the overall net effect of the Chesapeake Bay entrance is to cause a localized bias toward the north in the nearshore sediment transport.

The weir section of the southern jetty also tends to bias the sediment transport toward the north over an undefined but more localized zone south of the weir section. One effect of the inlet/jetty system on the south side of Rudee Inlet is to cause localized sheltering of these beaches from waves originating from the north. This results in a transport biased to the north in the shadow of the south jetty. Probably of greater importance to the transport bias toward the inlet is the relative elevation of the beach berm and the weir crest. The natural berm elevations to the south of Rudee Inlet are approximately 2 m MSL, whereas, as noted before, the greatest length of the weir crest elevation is at MSL. Figure 1C-3 presents profiles immediately south of the weir section (S-1) and 350 m south of the weir (S-10). Note that the profile nearest to the weir tends to conform to the weir profile. As a result of the weir section being lower than the natural berm crest, the weir tends to drain off sand from the south beach in much the same manner as a water weir functions. This effect, in terms of a gradual downward slope toward the weir, is quite evident in the field. A strong northerly flow of water over the weir is generally observable during periods of high tide. This flow is due in part to tidal effects and, in part, to a gradient in wave set-up along the beaches to the south.

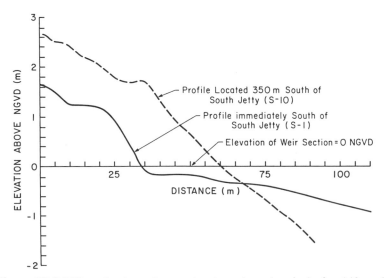

Figure 1C-3. Effect of weir section causing draw-down (erosion) of updrift profile.

1C.2.4 Beach Profile and Sediment Characteristics

The beach profile and sand size characteristics north of Rudee Inlet differ from those south of the inlet. In particular, the berm and foreshore sand sizes south of the inlet are coarser than to the north. Figure 1C-4 shows that the berm south of the inlet is higher and the foreshore slope significantly steeper than to the north. The associated sediment sizes are approximately 0.3 mm and 0.2 mm, respectively. Seaward of the foreshore, the profiles and sediment sizes are quite similar. The explanation for this discontinuity in sediment size and beach slope characteristics appears to be twofold. First, there is selective overpassing of sediment across the weir in the south jetty and settling in the deposition basin from which the sediment is transported to the north. As noted previously, the weir elevation is MSL and the material that tends to be transported across this weir is the finer fraction residing in the portion of the profile seaward of the foreshore. Secondly, nourishment projects have been carried out at Virginia Beach since 1952. The borrow areas have included bays and land sources. Thus the material present in the north beaches is a composite of that derived from bypassing around Rudee Inlet and that introduced during nourishment operations.

The seasonal sediment transport characteristics in the vicinity of Rudee Inlet are not well known. A directional wave gage has been in service off the south beach since July 1981. Based on the available wave data at Rudee Inlet, combined with measurements of impoundment in the deposition basin, both in our studies and by the Virginia Beach Erosion Committee, our estimate of net annual longshore sediment transport is 150,000 m^3 toward the north.

1C.2.5 Sand Bypassing at Rudee Inlet

The purpose of the weir/jetty system is to provide an impoundment basin in protected waters where the net transport from the south will accumulate for later transfer to the north beaches. In practice, this system appears to perform quite well, with only minor deposition across the seaward end of the entrance channel where a dredge is required to operate in unprotected waters. When the impoundment basin is filled to capacity, sedimentation will occur predominately along the southern side of the entrance channel to the west of the impoundment basin. Additionally, some deposition occurs on the north side of the channel which is believed to be due to a clockwise eddy that occurs due to water flow concentrated over the landward end of the weir. The transfer of sediment from the impoundment basin and the maintenance of the navigation channel are accomplished by two relatively small sand eductor systems located in the impoundment basin and a floating dredge with the mobility to service any area in which deposition is experienced. The sediment is transferred across the entrance channel by a submerged line (for the eductors) or a floating and/or submerged line (for the dredge) to the north side of the entrance channel where a permanent line is located along the beach. Discharge occurs at various

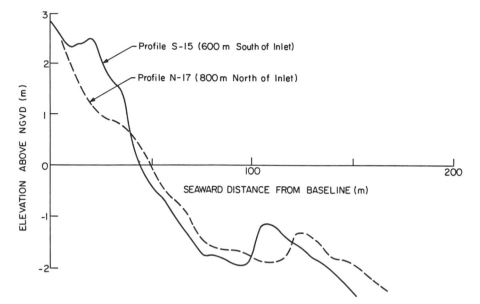

Figure 1C-4. Profiles north and south of Rudee Inlet showing steeper foreshore south of Inlet, November 5, 1981.

locations, depending on the condition of the beaches and the need for additional sand. An attempt is made to perform the transfer during periods when the tourist activity is low. The maintenance of the entrance channel and the transfer of sand is accomplished by the Virginia Beach Erosion Commission.

1C.3 References

Castel, D. and R. J. Seymour, 1982, Longshore sand transport report, February 1978 through December 1981, Report to U.S. Army Corps of Engineers and California Department of Boating and Waterways, IMR Ref. No. 86-2, 216 pp. (Reprinted March 1986).

Goldsmith, V., S. C. Strum and G. R. Thomas, 1977, Beach erosion and accretion at Virginia Beach, Virginia and vicinity, U.S. Army Coastal Engineering Research Center Miscellaneous Report No. 77-22, December.

Needham, B.H. and R.E. Johnson, 1972, Historical review of Rudee Inlet, Old Dominion University, Institute of Oceanography, Technical Report No. 9, October.

U.S. Army Corps of Engineers, 1982, Rudee Inlet, Virginia Beach, Virginia, Phase I, General Design Memorandum and Environmental Assessment, Norfolk District, North Atlantic Division, March.

Chapter 2

MEASURING THE INCIDENT WAVE FIELD

A. Torrey Pines Experiment

S. S. Pawka
Center for Coastal Studies
Scripps Institution of Oceanography

Wind generated surface gravity waves are the dominant driving force for surf zone dynamics on open coasts. A major thrust of the NSTS study was an investigation of the relationship between the incident waves and the processes that they drive (e.g., surf zone currents and sediment motion). A useful description of a linear wave field is obtained by the measurement of the frequency-directional spectrum. Particular attention is paid to certain moments of this spectrum. For example, S_{xy}, the onshore flux of longshore directed wave momentum, is an important wave parameter involved with longshore currents and resulting sediment transport. These wave field statistics were sampled with a linear array of pressure sensors in a mean depth of 9.6 m and by a pressure sensor and orthogonal-axis current meter in 5.7 m depth.

The directional spectra at Torrey Pines Beach are typically complicated (see Chapter 1A). High resolution (5-10 degrees) is required to define the narrow directional spreads created by the shadowing effects of the offshore islands. A linear array of length 363 m was used for the high resolution directional measurements. The sensors were placed in a 2-2-2-5 configuration with a spacing lag of 33 m. A shorter lag was originally planned but not implemented. The pressure sensors employed were Statham model PA 506-33 which are linear to a 0.1% of their range (less than 1 cm). The pressure signals were cabled to a central spar and transmitted by radio telemetry to the Shore Processes Laboratory at the Scripps Institution of Oceanography (Lowe *et al.*, 1972; Pawka *et al.*, 1976).

The linear array data were sampled continuously at 2.0 Hz during the daily experimental runs. The data were blocked in 17.1 minute segments for the routine analysis and the cross-spectra were averaged to yield 16 degrees of freedom with a bandwidth of 0.0078125 Hz. The usable range of the frequency spectrum is 0.02-0.25 Hz. The upper limit is set by the surface correction for attenuation of the pressure disturbances with depth. Some sample frequency spectra are shown in Figure 2A-1. The data on some of the days were contaminated with a high frequency noise that is suspected not to affect the frequency spectrum estimates in the usable range. However, the effects on the high resolution directional spectrum estimates are unknown.

The original directional spectrum analysis was performed using the Maximum Likelihood Estimator (MLE; Capon, 1969), a widely used high resolution technique. Figure 2A-2 (dashed line) is a plot of two typical directional spectra obtained from the linear array. The frequency range of well resolved directional spectra for this array is approximately 0.06-0.13 Hz. Although the MLE is a versatile estimator, it has proven to be deficient in the calculation of S_{xy} (Higgins *et al.*, 1981; Pawka, 1982). Therefore, two alternative methods were developed for the accurate estimation of S_{xy}. The first method (IMLE) achieves an improved directional spectrum by making iterative modifications to the MLE estimate. These modifications attempt to deconvolve the spectral smearing inherent in the estimate due to the finite nature of the array. Pawka (1982) describes the IMLE method and shows that the integration of these estimates yields accurate S_{xy} values throughout the 0.06-0.13 Hz range. Figure 2A-2 shows a comparison of MLE and IMLE estimates for two wave conditions. A second alternative method, the Moment Estimator (ME), makes direct estimates of S_{xy} by taking advantage of the spectral smearing. The spectral window is shaped to resemble the desired weighting function ($\sin \alpha \cos \alpha$ for S_{xy}) for the moment calculation. This method is derived in Pawka (1982) and shows similar accuracy to the IMLE method.

Many of the conditions during this experiment contained significant energy at frequencies above the usable range for the directional analysis of the 2-2-2-5 array (Figure 2A-1). Additional measurements were required for a complete description of the wave field. A system composed of a pressure sensor and a Marsh-McBirney #512 orthogonal axis current meter was located on the main range at 5.7 m depth. This system yields estimates of the frequency and directional spectrum for the range 0.05-0.30 Hz. Although the directional spectra are of low resolution, accurate estimates of several directional moments (including S_{xy}) are directly obtained from these data.

There was a relatively large uncertainty in the absolute orientation of the current meter. The estimated confidence in the current meter orientation was three to five degrees compared to a value of one to two degrees for the array lags. One check on the current meter orientation was obtained by comparing the mean direction (at a particular frequency) obtained from the current meter with the mode direction of the array directional spectrum. Refraction analysis was performed to convert the 10 m mode direction into a 5.7 m depth estimate. This comparison is only valid for very narrow unimodal directional spectra. A limited number of comparisons showed a consistent discrepancy of five degrees in the assumed orientation of the current meter. Due to the limited number of cases available for these comparisons an alternative approach was adopted.

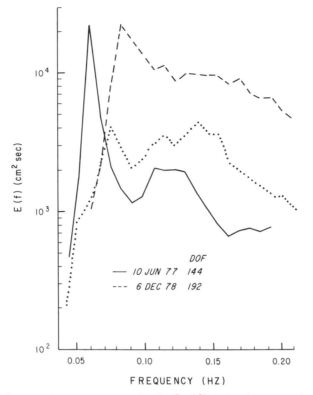

Figure 2A-1. Average frequency spectra for the $S_{xy}(f)$ estimation comparison data sets. The data were obtained from the 5.7 m depth pressure gage. DOF indicates the degrees of freedom resulting from the spectral averaging. Dotted line is for 18 November 1978 with 288 DOF.

IMLE estimates of $S_{xy}(f)$ were calculated for the frequency range 0.05-0.11 Hz and were refracted (with the use of the directional spectrum) to the 5.7 m depth. The current meter data were then rotated to achieve a best fit with these calculations. Data from five separate days were analyzed and yielded an optimal rotation that agreed within 0.5 degrees with the value obtained from the model comparisons. This best fit rotation angle was applied for the routine data analysis.

A comparison of $S_{xy}(f)$ estimates obtained from the rotated current meter data with refracted values from two array analysis methods for 18 November 1978 data is shown in Figure 2A-3. These data were not included in the set used for the current meter rotation. The IMLE estimates show excellent comparison with the current meter results in the range 0.05-0.13 Hz. The ME results appear to degrade at lower wave frequency (0.11 Hz) than the IMLE estimates. These

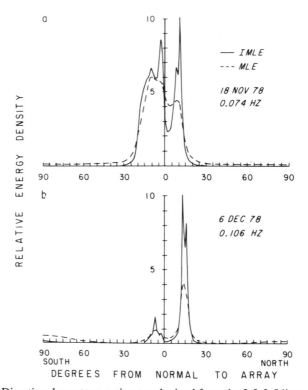

Figure 2A-2. Directional spectrum estimates obtained from the 2-2-2-5 linear array for a) the average of 18 consecutive 17.1 minute data segments on 18 November 1978 and b) 12 consecutive 17.1 minute segments on 6 December 1978. The spectra analyses with both the Maximum Likelihood Estimator (MLE) and the Iterative Maximum Likelihood Estimator (IMLE) are shown.

comparisons, which contain both southern and northern wave components, generate strong confidence in the analysis and current meter data rotation procedures.

A linear refraction model was used to shoal the offshore wave data into the 3 m depth contour. These calculations were made to facilitate the estimation of wave parameters at the wave breaking depth. The method of continuous directional spectrum refraction (Longuet-Higgins, 1957) was used to link specific sites on the 3 m contour to the offshore conditions. A numerical program developed by Dobson (1967) was employed for the calculation of the wave ray trajectories. The analysis was performed for five sites spaced by 40 m along the 3 m contour in the study area.

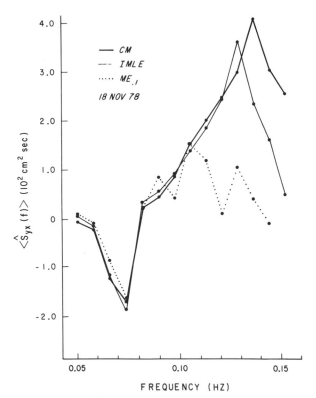

Figure 2A-3. Estimates of $S_{xy}(f)$ for 2 methods with the 2-2-2-5 linear array and the 2 axis current meter (CM). The data were averaged over 18 consecutive 17.1 minute segments. The subscript (.1) on the ME indicates the value of a noise rejection term which yields the best fit of the ME results with the current meter in the low frequency (0.059-0.074 Hz) range.

The refraction analysis produces linear filters which are used to transform the offshore directional spectra into the spectra at the 3 m depth sites. The linear array directional spectra were directly applied to the filters for the calculations. The 3 m depth spectra are then integrated to achieve the desired wave field parameters. It is recognized that linear theory breaks down as the waves near breaking and become strongly nonlinear. Linear theory was used, however, because of the lack of a finite-depth nonlinear shoaling model with demonstrated superiority (in the field) over linear shoaling.

The directional spectra obtained from a standard analysis of the pressure sensor and orthogonal-axis current meter (PUV) system (see e.g., Longuet-Higgins *et al.*, 1963) have inadequate resolution for the refraction calculations. Therefore, a unimodal distribution ($\cos^m \alpha_o$) was fit to the PUV cross-spectra.

The power, m, and the mode angle, α_o, are the fit parameters. This distribution was then applied to the spectrum transformation filters. A comparison was made of 3 m estimates made from linear array and PUV data and it showed good agreement of the energy and S_{xy} calculations. The PUV data were then used for the full frequency range (0.05-0.31 Hz) estimates at the 3 m depth sites. Further transformation of the wave parameters to the break point was estimated using monochromatic plane contour refraction.

2A.1 References

Capon, J., 1969, High-resolution frequency-wavenumber spectrum analysis, *Proceedings IEEE*, 57(8): 1408-1418.

Dobson, R. S., 1967, Some applications of digital computer to hydraulic engineering problems, Technical Paper No. 80, Department of Civil Engineering, Stanford University.

Higgins, A. L., R. J. Seymour and S. S. Pawka, 1981, A compact representation of ocean wave directionality, *Applied Ocean Research*, 3(3): 105-112.

Longuet-Higgins, M. S., 1957, On the transformation of a continuous spectrum by refraction, *Proceedings*, Cambridge Philosophic Society, 53: 226-29.

Longuet-Higgins, M. S., D. E. Cartwright and N. D. Smith, 1963, Observations of the directional spectrum of sea waves using the motions of a floating buoy, *Ocean Wave Spectra*, Prentice-Hall, Englewood Cliffs, N.J.: 111-36.

Lowe, R. L., D. L. Inman and B. M. Brush, 1972, Simultaneous data system for instrumenting the shelf, *Proceedings*, Thirteenth Coastal Engineering Conference, July 10-14, 1972, Vancouver, B.C., Canada, American Society of Civil Engineers, New York: 95-112.

Pawka, S. S., 1982, Wave directional characteristics on a partially sheltered coast, Ph.D. Dissertation, Scripps Institution of Oceanography, University of California, San Diego, 246 pp.

Pawka, S. S., D. L. Inman, R. L. Lowe and L. Holmes, 1976, Wave climate at Torrey Pines Beach, California, U. S. Army Corps of Engineers, Coastal Engineering Research Center, Technical Paper 76-5, 372 pp.

Chapter 2

MEASURING THE INCIDENT WAVE FIELD

B. Santa Barbara and Rudee Inlet Experiments

Richard J. Seymour
Scripps Institution of Oceanography
Institute of Marine Resources

Depth contours off Leadbetter Beach are neither straight nor parallel (Figure 1B-2). Therefore, a linear array, as described in the preceding section, could not be employed since it requires a constant relationship between wave numbers and frequencies along its length. Further, the contours suggest a significant longshore variation in the incident wave field. Two compact arrays were selected as the measurement scheme to provide some capability to measure longshore variation. Because these arrays are significantly smaller than the linear array, they reduce the problem of wave number variation. Each array consists of four bottom-mounted pressure sensors at the corners of a 6 m square. This arrangement allows the measurement of two components of local sea surface slope. These measurements provide an unbiased estimate of the longshore component of radiation stress as shown in Higgins *et al.* (1981). For beaches which have straight and parallel contours, this statistic can be used to estimate longshore transport as described in Seymour and Higgins (1978). At Santa Barbara, conservation of longshore radiation stress is not predicted by linear wave theory, and a more complicated procedure is required, as described below.

The arrangement of the two arrays at Santa Barbara is shown in Figure 1B-2. One of the arrays was installed approximately on the main instrument range and the second approximately 200 m to the east. Both arrays were at a depth of approximately 9 m. The angular orientation of the arrays is not controlled during installation, but is measured afterwards using an underwater magnetic gate compass with a digital readout. The compass is attached to the array frame by divers to determine the directional orientation. The accuracy of this device in determining the magnetic compass heading is approximately one degree. The horizontal position of the array was measured using an electronic surveying system. Two transponders were employed on the beach at known locations to triangulate the position of a small boat anchored over the array frame. The position error is estimated to be a few meters.

The wave measurement is made by the four solidstate pressure transducers. The pressure sensing elements are immersed in oil and are protected from sea water by a rubber diaphragm. Fouling of the diaphragm is inhibited by enclosing it with a thick layer of copper wool contained in a perforated plastic cap. The analog voltage output is converted at the transducer to a frequency modulated signal so that there will be no signal quality degradation over even very long transmission cable lengths. The transducer has a short term accuracy equivalent to a change in pressure appropriate to a 1 mm change in water column height. Because of temperature sensitivity, the mean value of pressure measured may deviate from the actual by several centimeters over periods of hours. The significance of this is that surface elevation changes (waves) can be measured within about 1 mm but water depth changes (tides and other causes) might have an error of 10 cm or more. An error in measuring the depth below mean sea level of the transducer yields a small error in estimating the wave energy resulting from the depth dependency term in transforming the pressure spectrum to an elevation spectrum.

The irregular depth contours at Santa Barbara, discussed previously, require that the directional information at the 9 m depth of the arrays be transformed shoreward to a depth at which the contours are essentially straight and parallel, which occurs at a depth of approximately 3 meters. The transformation was accomplished in the following way.

Longuet-Higgins, Cartwright and Smith (1963) describe a method (called hereafter the LCS method) for estimating a low-resolution directional spectrum from the motions of a buoy. The algorithm employs two components of sea surface slope as measured by a surface-following buoy and the surface elevation. The identical information can be obtained from the slope array. The LCS method is applied to the data at the array to obtain the directional spectrum. The LCS scheme also contains a method for estimating the central angle of approach for each frequency band, assuming that the incident energy is rather narrowly spread about some unimodal direction at each frequency. This approach is employed to condense the directional spectrum into a sequence of components each of which consists of a single angle and energy intensity and corresponds to a single frequency band. The components of this spectrum are then transformed to the 3 m depth contour using linear refraction techniques. The resulting spectrum is then considered to be the incident spectrum for surf zone processes.

The Santa Barbara arrays were sampled for 17-minute periods at six hour intervals continuously during the entirety of the trap experiment. These data were recorded and analyzed automatically at SIO using the wave network system described in Chapter 7. During the intensive experiment, the pressure signals from the two arrays were also recorded on the field system concurrently with all of the other NSTS instruments.

At the Rudee Inlet site, depth contours are reasonably straight and the problems that occurred at Santa Barbara were not anticipated. However, the successful employment of the slope arrays and the substantial reduction in installation costs for the compact array compared to a linear array, led to a decision to employ a single slope array off Croatan Beach, south of the trap site at Rudee Inlet (Figure 1C-1). This array was also sampled for 17-minute periods at six-hour intervals continuously through the experiment. The same method employed for the Santa Barbara data was used to calculate the incident wave field.

2B.1 References

Higgins, A. L., R. J. Seymour and S. S. Pawka, 1981, A compact representation of ocean wave directionality. *Applied Ocean Research*, 3(3): 105-112.

Longuet-Higgins, M. S., D. E. Cartwright and N. D. Smith, 1963, Observations of the directional spectrum of sea waves using the motions of a floating buoy, *Ocean Wave Spectra*, Prentice-Hall, Englewood Cliffs, N.J.: 111-36.

Seymour, R. J. and A. L. Higgins, 1978, Continuous estimation of longshore sand transport, *Proceedings*, Coastal Zone '78, Symposium on Technical, Environmental, Socioeconomic and Regulatory Aspects of Coastal Zone Management, March 14-16, San Francisco, CA., American Society of Civil Engineers, New York, 3: 2308-2318.

Chapter 3

MEASURING THE NEARSHORE MORPHOLOGY

A. Methods for Position Control and Beach Face Profiling

David G. Aubrey
Woods Hole Oceanographic Institution

Richard J. Seymour
Scripps Institution of Oceanography
Institute of Marine Resources

Accurate, repetitive surveying of the subaerial beach and shallow nearshore out to depths of about −1 m relative to Mean Sea Level (MSL) was performed as part of Task 4E of the Nearshore Sediment Transport Study (NSTS). This surveying was used to measure changes in beach shape and volume caused by cross-shore and longshore sediment transport during the inter-survey periods. The surveys also provided a daily measurement of the height of surf zone instrumentation above the sand bottom. Because accuracies of better than 5 cm were required for this task, the survey technique selected was similar to that described by Nordstrom and Inman (1975). A self-leveling engineers level and fiberglass, extendable survey rod were used for measuring vertical changes in the beach profile. The self-leveling level had a standard deviation of approximately ±2 mm over 1.6 km of double-run leveling. It was water resistant, with a magnification of about 32x and minimum focus distance of about 2 m. The leveling rods were graduated in 0.01 foot increments, with a linear accuracy of better than 1 in 4000. The fiberglass construction insured that the rod did not swell when wet and affect instrument accuracy. The survey line was a thin, plastic-sheathed steel line graduated at 3.0 meter increments. The construction of the line minimized stretching/contraction which would detract from survey accuracy. The maximum separation between level and rod was maintained at less than 60 m where possible; the primary exception to this occurred in the seaward portions of the profile when the level could not be relocated in the swash zone.

In addition to equipment limitations, other sources of survey error included the rod-holder wandering from the range line, natural small-scale bed roughness

on the order of 5 cm, scour beneath the rod in the swash and breaker zones, and
sudden changes in beach slope (scarps).

Each transect was distinguished by a fixed benchmark at its landward end.
Two range markers, previously located by standard surveying techniques and
separated by approximately 15 meters, defined the range. Vertical elevations
were referenced to a 2 m long metal pipe, driven into the sand, located at the
beginning of the range. Elevation readings were made at 3 m increments. The
level was normally moved once along the range to avoid exceeding the
maximum separation from the rod. The survey was discontinued when the rod-
holder could make no further progress through the surf zone, or was being
moved off line by longshore currents. Before moving to the next station, the
survey level was closed to the original elevation stake.

Longshore-directed transects were surveyed in a similar manner as that
described for cross-shore transects. These longshore transects were designed to
measure longshore variability in profiles, specifically when beach cusps and
other longshore-variable features were observed.

As an additional check on data quality, the survey data were entered into a
microcomputer and plotted soon after acquisition, then compared with previous
surveys. This technique gave quick survey data quality indicators, and provided
ample time for immediate resurvey in case of questionable data quality.

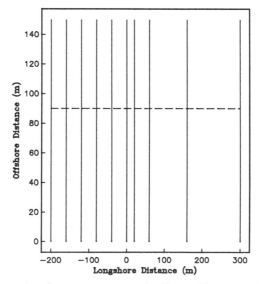

Figure 3A-1. Schematic of survey ranges at the Torrey Pines experiment, November,
1978. Dashed line is approximate location of mean sea level.

All horizontal and vertical control was referenced to well-documented, accurate survey monuments. All transects were set up based on these monuments; at each transect, a 2 m long, 2.5 cm steel pipe was inserted as a temporary benchmark for daily profiling. Vertical control between the primary and temporary benchmarks was transferred to within 0.01 foot.

The Torrey Pines experiment had ten shore-normal transects (Figure 3A-1; also Gable, 1979) with a bearing of 265 degrees TN. All transects were referenced to temporary benchmarks which were themselves referenced to a U. S. Coast and Geodetic Survey (USCGS) benchmark (see Gable, 1979, for details). Survey frequency can also be found in Gable (1979).

At the Santa Barbara experiment, five cross-shore transects were established (Figure 3A-2; also, Gable, 1981) with a bearing of 165 degrees TN. These were referenced to temporary benchmarks along each transect. In addition, eight longshore transects and five additional cross-shore transects were monitored over part of the experiment to document the three-dimensionality of the beach. Reference rods (0.15 cm diameter brass rods 1.3 meters long) were inserted along a shore-normal transect adjacent to the instrument range and monitored for a two week period prior to the major storm passage. This provided very accurate information on beach changes at a higher frequency sampling rate.

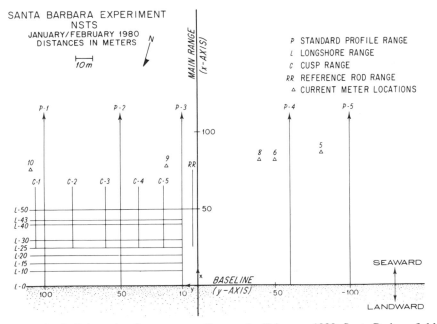

Figure 3A-2. Schematic of survey rangelines at the February, 1980, Santa Barbara field experiment.

Figure 3A-3. Tracked profiler.

The rod and level surveys described above were limited to approximately the −1 m contour under almost all conditions because of wave and current forces on the rod-holder. Surveys into deeper water were made using a boat and fathometer as described in the following section, which also delineates the problems of maintaining position and elevation accuracy and of achieving overlap with the rod and level portion of the transect. In an effort to eliminate some of the problems in obtaining long, continuous profiles by this composite approach, an experimental tracked vehicle profiler was developed. This profiler was described in detail in Seymour, Higgins and Bothman (1979) and is shown in Figure 3A-3. It consisted of a small electro-hydraulic tractor connected to shore by an umbilical cable carrying power and signals. A six wheel truck provided the shore support, including power generation, steering commands and data logging, as well as transport between range lines. The profiler measured inclination (tilt) along the track and offshore distance by an odometer sensing track movement. These parameters were sampled several times for each meter of travel and allowed computer integration to yield a finely sampled profile. The tractor was maintained on range by steering signals from the shore. A mast was provided that allowed the profiler to be controlled to a depth of −4 m. The system was capable of deployment under conditions of very strong longshore currents and breakers exceeding 2 m in height. Since the elevation was obtained by integration, a careful calibration program was required to maintain acceptable accuracy. Although almost 50 successful profiles into deep water were obtained at Santa Barbara, as described in Gable (1981), the maintenance and operation costs were considered too high to deploy the unit at Virginia Beach.

3A.1 References

Gable, C. G., Ed., 1979, Report on data from the Nearshore Sediment Transport Study Experiment at Torrey Pines Beach, California, November-December, 1978, University of California, San Diego, Institute of Marine Resources, IMR Reference Number 79-8.

_____. 1981, Report on data from the Nearshore Sediment Trnasport Study Experiment at Leadbetter Beach, Santa Barbara, California, January-February, 1980, University of California, San Diego, Institute of Marine Resources, IMR Reference Number 80-5, 314 pp.

Nordstrom, C. E. and D. L. Inman, 1975, Sand level changes on Torrey Pines Beach, California, U. S. Army Corps of Engineers, Coastal Engineering Research Center, Miscellaneous Paper No. 11-75, 166 p.

Seymour, R. J., A. L. Higgins and D. P. Bothman, 1978, Tracked vehicle for continuous nearshore profiles, *Proceedings*, Sixteenth Coastal Engineering Conference, August 27-September 3, 1978, Hamburg, Germany, American Society of Civil Engineers, New York, 2: 1542-1554.

Chapter 3

MEASURING THE NEARSHORE MORPHOLOGY

B. Offshore Surveys

Robert G. Dean
University of Florida
(Formerly University of Delaware)

3B.1 Introduction

The primary purpose of the offshore soundings conducted in conjunction with the NSTS studies at Santa Barbara, California and Rudee Inlet, Virginia was to determine sediment volumes in water depths too great for standard rod and level surveying procedures employed on the dry beach or in wading depths. Soundings are subject to particular types of errors which can lead to bias in the results and substantially misleading volume errors. These errors at Santa Barbara and Rudee Inlet included: data contamination by dredging, the horizontal position of the survey, the correct temporal and spatial tidal level, the fathometer calibration, the noise introduced into the data by waves and the requirement to survey a sufficient distance offshore to ensure that substantially all of the volumetric changes occurring are being measured. These individual error components and the means taken to minimize them are the subjects of later paragraphs in this section, followed by an overall error assessment viewed in light of the intersurvey volume changes measured. The errors will be discussed in the context of the Santa Barbara measurement program; however, they are similar for the Rudee Inlet study, except dredging occurred more frequently at the latter site. The survey plan for Santa Barbara, California is presented in Figure 3B-1 and comprised a total of 64 beach profiles and 62 sounding lines. Most (59) of the beach profiles and offshore sounding lines were colinear.

3B.2 Intersurvey Volume Changes

Table 3B-1 presents a summary of the intersurvey volume changes measured at Santa Barbara, California. Those intersurvey periods during which dredging occurred are noted as are the estimates of the dredged quantities which were added to the changes obtained from the surveys to obtain the totals. A value of ±35% is estimated as an upper limit of the error in estimating dredge

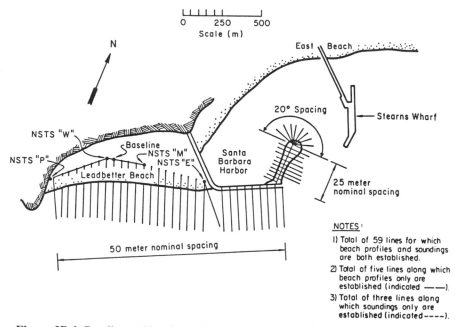

Figure 3B-1. Baseline and beach profiles and sounding lines measured during surveys.

quantities (which were based on surveys conducted by the dredging contractor, Great Western Dredging Company). It can be seen that during two of the intersurvey periods, the amount dredged exceeded by approximately a factor of three, the total volume change. Thus any substantial error in estimating dredged quantities would cause associated large percentage errors in the inferred total volume changes for these two intersurvey periods. The intersurvey volume changes represent the signal against which any survey errors must be compared.

3B.3 Error Components in Offshore Surveys

3B.3.1 Horizontal Postion of Survey Vessel

An attempt was made to maintain the survey vessel on the target sounding line by a transit positioned over the station and radio guidance to the boat operator by the transit person. The horizontal position of the survey vessel was determined by an Autotape microwave electronic positioning system. This system consists of a master unit on the survey vessel and two slave units which were positioned on known locations on the monumented baseline or, in some cases, known offsets therefrom. The master unit antenna was positioned directly above the fathometer transducer on the survey vessel to eliminate any requirement for translation of the position of the recorded depth information. At one-second intervals, the master unit interrogated the two slave units (each by a

Table 3B-1. Summary of Intersurvey Volume Changes
Including Effects of Dredged Volumes, Santa Barbara, California

Intersurvey Period	Measured Volume Change (m^3)	Estimated Dredging Contribution (m^3)	Total Volume Change (m^2)
October 13, 1979 to November 30, 1979	+32,830	0	32,820
December 1, 1979 to January 20, 1980	−101,500	166,570	65,070
January 21, 1980 to February 25, 1980	+76,380	6,430	82,810
April 11, 1980 to June 3, 1980	+10,290	0	10,290
June 4, 1980 to August 25, 1980	+22,220	0	22,220
August 26, 1980 to October 23, 1980	+38,760	0	38,760
October 23, 1980 to December 17, 1980	−78,900	115,540	36,640

distinct frequency signal); the slave units responded and the travel times from the two known locations were converted to distances and established a horizontal fix for the transducer location. These range data along with the sounding value obtained at the same one-second intervals were then digitized and stored on magnetic tape in an on-board recorder. The rated accuracy of the Autotape system is approximately 1 m with a range of approximately 60 km. Experience in the field indicated that the system was either quite accurate (limited checking was carried out on land where the distances could be measured) or, on occasions when the system malfunctioned, the inaccuracies were quite apparent by very erroneous readings. By far the greatest difficulties were encountered near the marina where many metal masts were present, apparently causing interference with the microwave signals. In most cases, this problem was alleviated by adjusting the antennas of the slave units so that the antenna was more oblique to the path of the transmitted signal. On a few occasions, it was necessary to delay soundings in a particular area until a later day. Due to forces of winds, currents or other reasons, the survey vessel track varied laterally from the target sounding line. Generally, with good sounding conditions, this deviation was kept reasonably small with a standard deviation on the order of 2 m. The deviations were generally smaller near shore where the depth gradients both along and transverse to the sounding lines, were greater. On occasion, the track deviated by a substantially greater amount than 2 m. The depth difference, Δh, on the actual and target sounding lines due to a lateral deviation from the target sounding line can be shown to be approximately

$$\Delta h \simeq \frac{\partial h}{\partial v} \Delta S_v \qquad (3B-1)$$

in which $\partial h / \partial v$ is the bottom slope in a local direction, v, perpendicular to the sounding line and ΔS_v is the lateral deviation from the sounding line. Referring to Figure 3B-1, it is seen that an attempt was made to orient the sounding lines perpendicular to the bottom contours such that $\partial h / \partial v \approx 0$. Thus in analyzing the data, it was possible to transfer sounding data laterally to the target sounding line without correction for the transverse component of bottom slope.

3B.3.2 Effects of Waves

The noise introduced into sounding data by waves can lead to substantial sources of errors in individual measurements and possible bias due to the set-down due to waves. These two effects will be discussed separately. First the oscillations introduced into the record by the waves is due to the combined heave, pitch and roll motions of the survey vessel. The survey vessel was a 7 meter Boston Whaler which is primarily subject to heave and it will be assumed in the analysis which follows that only the heave is important, although it is noted that whereas heave contributes unbiased noise, pitch and roll result in an upward bias to water depths. To minimize the wave-induced noise, each sounding line was surveyed three times successively during each survey. Depths were determined at 5 m intervals along the track line by averaging the results of the three successive soundings. It can be shown that if the standard deviation in depth associated with each measured point is σ, and the errors are unbiased, then the standard deviation σ_* associated with the sum of N measurements (the summing is required to obtain volumes) is

$$\sigma_* = \frac{\sigma}{\sqrt{N}} \tag{3B-2}$$

Since each line is at least 300 m long, depths are reported at 5 m intervals and three consecutive runs are made along each line and averaged, we find

$$N > 180 \tag{3B-3}$$

or

$$\sigma_* < 0.0745\sigma \tag{3B-4}$$

and for a wave height H, the root mean square heave displacement is

$$\sigma \approx 0.354H \tag{3B-5}$$

Since sounding conditions were generally characterized by waves less than 0.7 m height, we find the depth standard deviation, σ_*, associated with one line,

$$\sigma_* < 0.018 \text{ m} \approx 2 \text{ cm} \tag{3B-6}$$

Finally, when it is realized that the total volumes associated with an intersurvey period were determined from a total of at least 40 sounding lines, the overall standard deviation is on the order of 0.3 cm. The associated standard deviation in volume is approximately

$$\sigma_{\Delta V} = (0.003)\,(300 \times 50 \times 40) = 1800 \text{ m}^3 \tag{3B-7}$$

where the 50 represents the *maximum* spacing between lines. Hence, the estimate of 1800 m^3 yields an upper limit on the volume standard deviation due to wave height effects. Comparing this value to the *average* intersurvey volume change of 41,200 m^3 (averaging the last column, Table 3B-1), it is clear that, at most, the wave effects in an individual survey would be of minor importance. Fortunately this error (like all others) is not cumulative from survey-to-survey. Thus the confidence associated with the sum of all intersurveys volume changes is much better than with an individual survey.

Wave set-down can constitute a fairly important source of error since it can introduce a bias into the calculated volumes. The set-down, $\bar{\eta}$, outside the surf zone is given by Longuet-Higgins and Stewart as

$$\bar{\eta} = -\frac{1}{8}H_o^2 k_o \frac{\coth^2 kh}{2kh + \sinh 2kh} \tag{3B-8}$$

which, assuming shallow water waves, can be simplified to

$$\bar{\eta} = -\frac{H^2}{16h} \tag{3B-9}$$

Further, if Green's law is applied outside the surf zone, and considering

$$H_b \cong \kappa h_b \tag{3B-10}$$

where $\kappa \cong 0.8$,

$$\bar{\eta} \cong -\frac{H_b^2 \sqrt{h_b}}{16 h^{3/2}} \cong -\frac{\kappa^2 h_b^{5/2}}{16 h^{3/2}} \tag{3B-11}$$

Considering an effective beach slope, m, the areal error, ΔA, associated with the set-down is

$$A = \int_{x_b}^{x_m} \bar{\eta}(x)\,dx = -\frac{\kappa^2 h_b^{5/2}}{16\,m} \int_{h_b}^{h_m} \frac{1}{h^{3/2}}\,dh \tag{3B-12}$$

where h_b and h_m represent the breaking and maximum water depths, respectively.

$$\Delta A = +2\,\frac{\kappa^2 h_b^{5/2}}{16\,m}\left[\frac{1}{h_m^{1/2}} - \frac{1}{h_b^{1/2}}\right]$$

$$= -\frac{\kappa^2 h_b^{5/2}}{8 m h_m^{1/2}}\left[\left(\frac{h_m}{h_b}\right)^{1/2} - 1\right] \tag{3B-13}$$

As an example, consider $h_b = 1$ m, $h_m = 14$ m, $m = 0.025$

$$\Delta A = -2.3 \text{ m}^2 \tag{3B-14}$$

It is noted that additional contributions to this error in deeper water become very small. As $h_m \to \infty$,

$$\Delta A \to -\frac{\kappa^2 h_b^2}{8\,m} = -3.2 \text{ m}^2 \tag{3B-15}$$

which, of course, is not strictly valid, since these equations are based on shallow water approximations. An error of 3 m^3 per line is reasonably small especially when:

(1) compared to the approximate intersurvey area change per line of approximately 20 m^2,
(2) it is recognized that at Santa Barbara only the lines along Leadbetter Beach and the breakwater are subject to direct wave effects and most of the areal change occurred in the spit area,
(3) these errors are relative to the condition of no waves occurring and generally the waves were approximately the same for successive surveys, and
(4) the errors are not cumulative.

3B.3.2 Offshore Limit of Surveys

In general, all sounding lines were extended out to an approximate depth of 14 m which was judged to be well beyond the depth of effective transport. Confirmation was provided by the storm of February 1980 in which unusually large waves occurred, yet changes in bathymetry were limited to approximate depths of 5-6 m.

3B.3.3 Tidal Corrections

Possibly the greatest potential for bias is due to an error in the elevation of the instantaneous and local tidal plane which is used later to correct the soundings to some reference datum. Four sources of tidal elevations were available at Santa Barbara, including:

(1) a tide board located in the harbor near the root of the breakwater (this tide board was read visually during sounding operations and the values later interpolated),
(2) a filtered version of one of the four pressure gages on the slope array (this pressure includes barometric pressure contribution which should be taken into account),
(3) a NOAA-operated tide gage which was located near the seaward end of Stearns Wharf, and
(4) elevations taken by leveling to the sand/water interface inside the spit when the sounding operations were being conducted there.

In general, the corrections made using the tide data were effective as determined by:

(1) noting any apparent changes in elevations between successive surveys at the outer ends of each line, and

(2) always attempting to obtain overlap between the beach profiling and the sounding lines within a reasonably small time lag so that substantial profile changes would not occur between the two types of measurements. Finally, in accounting for the volume changes, the entire sounding profiles were used only out to a common depth (5 m) believed to be a good compromise between the errors that would occur by not carrying the volume calculations farther seaward and those that would occur due to small bias errors in the datum plane. Checks indicated that careful corrections using the tidal data and averaging three runs should yield offshore closure on a single profile with a standard deviation of approximately 4 cm. The volume error ΔV over an entire profile of an assumed width of 300 m is

$$\Delta V = 300 \times 0.04 \times 50 = 600 \text{ m}^2 \tag{3B-16}$$

The soundings usually lasted over the high tide portions of three days and there is no reason to expect that there is a net interline correlation of the errors. Thus the associated standard deviation in intersurvey total volume, by noting that these volumes were based on approximately 40 survey lines, is

$$\sigma_{\Delta V} = \left[\frac{600}{\sqrt{40}} \right] \times (40 \times 50) = 4100 \text{ m}^2 \tag{3B-17}$$

which is 10% of the average intersurvey volume change and 40% of the minimum intersurvey volume change.

3B.3.5 Fathometer Calibration

The fathometer was calibrated:

(1) at the commencement and completion of a sounding operation,

(2) after changes in chart paper,

(3) when a sounding area was changed, and

(4) at least every two hours. Calibration consisted of lowering, by a graduated line, a lead disc over the side of the vessel to various depths. The fathometer was adjusted if necessary to the known depth of the lead disc. Usually only very minor adjustments were necessary. These calibrations were always conducted in the area in which the soundings were being carried out and over the depth range of the soundings. If nonlinearities were present, they were accounted for in the data reduction.

3B.4 Summary

In summary, the average volume change for the seven intersurvey periods was 41,200 m^3. This discussion has shown that of the possible effects leading to errors in the calculated intersurvey volume changes, those two periods during which substantial dredging occurred are subject to large potential errors. With the possible exception of the tidal elevations, the remaining volume errors are on the order of 5 to 10%. The tidal elevations can lead to a bias in a single profile; however, there is no reason why the bias on one profile should be of the same sign as that on an adjacent profile. Although it is somewhat of a qualitative estimate, the errors due to the tidal elevations are estimated to amount to as much as ±10% of the average intersurvey volume and up to ±40% of the minimum intersurvey volume. In particular, the total volume change obtained by summing the five intersurvey volume changes, for which dredging effects were small in Table 3B-1, is 186,900 m^3. The error associated with this quantity is believed to be less than 5%.

3B.5 Reference

Longuet-Higgins, M. S. and R. W. Stewart, 1962, Radiation stress and mass transport in gravity waves, with application to 'surf beats,' *Fluid Mechanics*, 13: 481-504.

Chapter 4

MEASURING SURF ZONE DYNAMICS

A. General Measurements

R. T. Guza
Scripps Institution of Oceanography
Center for Coastal Studies

Edward B. Thornton
Naval Postgraduate School

4A.1 Introduction

The objective of the surf zone dynamics task was to characterize the horizontal velocity and sea surface elevation fields within the surf zone, and to relate them to the observed incident wave climate. Properties of the waves during shoaling, breaking, and run-up were measured with current meters, pressure sensors, wave staffs, and a run-up meter. This chapter is a brief description of the various sensors, and estimates of their accuracy. General considerations in the design of the experiments, and maps of typical sensor locations, are also given. More detailed information about the sensors (including manufacturers' specifications and close-up photographs) are given in the Torrey Pines and Santa Barbara experiment reports (Gable, 1979, 1981).

4A.2 Sensor Description -- Pressure Sensors

The pressure sensors were Statham Model PA506-33 absolute pressure, temperature compensated, transducers. Pressure sensors were calibrated by lowering into a salt water tank before and after both the Torrey Pines and Santa Barbara experiments. Changes in sensor gain were, with one exception, less than 2 percent. Changes in sensor offset were less than 3 cm for Santa Barbara, but substantially larger at Torrey Pines. Pre- and post-calibrations for each sensor are given in the experiment reports (Gable, 1979, 1981). A graph of typical calibration data (cm of water versus sensor output voltage) is given in Gable (1979, Fig. 4A-1) and demonstrates the good linearity of these sensors. Resolution of the pressure sensors, determined by the sensor dynamic range and the 12 bit data acquisition system, was generally 0.2 and 0.4 cm head of water for the Torrey Pines and Santa Barbara experiments, respectively. The pressure sensors functioned very well. The principal problem was bending of the pipes holding the sensors during the Santa Barbara storms. As a result, the vertical location of the sensors was sometimes not accurately known.

4A.3 Sensor Description -- Wave Staffs

Wave staffs were deployed throughout the surf zone in the Torrey Pines experiment, but only in the swash zone at Santa Barbara. The swash measurements are discussed separately later. The principal components of the Torrey Pines wave staffs were dual resistance wires stretched between mounting brackets secured to a 6.5 m long, 4 cm diameter fiberglass pole which was held vertical by jetting it about 1 m into the sand. The staffs had resolution on the order of 0.3 cm. Calibration by immersion in still water showed very linear sensor response. The principal problems with the wave staffs were wire stretching and breakage, and the entire staff pulling out of the sand. Wires which were sufficiently slack would occasionally short out, by touching each other. Gable (1979, Fig. 4A-4) shows a photograph of a wave staff installed next to an electromagnetic current meter at Torrey Pines Beach. A detailed discussion of the wave staff electronics is given by Flick *et al.* (1979).

4A.4 Sensor Description -- Run-up Meter

The run-up meter is a variation of the Scripps wave staff system and contains two parallel 0.016 cm diameter wires supported about 3 cm off the beach face. As the swash runs up the beach, more wire is submerged and the output voltage changes, resulting in an accurate representation of the up-rush. The downrush may occur rather more rapidly since percolation into the beach face may drop the water level below the wires. The run-up meter measures the most shoreward location where the depth is great enough to submerge the wires. The output of the run-up sensor is linear and relatively free of drift. Gain uncertainties are a few percent at most. However, work done since the NSTS experiment indicates that the magnitude of measured swash excursions depends rather critically on the distance between the sensing wires and the beach face. The wire heights in the NSTS experiments were not always kept at 3 cm, varying a few centimeters from this elevation. Tests with two side-by-side run-up meters indicate that these elevation differences introduce variance differences of order 20 percent. Hence, with 3 cm taken as a nominal wire elevation, there are roughly 10 percent uncertainties in swash amplitude.

4A.5 Sensor Description -- Current Meters

The current meters were Marsh-McBirney Model 512 OEM electromagnetic probes with 4 cm diameter spherical probes. The Marsh-McBirney electromagnetic current meter sensors operate on the principle of Faraday

Induction whereby a voltage is induced in a conductor moving relative to a magnetic field. In this case, seawater is the conductor, the magnetic field (AC) is set up by an electromagnet inside the probe and the electrodes measure the induced voltage to generate the output signal. Most sensors were calibrated three times (twice prior to, and once after the 1978 experiment) by towing at 100 cm/sec through still fluid. These tests showed gains close (±5 percent) to the manufacturer's specification of 100 cm/sec/volt and zero offsets. Therefore, the sensors were not calibrated specifically for the Santa Barbara experiment. Pre- and post-calibrations for each sensor used at Torrey Pines are given by Gable (1979).

A discussion of the performance of the current meters in steady and oscillatory flows is given in Cunningham et al. (1979) and elsewhere. These studies show the sensors to have a relatively flat frequency response, and to be reasonably linear devices. The calibration tests of Cunningham et al. (1979) lead to an estimated gain uncertainty of five to ten percent for these current meters. Some, and perhaps most, of this uncertainty is due to inaccuracies in the calibration procedure. This is somewhat irrelevant, however, since in practice a sensor is only as accurate as the calibration procedure. There are additional unanswered questions regarding the sensor response in flows with both steady and oscillatory components. It was beyond the scope of the NSTS experiment to completely resolve all the questions concerning current meters, and further work is needed to fully assess their accuracy. Sensor noise levels are less than 1 cm/sec rms, and resolution is 0.05 cm/sec. The output filter was three-pole with a 3 decible roll-off at 4 Hz.

The current meter (and other sensor) signals were degraded at times for a variety of reasons, so that care must be taken in deciding which data to analyze. When a current meter comes out of the water, for example in a wave trough, large spikes appear in the signal. These spikes have been decreased in magnitude by filtering and deglitching. During the period the current meter is out of the water, the signal is noisy and non-wavelike. Care should be taken to avoid misinterpreting these sections of data. The principal current meter failure modes at Santa Barbara were covering of the probe head with kelp mats and bending of the probe supports by kelp. Available visual observations of current meter condition (i.e., direction and amount of probe support bending) is given for each sensor and data run in Gable (1979, 1981). Generally, the stronger the longshore current, the worse the kelp problem. Thus, the runs with the strongest currents have the least reliable data.

A compass mounted on a special titanium (non-ferrous) jig was used to orient the slotted current meter bracket. The orientation technique can be used both above (at low tide) or underneath the water. Tests were conducted to insure negligible compass deviation due to ferrous material of the bracket and mounting pipe.

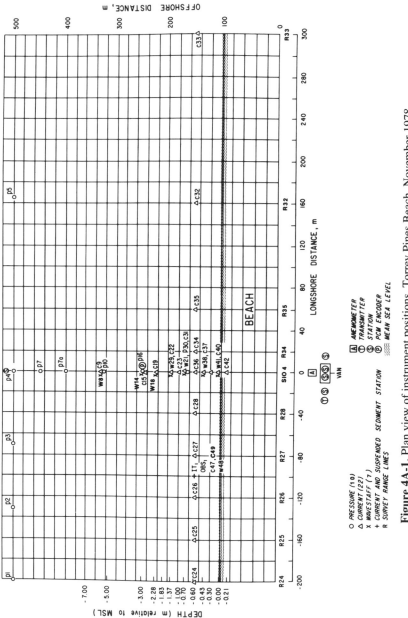

Figure 4A-1. Plan view of instrument positions, Torrey Pines Beach, November 1978.

4A.6 Sensor Description -- Surf Spiders (swash)

A "surf spider" consists of an electromagnetic current meter, wave staff, and electronic sediment measuring devices supported by a rigid triangular frame roughly 1.3 m on a side [Fig. 4A-6, Gable (1981)]. The purpose of the instrument is to simultaneously measure current water depth and sediment motion characteristics in the swash zone. The current meter was rigidly supported downward from the center of the frame; the probe height was adjustable and usually about 4 cm above the bed. The sediment monitoring device was offset towards the thinnest (1 cm diameter) support leg. The wave staff was about 50 cm seaward of the current meter and was of the resistance wire type. Surf spiders were deployed only at Santa Barbara. Since they were located in the swash, the current meters on surf spiders were frequently out of the water, and have sometimes been treated differently than the other current meters. When a functioning wave staff was less than 50 cm away, the time period when the current meter was out of the water could be identified. The current meter values during the intervals were set equal to zero. This is necessary because of the large spikes which occur in the current meter data when the probe is out of the water. Preliminary testing has shown that only a very short period of reimmersion is necessary for the current meters to function properly. However, the effect of cavitation around the probe head during strong backwash events is not known. Therefore, the current meter values corresponding to an immersion of the probe head of less than a few centimeters were also set equal to zero. It is believed that the signals from the surf spider current meters give at least a qualitative description of swash flow fields. Detailed calibrations with total water depths on the order of several probe diameters, and cavitation around the probe head, are needed for a quantitative interpretation of these data. The surf spider wave staffs sometimes had a very nonlinear response near the sand-water interface. This was not discovered until February 12, midway through the Santa Barbara experiment. Consequently, the wave staff data prior to this time were only useful for identifying time periods of current meter immersion and not for quantitative elevation measurements. The usefulness of the surf spider data is discussed, for each day they were deployed, in Gable (1981).

4A.7 Experiment Layout

The sensors were deployed in conceptually similar ways at Torrey Pines and Santa Barbara. The three basic elements in the experiment plans are:

(1) offshore (nominal depth 10 m) pressure sensor arrays used to monitor directional characteristics of the incident wave field;

(2) a densely instrumented cross-shore transect (extending from the offshore array to the beach face) designed to measure the transformation of wave properties during shoaling, breaking and run-up; and,

(3) a longshore transect of current meters, located roughly in mid-surf zone, used to detect rip currents and variability in the longshore current field.

Plan views of typical instrument deployments are given in Figures 4A-1 (Torrey Pines) and 4A-2 (Santa Barbara).

The instruments and electronics packages were mounted on one-inch steel pipes jetted into the sand. The pipes holding the current meter sensing element had 15 cm × 15 cm steel flanges, below sand level, to prevent rotation. The current meter probes fit into a mounting bracket attached to the top of a steel pipe. The mounting brackets allowed for three degrees of rotation so that the current meters could be leveled and properly oriented in the horizontal plane.

The instruments inshore of mean lower low water (MLLW) were installed on a dry beach at spring low tide. These sensors received power and transmitted signals via steel-jacketed cables connected to shore. Sensors seaward of MLLW were installed by divers. Signals from these sensors were telemetered (Lowe *et al.*, 1972) to shore at Torrey Pines, and directly cabled at Santa Barbara. Figure 4A-3 shows the on-offshore transect of instruments at Santa Barbara during spring low tide. The incident waves are very small. The longshore separation between pressure sensors and sensing elements of current meters, and between the E.M. sensing elements and instrumentation packages, is 1 m. The sensors are separated to avoid flow field distortion due to interference caused by adjacent mountings.

Although similar in overall structure and number of sensors, the two deployments differ in detail. Typical Santa Barbara beach face slopes are considerably steeper than Torrey Pines, and incident waves at Santa Barbara were anticipated to be rather small.

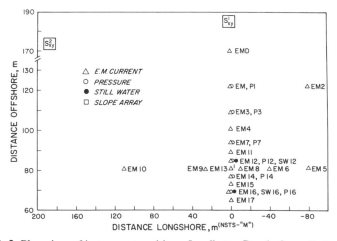

Figure 4A-2. Plan view of instrument positions, Leadbetter Beach, Santa Barbara, February 1980.

Figure 4A-3. Main cross-shore instrument range line. (1) EM sensing element, (2) EM electronics package, (3) pressure sensing element and electronics, (4) buoys locating offshore sensors.

The spacing of the instruments is dependent upon the temporal and spatial scales of the phenomenon to be measured. The minimum spacing of instruments must be correctly specified in order to avoid aliasing of spatial information. On the other hand, it is desirable to spread the instrumentation out as much as possible because of economy and the limited number of instruments available. The specification of the appropriate spacing of instruments within the surf zone is, in general, difficult because the scales of motion are not well understood, particularly in the longshore direction. An objective of the experiments was to obtain better information on space/time scales so that optimum instrument spacing may be determined for future experiments.

Classical models for longshore currents were used to determine the appropriate spacing of current meters on a transect perpendicular to the beach. Assuming steady state conditions and a plane beach, the alongshore momentum flux equation states that the gradient of the momentum flux (referred to as the radiation stress) of the unsteady velocity field (waves in this case) in the on-offshore direction (S_{xy}) is balanced by the bed shear stress directed alongshore

$$\frac{\partial S_{xy}}{\partial x} = -\tau_{by} \qquad (4A\text{-}1)$$

The radiation stress is the time averaged (overbar) covariance of the horizontal velocity components (u, v) integrated over depth

$$S_{xy} = \int_h^{\overline{\eta}} \overline{uv} \; dz \tag{4A-2}$$

The bed shear stress is a critical measurement for determination of sediment transport. Therefore, the accurate measurement of the radiation stress gradient, from which the bed shear stress can be inferred, is used as the criterion for instrument spacing.

The error involved in representing an arbitrary continuous function by discrete measurement points is examined using a Taylor series expansion for S_{xy}

$$S_{xy}(x+\Delta x) = S_{xy}(x) + \frac{\Delta x \partial S_{xy}(x)}{\partial x} + \frac{\Delta x^2}{2!} \frac{\partial^2 S_{xy}(x)}{\partial x} + \dots \tag{4A-3}$$

The true gradient as approximated by discrete points is then given by

$$\frac{\partial S_{xy}}{\partial x} = \frac{\Delta S_{xy}}{\Delta x} - \frac{\Delta x}{2} \frac{\partial^2 S_{xy}}{\partial x^2} + \dots \tag{4A-4}$$

The terms involving the second and higher order derivatives are a measure of the error of the approximation.

The percent error, R, in the radiation stress gradient measurement due to making point measurements Δx apart is approximately given by

$$R = \frac{\dfrac{\Delta x}{2} \dfrac{\partial^2 S_{xy}}{\partial x^2}}{\dfrac{\partial S_{xy}}{\partial x}} \tag{4A-5}$$

where derivatives higher than the second have been neglected.

For monochromatic shallow water linear waves inside the surf zone,

$$S_{xy} = E \sin \alpha \cos \alpha = AD^{5/2} \tag{4A-6}$$

where

$$E = \frac{1}{8} \rho g \, \gamma^2 \, D^2$$

$D = \overline{\eta} + h =$ total depth of water including set up
$\alpha =$ angle of crest to bottom contour
$A =$ complicated constant
Substituting (4A-6) into (4A-5), assuming a plane beach, and solving for Δx,

$$\Delta x = \frac{4}{3} R \frac{D}{D_x} \tag{4A-7}$$

Assuming for simplicity that $D \approx h$ (i.e., set-up is neglected) leads to

$$\Delta x = \frac{4}{3} R \frac{h}{h_x} \tag{4A-8}$$

where h_x is the beach slope. Note that the on-offshore spacing is only dependent on the spatial scales of the phenomenon being considered and is independent of time scales (i.e., wave frequencies). This is a consequence of the frequency nondispersiveness of shallow water waves; the group velocity depends only upon the depth.

Although the width of the surf zone decreases for increasing beach slope for a given breaker height, no economy in numbers of current meters is realized on steeper beaches. The surf zone width, ignoring set-up, is given by

$$x_b = \frac{h_b}{h_x} \tag{4A-9}$$

where h_b is the breaker depth. Combining (4A-8) and (4A-9), the number of current meters required is

$$N = \frac{x_b}{\Delta x} = \frac{3}{4R} \frac{h_b}{h} \tag{4A-10}$$

Equation (4A-10) states that the number of current meters is independent of the beach slope, h_x. As seen in (4A-8), increasing the beach slope requires a decrease in sensor spacing to maintain a constant relative error.

The criterion used in the derivation of (4A-8) is based on measuring gradients of S_{xy} with a constant relative error (4A-5), regardless of whether the gradient is large or not. It follows from (4A-6), however, that the largest gradients of S_{xy} occur at the breakpoint, where (4A-8) allows the largest instrument spacing. The smallest gradients of S_{xy} occur in very shallow water, where a good estimate requires very close instrument spacing. Since a very finite number of current meters were available, a constant spacing was used -- 14 m on the gently sloping Torrey Pines beach and 3 m on the steeper beach at Santa Barbara. Therefore, we have the greatest accuracy (for measuring gradients of S_{xy}) in the most energetic region just shoreward of the breakpoint, and the least accuracy in the inner surf zone.

The selection of sensor positions for the longshore line of current meters was very subjective. One objective was to detect rip currents which may make the measured behavior on the main range somewhat anomalous; for example, a strong seaward flowing rip current centered on the main range. Nearby current meters, at longshore locations not in the rip head, would presumably show rather different mean current structure. Since rip currents typically have longshore length scales equal to several surf zone widths, the sensors were spaced so as to have several current measurements across an anticipated circulation cell. Additional sensors (for example, EM 5 and EM 10, Figure 4A-2) were placed with the hope of verifying that the mean longshore current was homogeneous in the longshore direction. An inhomogeneous current field would indicate inhomogeneities in the incident wave field and would require corresponding additional terms in dynamic models.

4A.8 References

Cunningham, P. M., R. T. Guza, and R. L. Lowe, 1979, Dynamics calibration of electromagnetic flow meters, *Proceedings*, Oceans '79, IEEE: 298-301.

Flick, R. E., R. L. Lowe, M. H. Freilich and J. C. Boylls, 1979, Coastal and laboratory wavestaff system, *Proceedings*, Oceans '79, IEEE: 623.

Gable, C. G., Ed., 1979, Report on data from the Nearshore Sediment Transport Study Experiment at Torrey Pines Beach, California, November-December, 1978, University of California, San Diego, Institute of Marine Resources, IMR Reference #79-8.

_____. 1981, Report on data from the Nearshore Sediment Trnasport Study Experiment at Leadbetter Beach, Santa Barbara, California, January-February, 1980, University of California, San Diego, Institute of Marine Resources, IMR Reference #80-5, 314 pp.

Lowe, R. L., D. L. Inman and B. M. Brush, 1972, Simultaneous data system for instrumenting the shelf, *Proceedings*, Thirteenth Coastal Engineering Conference, July 10-14, 1972, Vancouver, B.C., Canada, American Society of Civil Engineers, New York: 95-112.

Chapter 4

MEASURING SURF ZONE DYNAMICS

B. Field Measurements of Rip Currents

Ernest C.-S. Tang and Robert A. Dalrymple

Ocean Engineering Group
Department of Civil Engineering
University of Delaware

4B.1 Purpose

The objectives of this task were to measure the characteristics of the nearshore circulation system at both field experimental sites and to develop a predictive model to:

(1) identify conditions where rip currents occur,
(2) provide information on the characteristics (spacing, strength, etc.) of these currents, and,
(3) determine their effects on the transport of sand along the beach.

4B.2 Summary of Experimental Equipment

The presence of rip currents was detected by several means. Rip current spacing and offshore extent were observed visually and photographically through the use of time lapse still photography in both experiments, coupled with fluorescent dye injections. The mean currents, velocities, directions and spatial periodicity were also observed with the large current meter array of Task I-A. Fine scale current measurements were measured through the use of one portable tripod at Torrey Pines and two at Santa Barbara. The tripods were equipped with a Marsh-McBirney current meter, a pressure sensor and a telemetry system to transmit the data ashore. Each tripod was capable of being moved by two divers and was anchored to the bottom with screw anchors. An adjustable arm permitted the current meter to be raised or lowered throughout the water column as well as turned 180° in the horizontal plane. This arm permitted the measurement of the velocity profile over the depth and the pressure sensor provided the wave and mean water level information. For the Torrey Pines experiment, a hydraulically filtered wave staff was used to examine the temporal variation of the mean (over a wave period) water level.

4B.3 Experimental Equipment

4B.3.1 Tripod -- Torrey Pines Experiment

The basic structure of the tripod system consisted of a center mast (3 inch I.D. steel pipe) threaded to a base tripod. Foot pads welded to each leg of the tripod were used for stability. The tripod could be anchored to the bottom through the use of three screw anchors. The height of the system was approximately eight feet.

A Marsh-McBirney (Model 512 OEM) electromagnetic current meter and a Viatran (Model 103) pressure sensor were mounted on the tripod. The current meter probe was held by a removable leveling bracket (Deep Ocean Work Systems) mounted on a three foot long pipe (1.5 inch O.D.) welded along with a reinforcing slant pipe on a 2.5 foot long sleeve, which was attached to the center mast with two half-inch diameter hexagon bolts through drilled holes on the mast. In order to make this arm system adjustable both vertically and horizontally, five sets of equally spaced bolt holes were drilled along the mast pipe, each set consisting of four holes distributed equally around the circumference of the pipe. The current meter electronic package and the pressure sensor were fastened to the legs of the tripod.

Data transmission was handled by a telemetry package consisting of a four-channel multiplexer and an FM transmitter, which transmitted the data to shore in analog form. The tripod carried a 24 volt battery supply for remote operation. The battery was stored in a waterproofed plexiglass box attached along with the telemetry package to the legs of the tripod. The antenna for the FM transmitter was fastened to the top of the center mast. A receiver onshore demultiplexed the FM signal into the four component signals and these were then input to the SAS system.

4B.3.2. Tripod -- Santa Barbara Experiment

The basic tripod assembly was redesigned for the Santa Barbara experiment. The second design was somewhat scaled down in size, with a central mast reaching to a height of six feet above the level of the foot pads, surmounted by a six-foot antenna mast. The tripod was constructed generally of 2 inch I.D. aluminum tubing, greatly reducing the overall weight of the system. The tripod is shown in Figure 4B-1. The typical set of instrumentation included a current meter (Marsh-McBirney Model 512 OEM) and a pressure sensor (Transducer, Inc.) with an operating range of 0-50 psia. Data transmission was handled again by a telemetry package consisting of a four-channel multiplexer and an FM transmitter, which transmitted the data to shore in analog form. The tripod carried a 24 volt battery supply for self-contained operation.

The two redesigned tripods proved light enough to be moved and positioned in gentle surf by two people. Due to anticipated low wave conditions at the Santa Barbara site, the tripods were equipped with only one sand screw for

Figure 4B-1. Tripod -- Santa Barbara Experiment -- showing pressure sensor and current meter on cantilever arm, and telemetry, battery pack and current meter package on the three legs. Screw anchors through the foot pads secure the tripod against bottom scour.

anchoring, which unfortunately proved to be inadequate during the majority of the test period.

4B.3.3 Current Meter

The current meter used on the tripod systems was a Marsh-McBirney electromagnetic meter, Model 512 OEM. The meter produces two output signals corresponding to a bi-directional current measurement with an output signal strength of ±5 volts, corresponding to velocities of ±3 meters per second. The time constant of the meter is 0.04 second. This meter is the same as used in Task I-A and was calibrated at SIO prior to the experiments. The probe was mounted to the instrument arm of the tripod using a Scripps-designed, Deep Ocean Work Systems break-away connector designed to protect the probe shaft.

4B.3.4 Pressure Sensor -- Torrey Pines Experiment

The pressure sensor (Viatran Corporation, Model 103) used in the Torrey Pines experiment has four strain gages arranged in the form of a Wheatstone bridge bounded to a sensing element. The signal from the bridge is amplified by a solid state signal conditioning package in the transducer, which was waterproofed by a plexiglass housing. The sensor has a pressure range of 0-25 psia. The output voltage ranges from 0-6 volts.

4B.3.5 Pressure Sensor -- Santa Barbara Experiment

The pressure sensor mounted on the tripod consisted of a bonded strain gage sensor (Transducer, Inc. Model JAP-69F-50) with an integral amplifier mounted in an oil-filled watertight housing. The sensor had an operating range of 0-50 psia with a frequency response of 0-1000 Hz. The output voltage ranges from 0-5 volts.

4B.3.6 Telemetry System

The telemetry system used at both Torrey Pines and Santa Barbara consisted of a tripod-mounted transmitting unit and a shore-base receiving unit. The transmitting unit consisted of a four-channel multiplexer which took output signals from the instruments and produced a single signal consisting of four modulated frequencies in the audible range. (Note that only three were used for data transmission.) This audible signal was then transmitted to shore by a FM transmitter operating in the high FM range. Each tripod operates at a separate broadcast frequency, allowing for the separate reception of data from several tripods simultaneously. The onshore receiving system consisted of a mirror image system to the transmitter, which then fed the four original output signals in analog form to the Scripps SAS system. The telemetry system was constructed by Deep Ocean Work Systems, Inc.

4B.3.7 Still Water Level Gage -- Torrey Pines Experiment

A Scripps Institution of Oceanography dual resistance wire analog wavestaff supported on a fiberglass pole (92 inches long) was fixed inside a ten-foot-long steel pipe (2-3/8 inch I.D.). The pipe was jetted into the sand several feet and guyed by screw anchors for stability. The bottom end of the pipe, buried in the sand, was open such that water was allowed to flow into the pipe, but the high frequency components of the water level variation inside the pipe would be filtered out by the sand. To prevent water from spilling into the instrument housing from the top, a siphon-shaped tube was threaded on the top of the pipe. This permitted air to enter the gage and allowed the water level inside to respond only to the hydraulic forcing.

4B.3.8 Time Lapse Camera System

A self-controlled time lapse Nikon camera system was mounted in a weatherproof box at both the Torrey Pines and Santa Barbara experiments. At Torrey Pines, the camera box was mounted on the cliffs, while at Santa Barbara the box was mounted on the anemometer tower. The camera provided a visual record of the study area, principally the main instrument range, at controlled intervals during periods of testing. The resulting photographs document visible complex flow patterns, such as rip currents present in the instrumented area during tests.

The camera system consisted of a battery-powered, motor-driven Nikon 35-mm SLR camera with an extended back which held 250-foot film cassettes. The interval between photographs was controlled by a Nikon intervalometer, which was generally set at eight minutes. The resulting photographs were Ektachrome 35-mm slides.

Chapter 4

MEASURING SURF ZONE DYNAMICS

C. Measurement Errors for Electromagnetic Current Meters

David G. Aubrey
Woods Hole Oceanographic Institution

4C.1 Introduction

The Nearshore Sediment Transport Study (NSTS) field experiments relied on Marsh-McBirney electromagnetic current meters to provide estimates of currents in the surf zone, for modeling of both hydrodynamics and sediment transport. These meters were chosen because of their history of successful use in previous surf zone studies, relying on their compactness and durability in this rough environment. The present work was motivated by examination of the NSTS experimental data from Santa Barbara, California, in which the higher-order velocity moments calculated from current meter time series showed a time-variability not obviously related to time scales of change of forcing (e.g., wave groupiness, infragravity waves), and by field studies which show persistent offshore near-bottom flow which is not yet satisfactorily explained by nearshore circulation theories. Because of the importance of these quasi-steady flows and higher order velocity moments to sediment transport in the nearshore, the present study examined under carefully controlled laboratory conditions the dynamic response of electromagnetic current meters (EMCM) typically used for field experimentation.

A long history of theory and calibration of EMCM's was reviewed by Aubrey *et al.* (1984), in which the present study is described in exhaustive detail. Pertinent to this study was empirical work performed on current meter response by NOAA National Oceanographic Instrumentation Center (e.g., Bivins and Appell, 1976; Mero *et al.*, 1977; and Kalvaitas, 1977). Cushing (1961, 1965, 1974 and 1976) presented a series of papers mixing theory and observations of the behavior of EMCM's. Some limited unsteady testing of these current meters has been performed (Appell, 1977a,b; Kalvaitas, 1977; Cunningham *et al.*, 1979), showing that unsteady superimposed motions affect the steady response of these instruments, particularly when the unsteady component is large compared to the steady component. McCullough's (1978) summary graphs of gain versus steady/oscillatory ratio show this dependence well, when the unsteady flow has a significant vertical component. These studies all showed that EMCM response varies with flow regime in a complex fashion. None of the above studies systematically examined the dynamic response of these meters to steady/oscillatory flow, which is a major aspect of this study (Aubrey *et al.*, 1984; Aubrey and Trowbridge, 1985).

Flow around a sphere is complex and varies with the nature of the incident flow and geometry of the sphere exterior (Aubrey *et al.*, 1984). EMCM's used in NSTS have hydrodynamically smooth insulating exteriors, which are interrupted by protruding electrodes and mounting stings extending from the top and bottom of the sensor (Figure 4C-1). Besides the wakes generated by each of these protrusions, the three-dimensional roughness elements affect the position of boundary layer separation on the sphere. Separation on spheres is also affected by ambient turbulent intensities and scales (e.g., Fernholz, 1978; Nakamura and Tomonari, 1982). Since turbulence intensities and scales vary markedly in the surf zone, this effect could be critical to EMCM behavior. For instance, Heathershaw (1976) measured turbulent intensities of up to 18% in the Irish Sea at 1 m off the bottom. W. D. Grant (personal communication, March 1985) typically finds turbulent intensities of 10% or more in water depths of 20 to 40 m on the continental shelf. Gross and Nowell (1983) present data indicating turbulent intensities of 20% at a distance of 70 cm from the bottom in a tidal flow in Puget Sound. In the surf zone, shear-generated turbulence combines with turbulence associated with breaking waves to provide high levels of turbulence over much of the surf zone. No accurate measurements of surf zone turbulent intensities are available in the literature, however. This combination of high levels of turbulent intensity and the sensitivity of flow around a sphere to this turbulence suggests the possibility of a strongly flow-dependent current meter sensitivity. Some previous work on the effect of turbulence on sensor behavior was performed by Bivins (1975), Bivins and Appell (1976), and Griffiths (1979).

Sixteen Marsh-McBirney current meters (Figure 4C-1) were examined in either the Ralph Parson's Laboratory ship model tow tank (at Massachusetts Institute of Technology) or at the tow tank/flume facility at the Woods Hole Oceanographic Institution. Four Marsh-McBirney MM551 current meters, seven MM551M current meters, three MM5512/OEM current meters (these are the meters most extensively used in NSTS), and one each of Sea Data Corporation models 635-9 and 635-12F directional wave sensors were calibrated in this study.

All data were examined using regression equations based on linear models with one or more independent variables (x), where voltage is the dependent variable (y). Independent variables are generally dynamical terms from the equations of motion (derived from dimensional analysis). The basic model is

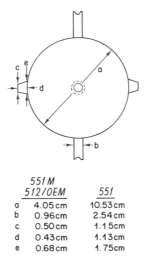

	551M 512/0EM	551
a	4.05 cm	10.53 cm
b	0.96 cm	2.54 cm
c	0.50 cm	1.15 cm
d	0.43 cm	1.13 cm
e	0.68 cm	1.75 cm

Figure 4C-1. Schematic of the large diameter and small diameter Marsh-McBirney electromagnetic current sensors, showing primary dimensions and location of roughness elements.

$$y = \alpha + \sum_{i=1}^{m} \beta_i x_i + \varepsilon \qquad (4C\text{-}1)$$

where m is the number of independent variables, α and β are coefficients, and ε is an error with zero mean and variance σ^2. β represents the sensitivity (units: volts/m/sec) of the instrument, in this usage, as distinguished from the gain, which is the inverse of the sensitivity. The offset (or intercept) consists of two parts. Electronic offset is the measured voltage derived from an immersed sensor with no water flowing past it. Numerical offset refers to the value of α calculated from the data after removing the electronic offset; it reflects the nonlinear response of the instrument to any given dynamical variable. For a linear instrument, the numerical offset would be zero. Estimates of α and β are derived using standard least-squares procedures. Either the F-test or the t-test is used to determine error bounds on sensitivity and offset.

Dimensional analysis defines the parameters with which to evaluate current meter response. For cases of pure steady, pure oscillatory, and combined steady and oscillatory flow, the appropriate non-dimensional ratios were formed. Details are contained in Aubrey *et al.* (1984).

4C.2 Results

Current meter response was evaluated for five categories of flow: pure steady, pure oscillatory, combined steady/oscillatory, horizontal cosine response, and grid turbulence. In summary form, the results are as follows:

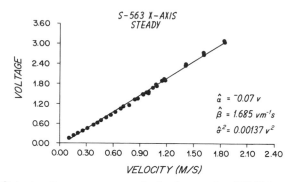

Figure 4C-2. Output voltage versus tow velocity for x-axis of S563 for pure steady conditions. As in the following figures, regression estimates of fit are indicated on the figure; specific information can also be obtained from tables in Aubrey *et al.* (1984).

4C.2.1 Pure Steady Flow

Twenty-one calibrations under steady flow conditions (Figure 4C-2) provided the following results:

a) Non-linear meter response results in a root-mean-square (rms) deviation of 0.042 volt in numerical offset, α, between the manufacturer's specifications and observed values (corresponding approximately to 2.5 cm/sec in velocity). This offset reflects nonlinear dependence of sensitivity on flow speed.

b) Observed sensitivities of current meters differed from manufacturer's specifications by an average of 5.3% (0.086 volts/m/sec root-mean-square sensitivity). Maximum deviation, observed on four sensors, was 11%.

c) Changes of 25% in sensitivity can result from biofouling, but magnitude of change depends on type of biofouling. This effect appears to be hydrodynamic, not electromagnetic.

d) Sensitivities of two orthogonal axes of the same current meter differed from each other an average of 4.8% from the expected sensitivity.

e) A two-segment least-squares fit to the data consistently resulted in lower error variance than a single straight line fit. Lack of consistency in the position of the intersection of the two line segments suggest that this improvement in fit is statistical, not dynamical (such as would be expected in transition from laminar to turbulent flow).

4C.2.2 Pure Oscillatory Flow

Four sensors were calibrated in pure oscillatory flow (Figure 4C-3), with periods ranging from 1 to 12 seconds, and double amplitude ranging up to 0.5 m. These tests documented:

a) Sensitivity is approximately the same (within 2%) in oscillatory flow as in pure steady flow.

b) Because of nonlinearity in sensitivity, numerical offsets are high for oscillatory flow.

c) Unsteady results show a sensitivity dependence on the ratio of orbital amplitude to sphere diameter (A/d) and on the oscillatory Reynolds number (Ad/vT), which has no obvious physical explanation.

4C.2.3 Combined Steady/Oscillatory Flow

Two current meters were subjected to combined steady/oscillatory flow tests. These tests revealed the following response characteristics:

Steady response to steady/oscillatory flow (Figure 4C-4):

a) Steady sensitivity was 7.1% and 10% lower in combined steady/oscillatory flows than in pure steady flow.

b) Decrease in steady sensitivity is qualitatively consistent with complex wake structure set up in the combined steady/oscillatory tests.

c) Error variance for regression model fits was increased by 1.5 to 7 times that for pure steady flow, suggesting the possibility of biasing steady results.

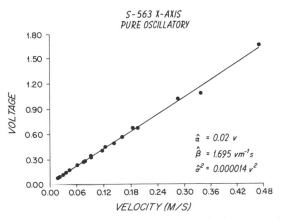

Figure 4C-3. Output voltage versus peak orbital velocity for pure oscillatory flow.

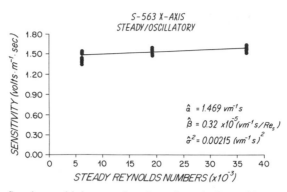

Figure 4C-4. Steady sensitivity as a function of steady Reynolds number for combined steady/oscillatory flow conditions. Sensitivity increases with Re.

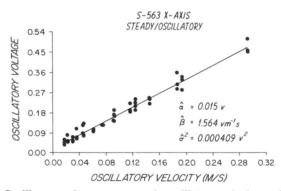

Figure 4C-5. Oscillatory voltage versus peak oscillatory velocity under combined steady/oscillatory flow conditions. Although the regression is excellent, the sensitivity is lower than in the pure oscillatory case.

Oscillatory response to steady/oscillatory flow (Figure 4C-5):

a) Oscillatory sensitivity in combined steady/oscillatory flow is consistently less than for pure oscillatory flow, by 8.5% to 14.5%. This result is consistent with the steady response characteristics seen above.

b) Oscillatory sensitivity is not significantly correlated with the ratio of steady to oscillatory velocity (UT/A), contrasting with results of other investigators (e.g., McCullough, 1978). This lack of dependence is not easily explained.

4C.2.4 Horizontal Cosine Response

Horizontal cosine response was determined for two sensors, with the following results (Figure 4C-6):

a) Average sensitivity (over all angles) for each of the axes and at each of three Reynolds numbers is dependent on Reynolds number, with sensitivity increasing with Reynolds number. Sensitivity is lower than comparable sensitivity for on-axis steady tow results, consistent with an inter-cardinal undersensitivity in horizontal flows, with greatest effects at low Reynolds number.

b) Because of their non-linear sensitivity, response to off-axis flows near 90° shows large errors in normalized sensitivity.

c) There is a shoulder-like structure in the cardinal sensitivity, particularly for lower Reynolds number. Near 0° flow, normalized sensitivity is low. The sensitivity rises as the relative angle increases to about 20°, finally turning to undersensitivity as angle increases. This near-cardinal behavior is likely due to asymmetries in sensor geometry as the protruding electrode is slowly rotated away from the forward stagnation point.

d) Root-mean-square error in horizontal cosine response varies as a function of axis and Reynolds number. Errors increase with Reynolds number, from a value of 0.01 m/sec up to 0.03 m/sec.

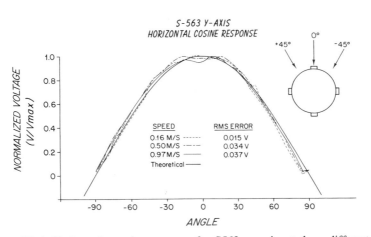

Figure 4C-6. Horizontal angular response for S563, y-axis, at three different tow speeds. Note "shoulder" about the 0° flow direction indicative of asymmetry of roughness distribution on sphere at low angle of attack.

4C.2.5 Grid Turbulence Tests

As a pilot study only, the effects of scales and intensities of free-stream turbulence (generated by grids) on current meter response were examined. Tests were not adequately controlled to verify the blockage effect behind the test grids, so theory had to be developed to estimate blockage effects (Aubrey *et al.*, 1984). The following results are presented only to indicate the potential magnitude of the turbulence dependence of these current meters, and to encourage future, more careful studies. Tests include:

a) Sensitivity of current meters decreased by 24% and 45% for the two grids compared with steady tow results (Figure 4C-7). The grids had different geometries, mesh size, solidity, and diameter.

b) A simple model of blockage behind grids shows Grid 1 results can be explained by flow blockage, even though experimental results where blockage occurred were not used in the analysis. However, Grid 2 results cannot be explained by the simple blockage model, indicating either that the sensor was not working or that increased intensity of turbulence behind Grid 2 caused hydrodynamic effects registered as a large reduction in sensitivity.

c) Error variances for Grid 1 tests were significantly higher than for pure steady tests, but error variances for Grid 2 were lower than under pure steady conditions. This indicates that the sensitivity change may be due to hydrodynamic effects.

4C.3 Implications

Experimental results on electromagnetic current meter behavior reveal a complex response to varying hydrodynamic conditions, including steady, oscillatory, combined steady/oscillatory, and turbulent flows. These findings suggest that the meters be used with care in situations demanding great accuracy, their use following a careful error analysis to assure that the instruments meet accuracy specifications for a particular use. This complex response to varying flow conditions is not unexpected given the complicated wake and boundary layer around a hydrodynamically rough sphere. Implications of these test results on NSTS data are profound, since these data were obtained in an environment with broad spectra of wave motions, quasi-steady currents at right angles to this motion, and high levels of ambient turbulence with scales and intensities varying widely. Some of these implications are discussed below.

a) Estimation of velocity for kinematic purposes: Measurement of the mean or oscillating component of velocity is often used to verify theory. For instance, examination of low frequency, infragravity waves in the surf zone requires knowledge of the sea surface or velocity distribution throughout the surf zone, an example of which is the recent work by Huntley *et al.* (1981) determining the dispersion relationship of infragravity waves from current

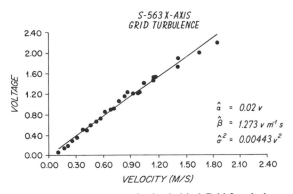

Figure 4C-7. Output voltage versus tow velocity behind Grid 2 turbulence. Scatter is relatively low, while sensitivity is much decreased compared to pure steady conditions. This decrease cannot readily be explained by flow blockage.

meter records. Alternatively, velocity records might be useful for examining the quasi-steady longshore current responding to onshore gradients in radiation stress.

If the electromagnetic current meters are used in pure steady or pure oscillatory flow, they can provide accurate estimates of the velocity field. Standard deviations of one to several cm/sec are common for these measurements. Errors can be reduced by careful calibration, applying a non-linear calibration curve, and continual attendance to current meters in the field. Under these conditions, rms errors of less than 1-2 cm/sec should be possible for limited Reynolds number ranges. Care must be taken to obtain good zero offsets in the field, on a frequent basis.

Combined steady/oscillatory flow, such as that found in the surf zone, considerably degrades sensor response, with errors of 10% or so in gain. The rms deviations in calibrations are higher than in pure steady or pure oscillatory flow. The possible dependence of current meter sensitivity on ambient scales and intensities of free stream turbulence may also affect their use in the surf zone.

This behavior depends on careful (and frequent) calibration, lack of significant biofouling, and maintaining the sensor several sensor diameters from any material with different electrical conductivity. Regardless of the inherent accuracy of these sensors, violation of any of the above conditions can reduce the sensor's accuracy to unacceptable levels.

b) Estimation of higher order velocity moments for sediment transport calculations: Since direct in situ measurement of sediment transport is difficult with existing technology, and certainly not possible over broad areas, a common way to estimate sediment transport is by applying a theoretical or empirical model to measured currents. Most models assume sediment transport is proportional to the third order velocity moment, while the Einstein-Brown formulation for waves (modified by Madsen and Grant, 1976) has sediment transport proportional to the sixth moment of the velocity. Since errors in velocity propagate rapidly as higher order moments are calculated, this use of the velocity measurements is highly sensitive to sensor response.

As an example, a simple model of velocity can be used

$$u = u' + \varepsilon \tag{4C-2}$$

where u' is the true velocity, u is the observed velocity, and ε is the error in the observation. Higher order moments become

$$<u^2> = u'^2 + 2u'\varepsilon + \varepsilon^2 \tag{4C-3}$$

and

$$<u^3> = u'^3 + 3u'^2\varepsilon + 3\varepsilon^2 u' + \varepsilon^3 \tag{4C-4}$$

In this latter case, the relative error is approximately $3\varepsilon(u')^2/(u'^3) = 3\varepsilon/u'$, whereas the error in u is ε/u'. By incorporating the error into higher order moments, the error is more than tripled. For a ratio of ε/u' of 0.1 (a 10% error in gain), relative error in $<u^3>$ is 0.33, an unacceptable error in many situations. For an error ratio $\varepsilon/u' = 0.45$ (characteristic of the error observed for a dirty probe calibration and in a free-stream turbulence test), the relative error in $<u^3>$ is greater than 2.05, for a signal-to-noise ratio of less than one-half.

This simple error analysis indicates the magnitude of the problem of cascading errors when estimating dynamical quantities from imperfect kinematic observations. A more complete analysis must properly account for the many sources of variability in these sensors, both electronic and hydrodynamic.

The two examples presented above illustrate the possible errors resulting from use of electromagnetic current meters in surf zone experiments. Use of these current meters must reflect a trade-off between allowable errors, ability to calibrate periodically, availability of alternative sensors, and cost. Since the surf zone environment is physically harsh, few sensors exist which will withstand continual wave pounding, sand abrasion, and kelp fouling. The electromagnetic current meter was selected for use in the NSTS experiments because of its ruggedness, ease of installation, and relatively good steady flow response. Errors documented in the present calibration study show the sensor is generally accurate within 10% (for both steady and oscillatory motions), with decreased accuracy when biofouling occurs or when the sensor head is closer than several probe diameters from the bed. The possibility of contamination by free stream turbulence is frightening, particularly given the possible magnitude of this effect (45% in this study). Since the list of alternative sensors is limited for energetic surf zone environments, the electromagnetic sensor is still a reasonable choice. Its use must be accompanied by a careful error analysis, to assure that the types of information inferred from the data are within acceptable error limits. These calibration results notwithstanding, the electromagnetic current meter is an appropriate instrument for surf zone use, although the scientists and engineers working in this environment must continue to search for higher quality instruments to continue answering some basic questions about surf zones and associated sediment transport.

4C.4 Acknowledgements

This work was initiated with funding from the NOAA National Office of Sea Grant, grant number NA80-AA-D-00077, as part of the Nearshore Sediment Transport Study. Most calibration runs and analysis were completed with funding from the U. S. Army Corps of Engineers, Coastal Engineering Research Center, under contract number DACW/2-82-C-0014. Some support came from the Woods Hole Oceanographic Institution's Coastal Research Center.

Mr. S. T. Bolmer performed many of the calibrations and reduced much of the analog data. Abigail Ames Spencer helped in the laboratory calibrations during our initial tests. Pam Barrows typed the report. Mr. Y. Meija of the MIT ship model test facility provided needed expertise in experimental work at his facility. His participation assured smooth operations at MIT. Mr. W. D. Spencer performed many of the calibrations, reduced much of the data, and designed some of the experimental hardware. Professor J. Trowbridge jointly conducted some of this work with the author.

4C.5 References

Appell, G. F., 1977a, Performance Assessment of advanced ocean current sensors, *Proceedings*, Oceans '77, October 17-19, Los Angeles St. Bonaventure Hotel, IEEE and MTS, 2: 30E-1.

_____. 1977b, Current meter performance in a near surface simulated environment, *Exposure*, 5(2): 5-9.

Aubrey, D. G., W. D. Spencer and J. H. Trowbridge, 1984, Dynamic response of electromagnetic current meters, Woods Hole Oceanographic Institution Technical Report Number 84-20, 150 pp.

Aubrey, D. G. and J. H. Trowbridge, 1985, Kinematic and dynamic estimates from electromagnetic current meter data, *Journal of Geophysical Reserarch*, September 20, 90(C5): 9137-9146.

Bivins, L. E., 1975, Turbulence effects on current measurement, MS thesis, University of Miami, Coral Gables, FL, 104 pp.

Bivins, L. E. and G. F. Appell, 1976, Turbulence effects on current measuring transducers, *Exposure*, 3(6).

Cunningham, P. M., R. T. Guza and R. L. Lowe, 1979, Dynamic calibration of electromagnetic flow meters. *Proceedings*, Oceans '79, IEEE and MTS: 298-301.

Cushing, V., 1961, Induction flowmeter, Rev. Sci. Instr., 29: 692.

———. 1965, Electromagnetic flowmeter, Rev. Sci. Instr., 36: 1142-1148.

———. 1974, Water current measurement, *Geoscience*, Wolff and Mercanti (eds.), Chapter 4, John Wiley and Sons, New York.

———. 1976, Electromagnetic water current meter, *Proceedings*, Oceans '76, IEEE and MTS: 25C-1 to 25C-17.

Fernholz, H.-H., 1978, External flows, In: P. Bradshaw (ed.), *Turbulence*, Chapter 2: 45-107, Springer-Verlag, New York, 339 pp.

Griffiths, G., 1979, The effect of turbulence on the calibration of electromagnetic current sensors and an approximation of their spatial response, Institute of Oceanographic Sciences Report No. 68, Wormley, 14 pp. plus figures.

Gross, T. F. and A. R. M. Nowell, 1983, Mean flow and turbulence scaling in a tidal boundary layer, Cont. Shelf Res., 2: 109-126.

Heathershaw, A. D., 1976, In: *The Benthic Boundary Layer*, pp. 11-31, I. N. McCave (ed.), Plenum Publ. Co., New York.

Huntley, D. A., R. T. Guza and E. B. Thornton, 1981, Field observations of surf beat, 1. Progressive edge waves, *Journal of Geophysical Research*, 86: 6451-6466.

Kalvaitas, A. N., 1977, Current sensor dynamic testing, *Polymode News*, no. 28 (unpublished manuscript).

Madsen, O. S. and W. D. Grant, 1976, Sediment transport in the coastal environment, MIT Ralph Parsons Lab Report 209, 105 pp.

McCullough, J. R., 1978, Near-surface ocean current sensors: problems and performance, Proceedings of a working conference on Current Measurement, Technical Report DEL-SG-3-78, College of Marine Studies, University of Delaware, Newark, DE 19711: 9-33.

Mero, T., G. Appell and R. S. McQuivey, 1977, Marine dynamics and its effect on current measuring transducers, NBS flow measurement symposium, February 23-25, Gaithersburg, MD.

Nakamura, Y. and Y. Tomonari, 1982, The effects of surface roughness on the flow past circular cylinders at high Reynolds numbers, *Journal Fluid Mechanics*, 123: 363-378.

Chapter 5

MEASURING SEDIMENT DYNAMICS

A. Discrete Sampling of Bedload and Suspended Load

James A. Zampol and B. Walton Waldorf
Scripps Institution of Oceanography
Center for Coastal Studies

5A.1 Introduction

This aspect of the program included the development of fluorescent sand tracer and suspended sediment sampling devices and procedures for their use in measuring the transport rate of sand. These devices and procedures were employed at Torrey Pines Beach and at Leadbetter Beach.

5A.2 Sampling Tracer Distribution

The measurements required to estimate longshore transport are the tracer advection rate and the thickness of the moving bedload layer (Inman *et al.*, 1968). In the past, both irradiated sand and sand tagged with fluorescent dye have been used as tracer for measuring sediment transport. Crickmore (1967) used scintillation counters *in situ* to measure the surface distribution of irradiated tracer. Inman and Chamberlain (1959) used greased card samplers and cores of the sand bed to sample the distribution of irradiated grains, where the concentration of irradiated grains was determined by exposing photographic film to the samples. The horizontal distribution of fluorescent tracer particles has been sampled with the greased card procedure (Ingle, 1966) and with cores and surface volume samplers designed to sample into the surface layer (Inman *et al.*, 1968; Komar and Inman, 1970).

The thickness of the moving bedload layer has been estimated by burying a plug of dyed sand on the exposed beach at low tide and excavating the site at the following low tide. The thickness of dyed sand replaced by undyed sand was assumed to be the thickness of sand in the moving layer (King, 1951; Williams, 1971). Others have estimated this from the vertical distribution of tracer grains within cores of the sand bed. Crickmore (1967) and Komar and Inman (1970) used hand-held core tubes to obtain core samples. More recently, investigators have designed more sophisticated sampling devices to aid in core sample collection and to minimize disturbance of the core sample (Gaughan, 1978; Inman *et al.*, 1980; Kraus *et al.*, 1981).

Experiments to measure transport rates were conducted during the program using fluorescent dyed, native beach sand introduced along a line across the surf zone. Two methods were used to sample the distribution of tracer grains. The first, spatial sampling, utilized surface volume grab samplers and core samplers to measure the areal distribution of tracer along the beach. The second sampling method, temporal sampling, consisted of core samples collected repeatedly with time along a fixed cross-shore line (Inman *et al.*, 1980; Kraus *et al.*, 1981; Chapter 6B).

The grab sampler is designed for rapid collection of a volume of surface sand from the nearshore zone. It consists of a 7.5 x 12.5 x 2 cm deep sample compartment and a sample retainer lid; each piece is mounted on one end of a pair of arms which are hinged together in a scissors-like arrangement (Figure 5A-1). The sample compartment is forced into the bed and the retainer brought into the closed position by bringing the handles together, trapping the sand in the compartment.

The nearshore sand coring device is designed to facilitate recovery of an undisturbed vertical core of the bed (Figure 5A-2). Clear butyrete tubing (4.13 cm inside diameter, 0.16 cm wall thickness) is used to collect samples up to 15 cm in length. Clear tubing allows visual inspection of the amount of sample retained.

Figure 5A-1. Photograph of the surface volume grab sampler in use.

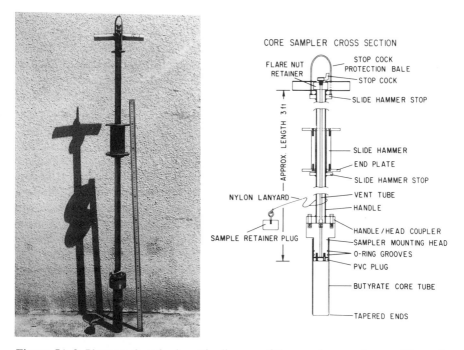

CORE SAMPLER CROSS SECTION

Figure 5A-2. Photograph and schematic diagram of the nearshore sand corer. The scale in the photograph is one meter in length.

The core tube is mounted on the end of the coring device over a double O-ring seal which provides the necessary friction to retain the core tube on the sampler head during withdrawal from the sand, and provides a water tight seal to the mounting head (Figure 5A-2). During penetration of the core tube into the sand, a vent provides a path for the removal of the water trapped above the sand in the core tube. A knurled PVC plug at the base of the sampler mounting head prevents sand from entering the vent while allowing water to pass. After penetration of the core tube, closing the vent creates a sufficient pressure differential to retain the sample within the core tube as it is withdrawn from the bed. Following removal of the sample from the bed, a retainer plug is inserted into the bottom of the core sample. The plug displaces water from the core tube below the bottom of the sample and supports the core's contents. Water must be displaced or it will migrate upwards, disrupting the vertical structure within the core sample.

Spatial samples were collected at predetermined longshore and cross-shore locations. The longshore location of each spatial sampling transect was indicated by a pair of flags aligned in the cross-shore direction. A marked line, stretched through the surf zone normal to the beach at each sampling transect, indicated the cross-shore location of samples. Immediately following collection each grab sample was transferred to a plastic bag and a labeled tag inserted to specify longshore and cross-shore location, and time of collection.

Temporal core samples were collected at cross-shore positions downcurrent of the tracer injection site; the direction and distance from the tracer injection line was chosen based on dyed water measurements of the longshore current velocity. Markers attached to a line which spanned the surf zone indicated the cross-shore sampling locations. Sampling personnel remained slightly downcurrent of the sample range until a signal from the beach was given indicating sampling should commence. Samples were collected, carried downcurrent of the sample site, and brought to the beach. This procedure was used to minimize disturbance of the bed between the tracer injection rangeline and temporal core sample sites.

Core samples were removed from the sampling device using a vise modified to encompass the diameter of the core tube. During sample removal, water filling the vent tube was drained by piercing the core tube between the PVC plug and the mounting head and opening the stop cock. Core samples were capped at both ends and stored vertically in a partitioned box which was labeled with locations and times of collection.

The concentration of fluorescent tracer in grab and core samples was determined by visually counting, under ultraviolet illumination, the number of dyed grains in subsamples of known mass. The grab samples were transferred from the plastic bags to evaporating dishes. Samples were then rinsed and decanted three times with de-ionized water to remove seawater in order to prevent the precipitation of salt during drying which cements the sand into clumps. The samples were evaporated to dryness at 50°C in a drying oven. Dried samples were split into subsamples of approximately 10 grams. A subsample was weighed on a precision balance and then spread onto a black counting grid to a single grain layer thickness. The sand was illuminated with a longwave ultraviolet light source (Black-Ray Model B100-A) in an otherwise completely darkened room and the fluorescent grains counted. The reliability of this analysis technique was determined by having personnel analyze a series of standard samples of known tracer content that were exposed to the same laboratory techniques as the samples collected in the field (Table 5A-1). The tracer concentrations of standard samples were not observed to decline significantly with time or by abrasion as simulated by placing samples in a rock tumbler for 24 hours.

Tests using stratified beach sand showed that the only measurable disturbance during core sampling occurred adjacent to the wall of the core tube. A sample extruder (Figure 5A-3) that slices away the peripheral grain layers as each increment of the core sample is extruded was used to remove core samples from the core tubes. This procedure removed tracer that may have been smeared due to drag along the walls of the core tube. The core samples were extruded in increments ranging from 0.25 to 2.0 cm depending on the vertical

Table 5A-1. Precision of Tracer Concentration by Visual Counting in Number of Dyed Grains per gram of Sample (The number of counters were 9 and 11 for Torrey Pines and Santa Barbara, respectively.)

Torrey Pines Tracer			Santa Barbara Tracer				
Green				Red		Green	
Sample	Mean	Standard Deviation	Sample	Mean	Standard Deviation	Mean	Standard Deviation
1	13.2	1.5	1	56.4	1.7	46.0	2.7
2	39.3	3.9	2	54.5	1.1	44.7	3.9
3	17.2	0.7	3	59.0	1.3	49.4	2.3
4	13.9	1.3	4	55.0	1.0	53.3	2.7
-	----	---	5	57.5	1.3	51.8	2.8

resolution desired. Each interval of the core sample was rinsed, dried, and counted using the same techniques as described for analysis of the grab samples.

5A.3 Suspended Sediment Sampling

The suspended sediment experiments were designed to give the distribution of suspended load across the surf zone so that the suspended load transport rate could be estimated. The distribution of suspended load in conjunction with measurements of the velocity field of the longshore current provide an estimate of the suspended load longshore transport rate.

The distribution of sand-sized sediment in suspension in the nearshore environment has been investigated both by direct in situ collection of samples and by indirect measurement of properties of the water and sediment that are assumed to be functions of sediment concentration. Direct methods include systems which average over times greater than wave periods such as permeable sample bags (Inman, 1948; 1978), bamboo samplers (Fukushima and Mizoguchi, 1958), pumping systems (Watts, 1953; Fairchild, 1965; Jensen and Sorensen, 1972; Kilner, 1976; Thornton and Morris, 1977; and Coakley et al., 1978), or suction devices (Nielsen et al., 1982). Other direct samplers collect a nearly instantaneous discrete bulk sample (Kana, 1976; Inman et al., 1980). Indirect methods which have been used in the field include gamma ray absorption (Basinski and Lewandowski, 1974), light attenuation (Brenninkmeyer, 1974), light backscattering (Thornton and Morris, 1977; Downing et al., 1981), and acoustical properties (Wenzel, 1974).

Mechanical water core samplers which collect a vertical array of instantaneous in situ discrete samples of water and sediment were used in the Nearshore Sediment Transport Study at Torrey Pines and Santa Barbara. The samplers consist of an 8.9 cm inside diameter clear acrylic tube which is internally divided into chambers by PVC disks mounted on a central rod (Figure 5A-4). This tube is moved relative to the inner column of chambers by means

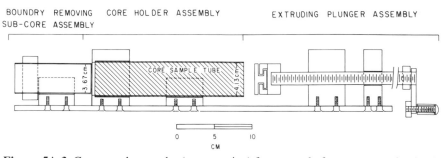

BOUNDRY REMOVING   CORE HOLDER ASSEMBLY   EXTRUDING PLUNGER ASSEMBLY
SUB-CORE ASSEMBLY

Figure 5A-3. Core sample extruder (cross section) for removal of core contents by depth increments. Boundary removing sub-core assembly eliminates tracer smeared by penetration of the core tube into the sand bed.

Figure 5A-4. Photograph of the water core suspended sediment sampling devices in open (left) and closed (right) position.

of a spring at the top of the sampler. The outer tube is ported such that, in the cocked position, water and sediment are free to flow through the sample chambers. When the sampler is tripped, the tube is forced downward, retaining the contents of the sample chamber. Since a sample chamber is occasionally partially above the water surface or may sample air bubbles within the water column, clear tube material is used so that the sample volume (if less than full) can be measured.

The water core samplers were designed to provide relatively undisturbed in situ samples of the distribution of suspended sediment with depth. A number of samples employing different principles were tested, including Van Dorn type samplers and slurp type samplers in which the inner partitions move relative to a stationary outer tube drawing water into a chamber. Motion pictures of the samplers in operation, using confetti to trace the water motion, showed that the water core sampler caused the least disturbance of the water column (Figure 5A-5).

The water core suspended sediment samplers that were used for the Torrey Pines Beach experiments used disks which were attached to the central rod. An O-ring around each disk provided the seal between the disks and the tube wall. A revised version of the sampler design was employed for the Leadbetter Beach experiments in which the fixed disks were replaced by a series of spools stacked on the central rod with O-rings between the spools. Half of the spool ends are tapered such that, as the spool stack is compressed by the closing spring, the O-rings are squeezed against the tube wall. This provides a tight seal when the sampler is closed but reduces the sliding friction during closure. The later versions of the samplers were also designed to sample closer to the bed and at more closely spaced vertical intervals. The sample chamber volumes range from 300 to 800 ml and the chamber elevations range from 14 to 93 cm above the bed (at the center of the chamber).

For the Torrey Pines and Leadbetter Beach experiments discrete suspended sediment samples were collected along a line across the surf zone. The range flags marking the spatial sampling grid were used for longshore positioning. A marked line or known surf zone instrument locations were used, as available, to position the samplers in the cross-shore direction.

Prior to sampling, the sampler was oriented so that water flow was through the chambers. The sampler was tripped when the foot of the sampler was placed on the bed to prevent bed scour from contaminating the suspended sample. In addition, the operator was positioned downcurrent of the sampler in order to prevent operator-induced suspended sediment from reaching the sampler. The water depth at time of passage of the bore crest and trough were measured with a hand-held wave staff.

To empty the sampler, the full sampler was placed on a rack which holds it at a slight angle from the horizontal (Figure 5A-6). The contents of the chambers were drained simultaneously through funnels into numbered bottles.

Figure 5A-5. Frames from motion pictures of the water core suspended sediment sampler (top: three-quarter closed) and a slurp type sampler (bottom, closing in water tagged with confetti). Water motion is shown by elongated path of confetti.

The chambers and funnels were rinsed carefully with a wash bottle to insure that all of the sediment was recovered.

The samples were decanted to a final volume of approximately 200 ml. A two-stage decantation procedure was used to guard against the accidental loss of sediment. The sediment and remaining water were washed into filter funnels and either gravity or vacuum filtered. The sediment was washed during final filtering with fresh water to prevent the precipitation of salt. The filtered sediment was dried in a low temperature oven, removed from the filter, and accurately weighed. The sample mass has an estimated uncertainty of 1 to 2 mg. The sample volume generally has a relative error of less than 10% caused

Figure 5A-6. Rack for emptying and rinsing the suspended sediment sampling devices. The rack holds a row of funnels which drain the water and sediment into numbered bottles.

by the dimensional variability of the samplers and by errors in height measurement when the sample chamber is not full. However, for cases in which the sample filled less than half of the sample chamber the relative error can exceed 10%, and it becomes large for very small sample volumes.

In addition to the water core samplers, a plastic baggy sampler was used on some occasions to sample within the swash zone. This sampler consists of a plastic bag mounted between a pair of jaws. The open jaws are oriented into the flow until the bag fills, then closed to seal the bag.

5A.4 References

Basinski, T. and A. Lewandowski, 1974, Field investigations of suspended sediment, *Proceedings*, Fourteenth Coastal Engineering Conference, June 24-28, 1974, Copenhagen, Denmark, American Society of Civil Engineers, 2: 1096-1108.

Brenninkmeyer, B. M., 1974, Mode and period of sand transport in the surf zone, *Proceedings*, Fourteenth Coastal Engineering Conference, June 24-28, 1974, Copenhagen, Denmark, American Society of Civil Engineers, New York, 2: 812-827.

Coakley, J. P., H. A. Savile, M. Pedrosa, M. Larocque, 1978, Sled system for profiling suspended littoral drift, *Proceedings,* Sixteenth Coastal Engineering Conference, August 27-September 3, 1978, Hamburg, Germany, American Society of Civil Engineers, New York, 2: 1764-1775.

Crickmore, M. J., 1967, Measurement of sand transport in rivers with special reference to tracer methods, *Sedimentology*, 8: 175-228.

Downing, J. P., R. W. Sternberg and C. R. B. Lister, 1981, New instrumentation for the investigation of sediment suspension processes in the shallow marine environment, *Marine Geology*, 42: 19-34.

Fairchild, J. C., 1965, A tractor-mounted suspended sand sampler, *Shore and Beach*, 33(2): 31-34.

Fukushima, H. and Y. Mizoguchi, 1958, Field investigation of suspended littoral drift, *Coastal Engineering in Japan*, 1: 131-134.

Gaughan, M. K., 1978, Depth of disturbance of sand in surf zones, *Proceedings*, Sixteenth Coastal Engineering Conference, August 27-September 3, 1978, Hamburg, Germany, American Society of Civil Engineers, 2: 1513-1530.

Ingle, J. C., Jr., 1966, The movement of beach sand, *Developments in Sedimentology*, Elsevier, New York, 5: 221 pp.

Inman, D. L., 1948, Observations of nearshore sand transport by waves at Scripps Institution of Oceanography, La Jolla, *Geological Society of America Bulletin*, 59: 1374.

_____. 1978, Status of surf zone sediment transport relation, *Proceedings*, Workshop on Coastal Sediment Transport with Emphasis on the National Sediment Transport Study, DEL-SG-15-78, Sea Grant College Program, University of Delaware, Newark, DE: 9-20.

Inman, D. L. and T. K. Chamberlain, 1959, Tracing beach sand movement with irradiated quartz, *Journal Geophysical Research*, 64(1): 41-47.

Inman, D. L., P. D. Komar and A. J. Bowen, 1968, Longshore transport of sand, *Proceedings*, Eleventh Conference on Coastal Engineering, London, England, American Society of Civil Engineers, New York, 1: 298-306.

Inman, D. L., J. A. Zampol, T. E. White, D. M. Hanes, B. W. Waldorf and K. A. Kastens, 1980, Field measurements of sand motion in the surf zone, *Proceedings*, Seventeenth Coastal Engineering Conference, March 23-28, 1980, Sydney, Australia, American Society of Civil Engineers, New York: 1215-1234.

Jensen, J. K. and T. Sorensen, 1972, Measurement of sediment suspension in combination of waves and currents, *Proceedings*, Thirteenth Coastal Engineering Conference, July 10-14, 1972, Vancouver, B.C., Canada, American Society of Civil Engineers, New York, 2: 1097-1104.

Kana, T. W., 1976, A new apparatus for collecting simultaneous water samples in the surf zone, *Journal of Sedimentary Petrology*, 46(4): 1031-1034.

Kilner, F. A., 1976, Measurement of suspended sediment in the surf zone, *Proceedings*, Fifteenth Coastal Engineering Conference, July 11-17, 1976, Honolulu, Hawaii, American Society of Civil Engineers, New York, 2: 2045-2059.

King, C. A. M., 1951, Depth of disturbance of sand on sea beaches by waves, *Journal Sedimentary Petrology*, 21: 131-140.

Komar, P. D. and D. L. Inman, 1970, Longshore sand transport on beaches, *Journal of Geophysical Research*, 75(30): 5914-5927.

Kraus, N. C., R. S. Farinato and K. Horikawa, 1981, Field experiments on longshore sand transport in the surf zone, *Coastal Engineering in Japan*, v24.

Nielsen, P., M. O. Green and F. C. Coffey, 1982, Suspended sediment under waves, Coastal Studies Unit Technical Report No. 82/6, Department of Geography, The University of Sydney, Sydney, N.S.W. 2006, 157 pp.

Thornton, E. B. and W. D. Morris, 1977, Suspended sediments measured within the surf zone, *Proceedings*, Coastal Sediments '77: 655-668.

Watts, G. M., 1953, Development and field tests of a sampler for suspended sediment in wave action, Beach Erosion Board Technical Memo No. 34, U. S. Army Corps of Engineers, 41 pp.

Wenzel, D., 1974, Measuring sand discharge near the sea-bottom, *Proceedings*, Fourteenth Coastal Engineering Conference, June 24-28, 1974, Copenhagen, Denmark, American Society of Civil Engineers, New York, 2: 741-755.

Williams, A. T. 1971, An analysis of some factors involved in the depth of disturbance of beach sand by waves, *Marine Geology*, 11(3): 145-158.

Chapter 5

MEASURING SEDIMENT DYNAMICS

B. Continuous Bedload Sampling

Robert L. Lowe
Scripps Institution of Oceanography
Center for Coastal Studies

5B.1 Introduction

An acoustic bedload sensor was designed and developed as a part of the program. Its operation is based on the fact that sand at rest must dilate before moving and remain dilated in motion. Thus, it is possible to obtain reflections from the surface of the moving sand as well as from the layer at rest due to the different acoustic properties of the stationary and dilated sand.

The instrument employs two acoustic devices (Figure 5B-1). One device transmits a vertically incident acoustic beam which provides a measurement of the thickness of the moving sand layer. The other device measures the velocity of the sand layer using Doppler sonar techniques. The sensor shown in Figure 5B-1 is described in detail in Lowe (1980).

A trial instrument was deployed during the Torrey Pines experiment, but gave no useful data on bedload thickness because of design deficiencies. However, information leading to design improvements was obtained. The improved prototype acoustic sensor, potentially capable of continuously measuring the position of the water-sand interface and the thickness of the moving bedload layer, was deployed for a total of 14 hours during the Santa Barbara field experiment. In addition, a series of laboratory flume calibration runs of the bedload thickness sensor have been conducted. The flume tests indicated that the instrument was capable of accurately measuring the surface bed features (ripples) that migrated past the sensor in unidirectional flow of varying magnitudes. Lack of independent methods to accurately measure the instantaneous bedload thickness have so far frustrated direct calibration efforts.

A computer program was developed which objectively analyzed the return data to estimate bed thickness. The criteria used to determine the upper and lower boundaries of the moving layer are a maximum derivative of the return signal associated with a minimum amplitude.

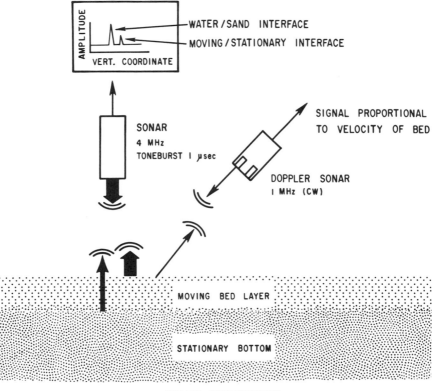

Figure 5B-1. Bedload instrumentation.

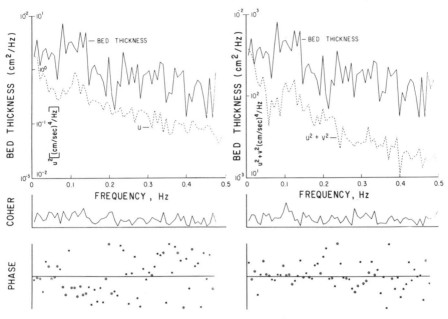

Figure 5B-2. Cross-spectral analysis of bed thickness
observations and bottom stress for two different conditions.

Deployment of the device in the surf zone in approximately one meter depth, indicated an apparent thickness signal that was correlated with the square of the magnitude of the horizontal water velocity near the bed. This parameter is generally held to be proportional to bed stress (Figure 5B-2). Some useful information about the bed motion was obtained at Santa Barbara, but no bedload transport estimates were made. More development work on the sensors themselves is needed, as well as a definitive method for calibration.

5B.2 Reference

Lowe, R. L., 1980, Acoustic bedload sensor, *Abstracts*, Seventeenth International Conference on Coastal Engineering: 215-216.

Chapter 5

MEASURING SEDIMENT DYNAMICS

C. Continuous Suspended Load Sampler

Richard W. Sternberg
Contribution No. 1701 of the School of Oceanography
University of Washington

Knowledge of the response of sediment particles to the hydraulic forces produced by shoaling waves is essential in studies of both large and small scale sedimentation patterns along sandy coasts. While accurate fluid-flow measurements are possible in even the most energetic nearshore waters because of technological advances in commercially available current meters, the technology for measuring sediment response is far less satisfactory (Downing, 1978). Direct samplers such as pumps (Watts, 1953; Fairchild, 1973) and water corers (Kana, 1979; Inman *et al.*, 1980) can provide accurate data but are limited by spatial and temporal resolution. Optical instruments previously used (Brenninkmeyer, 1974, 1976) had excellent frequency response but lacked dynamic range; they also were subject to interference from gas bubbles entrained by breaking waves. In order to integrate sediment dynamics studies into the surf zone dynamics aspects of NSTS, it was necessary, therefore, to design and develop new instrumentation to measure the spatial and temporal characteristics of the suspended sediment concentration field.

5C.1 Optical Instrumentation

Suspended sediment concentration was measured with an optical backscatter sensor (OBS) recently developed at the University of Washington (Downing *et al.*, 1981). An OBS sensor consists of a stainless steel housing 19 mm in diameter by 11 mm in length (Figure 5C-1). It supports:

(1) an infrared emitting diode (IRED) with peak radiant intensity at 950 nm,
(2) a low capacitance silicon photo-voltaic cell with peak spectral response at 900 nm, and
(3) a KODAK WRATTAN R filter with transmittances of 0.55 percent below 790 nm and 83.2 percent above 950 nm. The optical components are encapsulated in the recessed face of the sensor head with crystal-clear epoxy resin.

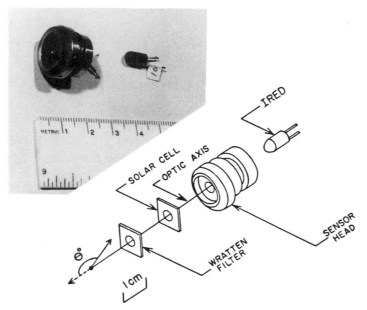

Figure 5C-1. Photograph and expanded sketch of the OBS sensor.

The principal of operation is: A conical sample volume, 1.3 cc, is irradiated through a 5.6 mm aperture at the geometric center of the photo-detector by an IR-beam with a half-cone angle of 14 degrees. High-angle (110-165 degrees) backscattered radiation is converted to photocurrent by a large area detector oriented in a plane normal to the emitter beam axis and located close to it.

There are a number of advantages associated with this sensor design. The small sample volume permits excellent spatial resolution of the concentration field and minimizes multiple scattering. The compact arrangement of the emitter and detector minimizes losses due to scattering and absorption because the optical path length between emitter and detector is short (≤ 2 cm). As a result the sensor has an excellent signal-to-noise ratio and only conventional low-gain electronics is required for signal conditioning. The output of the sensor electronics is bipolar (± 5VDC) and was fully compatible with the NSTS data system.

The specifications of the OBS instrumentation are summarized in Table 5C-1.

5C.2 Sensor Calibration

A laboratory test tank that produces strong currents and steady sediment suspensions in a space large enough to accommodate the OBS instrumentation was used for all calibrations (Downing *et al.*, 1981). Figure 5C-2A shows a representative calibration curve for an OBS sensor using a beach sand with median diameter of 0.165 mm. The response of the instrument is linear and in

Table 5C-1. Instrument Specificiations
Optical Back-scatterance Instrument (OBS)

Supply Voltage	+27 VDC
Current Demand	185 ma
Output	±5V full scale
Low-pass Cutoff Frequency (− 3dB)	10 Hz
Output Impedance	<1.0 ohm
Dynamic Range	0.1 − 150.0 ppt (by weight)
Resolution	0.1 ppt

this particular calibration a concentration range of 0.1 to 100 ppt by weight was encompassed. The response of the instrument to air bubbles injected into the test tank produced a minimal response with both low and high sediment concentrations (indicated on Figure 5C-2A by arrows).

The size and refractive indices of sediment particles strongly influence their light scattering characteristics. Consequently, each optical sensor was individually calibrated with Leadbetter Beach sediment before and after the Santa Barbara experiment. The Leadbetter Beach sediment is a fine-to-medium sand and its textural characteristics are summarized in Table 5C-2. Typical pre- and post-calibration curves are shown in Figure 5C-2B. The instrument response is linear in the concentration range from 0.1 ppt (the threshold of detection) to 150 ppt.

Of the twenty-five sensors deployed, pre- and post-calibration gains were within 6 percent for 19 sensors, between 6 and 50 percent for 3, and greater than 50 percent for the remaining 3. The major cause of calibration changes is thought to be physical damage from moving gravel and cobbles in the surf zone. Cobbles with diameters in excess of 20 cm were exposed by beach erosion in response to the major storm that occurred during the experiment and collisions with some of the sensors are thought to have caused cracks in active sections of the photo-voltaic detectors.

Table 5C-2. Textural Characteristics
of Leadbetter Beach Sand

Median	=	0.227 mm
Mean	=	0.236 mm
Sorting	=	0.40
Phi Skewncss	=	− 0.13
2nd Phi Skewness	=	0.11
Phi Kurtosis	=	0.67

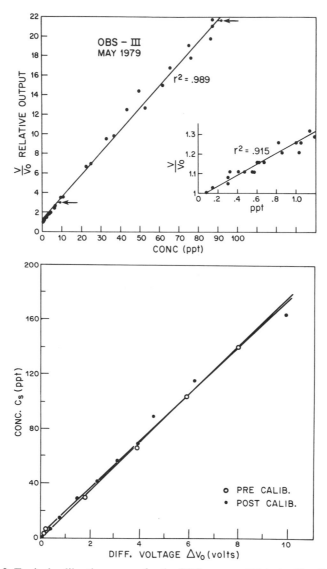

Figure 5C-2. Typical calibration curves for the OBS sensor. (A) test calibration with the sensor output normalized to the output from the water in the test tank prior to the addition of sand ($D_{50} = 0.165$ mm); the two test values indicated by arrows show the influence of air bubbles on the sensor calibration. (B) sensor calibrations on Leadbetter Beach sand collected before and after the field experiment.

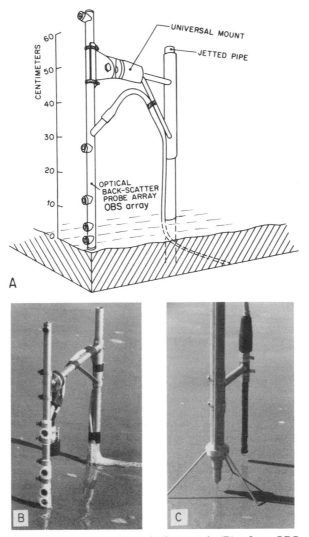

Figure 5C-3. Schematic drawing (A) and photograph (B) of an OBS sensor array installed on the beach face. The white covers around the lowest 4 sensors are protective devices. (C) is a MOBS sensor array installed on the beach face.

Table 5C-3. Sensor Array Status and (OBS and MOBS) Location. X indicates operation sensor array. Locations are referenced to control marker NSTS "M" in meters -- X coord./Y coord.

DATE	AO	loc.	BO	loc.	CO	loc.	AMO	loc	BMO	loc.
07 Feb	X	72.0/−6.0	X	78.0/−6.0	--		--			
08 Feb	X	72.0/−6.0	X	78.0/−6.0	--		X	60.7/−1.6	--	
09 Feb	X	72.0/−6.0	X	78.0/−6.0	--		X	63.7/−1.6	X	66.8/−1.6
13 Feb*	X	72.0/−6.0	X	78.0/−6.0	X	63.0/−6.0	--		--	
13 Feb	--		X	78.0/−6.0	X	63.0/−6.0	--		--	
14 Feb	--		X	78.0/−6.0	X	63.0/−6.0	X	58.1/−1.5	X	55.0/−1.6
14 Feb*	--		X	78.0/−6.0	X	63.0/−6.0	--		--	
15 Feb	--		X	78.0/−6.0	X	63.0/−6.0	X	58.2/−1.6	X	54.9/−1.6
15 Feb*	--		X	78.0/−6.0	X	63.0/−6.0	--		--	
16 Feb	--		X	78.0/−6.0	X	63.0/−6.0	--		--	
16 Feb*	--		X	78.0/−6.0	X	63.0/−6.0	--		--	
17 Feb	--		X	78.0/−6.0	X	63.0/−6.0	--		--	
18 to 21	--	--------	--	STORM--------	--	--------------	--	--	--	--------
22 Feb	X	63.4/−8.2	X	56.8/−8.3	--		--		X	18.8/−6.7
23 Feb	X	63.4/−8.2	X	56.8/−8.3	--		X	40.2/−6.6	X	23.7/−6.7
24 Feb	X	63.4/−8.2	X	56.8/−8.3	--		X	40.2/−6.6	X	23.7/−6.7

*Second run of day.

5C.3 Sampling Techniques

To meet the program objectives, vertical arrays of OBS sensors were placed on a transect across the surf zone. In all cases sensor arrays were cantilevered across the flow (upcoast) from mounting pipes jetted into the seabed. Each vertical array consisted of five sensors. The arrays were constructed in two sizes (OBS and MOBS) for use in the outer and inner surf zone, respectively. The sensors in an OBS array are mounted in a steel tube with a logarithmic spacing between sensors (Figures 5C-3A, B). Total length of the array is 50 cm and the spacings between sensors is 3, 6, 12, and 30 cm. This configuration was used in the outer surf zone where water depths were greater than 0.5 m. PVC plastic shields were fitted on the lower sensors at Leadbetter Beach to minimize damage from moving cobbles in the breaker zone.

MOBS (mini-optical backscatter sensor) arrays were used in the inner surf zone where water depths were generally less than 0.5 m. Each MOBS consists of five OBS sensors mounted on a structural rod that is encapsulated in plastic (Figure 5C-3C). The total length of the MOBS array is 16 cm and the spacings between sensors is 3, 3, 5, and 5 cm.

Diver observations were used to determine how close to the seabed the OBS and MOBS arrays could be placed without causing appreciable disturbance to the sediment. During numerous observations, with waves as large as 90 cm, slight sediment scour was induced by the mounting pipes jetted into the seabed but the OBS and MOBS arrays cantilevered upcoast from these pipes could be placed as close as 3.0 cm without causing erosion.

The OBS and MOBS sensor arrays were deployed in a shore-normal transect in conjunction with current meters, pressure gages, and wave staffs used for the surf zone dynamics aspects of the NSTS. Sensor electronics were buried in the seabed beneath each OBS sensor array and connected to the shore laboratory with a multiconductor cable. Continuous output signals from each of the five OBS sensors in an array were transmitted back to the shore laboratory.

In shallow water (< 0.5 m) the MOBS sensor arrays were attached to one leg of a tripod structure jetted into the beach face. Each tripod served as a stable platform on which a current meter and pressure gage or wave staff also were mounted. The MOBS sensor array electronics were strapped to a cross member of the tripod structure well above the sea surface and the multiconductor cable connecting the sensor electronics to the shore laboratory was led down one of the tripod legs, then buried.

All sensor output signals (OBS and MOBS arrays) were transmitted continuously to the shore station and subsequently scanned and recorded in the NSTS data system. Each sensor array was always located in proximity to a current meter and pressure sensor or wave staff. The specific locations of each sensor array relative to the main instrument network and a sampling schedule for the Santa Barbara experiment are shown in Table 5C-3.

5C.4 References

Brenninkmeyer, B. M., 1974, Mode and period of sand transport in the surf zone. *Proceedings*, Fourteenth Coastal Engineering Conference, June 24-28, 1974, Copenhagen, Denmark, American Society of Civil Engineers: 812-827.

———. 1976, In-situ measurements of rapidly fluctuating high sediment concentrations. *Marine Geology*, 20: 117-128.

Downing. J. P., 1978, Sediment transport measurement in the nearshore environment: a review of the state of the art. *Proceedings*, Nearshore Sediment Transport Study Workshop on Instrumentation for Nearshore Processes. California Sea Grant Publ. 62: 58-83.

Downing, J. P., R. W. Sternberg and C.R.B. Lister, 1981, New instrumentation for the investigation of sediment suspension processes in the shallow marine environment. *Marine Geology*, 42: 19-34.

Fairchild, J. C., 1973, Longshore transport of suspended sediment. *Proceedings*, Thirteenth Coastal Engineering Conference, July 10-14, 1972, Vancouver, B.C., American Society of Civil Engineers: 1069-1088.

Inman, D. L., J. A. Zampol, T. E. White, B. W. Waldorf, D. M. Hanes and K. A. Kastens, 1980, Field measurements of sand motion in the surf zone, *Proceedings*, Seventeenth Coastal Engineering Conference, March 23-28, 1980, Sydney, Australia, American Society of Civil Engineers: 1215-1234.

Kana, T. W., 1979, Suspended sediment in breaking waves. Coastal Research Division, Technical Report CRD-18, University of South Carolina, 153 pp.

Watts, Q. M., 1953, A study of sand movement at Lake Worth Inlet, Florida. Beach Erosion Board Technical Memo 42, 24 pp.

Chapter 6

TRANSPORT DETERMINATION BY TRACERS

A. Tracer Theory

Ole Secher Madsen

Ralph M. Parsons Laboratory for Water Resources and Hydrodynamics
Massachusetts Institute of Technology, Cambridge, Massachusetts

6A.1 Introduction

An important aspect of environmental engineering is the prediction of the behavior of a foreign substance introduced into the environment. In environmental fluid mechanics a knowledge of the environmental conditions (the transporting system) is generally assumed and the problem is that of predicting the behavior of an ensemble of contaminated (marked) fluid particles released into the transporting system. The fundamental idea behind the use of tracers for the determination of transport is essentially the inverse problem, i.e., one deliberately marks a finite number of particles within the transporting system with the expectation of gaining information about the characteristics of the transporting system itself by monitoring the behavior of the marked particles (the tracers).

A classical example of the use of tracer theory is the experiment in which a finite amount of dye is injected into a pipe flow. By monitoring the dye concentration at stations downstream of the injection point, e.g., measuring the time interval, Δt, between peak concentration at two stations, Δx apart, the average fluid velocity is found as $\bar{U} = \Delta x / \Delta t$ and the discharge is obtained from $Q = \bar{U} A$, where A is the pipe cross-sectional area. The basic assumption behind the use of tracer theory in the pipe flow experiment, and behind the use of tracer theory for the determination of discharges in general, is that advection (the downstream transport of tracers associated with the flow in the transport system) dominates diffusion (the spreading of the tracers due to small scale random motions, molecular or turbulent, in the transport system) and dispersion (the spreading of the tracers by the combined effect of diffusion and a non-uniform advective velocity across the flow area of the transport system). The pipe flow experiment is generally quite successful when its result is compared to a more convincing measurement of the discharge, e.g., by timing the discharge of fluid into a bucket at the end of the pipe.

There exists a direct analogy between the behavior of marked fluid particles in a turbulent pipe flow and the behavior of tracer particles in the longshore sediment transport system (the surf zone). The analog to the turbulent diffusion in the pipe is the stirring action of breaking waves in the surf zone while the advection and dispersion in the pipe have their counterparts in the effects of the varying wave intensity and longshore current on the longshore sediment transport. There are, however, some very important differences between the two transport systems. Only for the pipe flow system do we (1) know the cross-sectional area of the transport system, (2) have assurance that no tracer is lost, (3) have alternative methods for measuring the discharge thereby assessing the accuracy of the results and the validity of the assumptions underlying the application of tracer theory.

From the pipe flow analogy it follows that the use of tracer theory will provide, at best, information about gross features of the transport system (discharge, not the velocity distribution). The lack of alternative methods for determining the longshore sediment transport rate makes it imperative that one examines, in detail, the basic assumptions underlying the use of a particular tracer methodology and evaluates the extent to which these assumptions are violated for the longshore sediment transport system. In the following sections a conceptual model for the behavior of tracer particles in the surf zone is proposed and this model is used as the basis for the examination of various tracer methodologies with the emphasis on the assumptions and possible sources of errors associated with each methodology.

6A.2 Conceptual Model for Tracer Behavior in the Surf Zone

As a simple conceptual model of the longshore sediment transport system, one might view it as consisting of two layers. The upper layer is the transporting layer in which sediment is advected in the longshore direction at a velocity V_s. The lower layer of thickness Z_0 consists of sediment at rest. Due to the stirring action of the breaking waves and the natural tendency of the sediment in transport to settle out, there is an exchange of particles across the interface between the two layers. This exchange rate is expressed in terms of the probability per unit time per unit area of a particle in the lower layer making the transition to the upper layer and vice versa. This simple conceptual model based on a physical interpretation of the transport system turns out to yield identical results to the model of tracer behavior developed by Einstein (1936) and later used by Hubbel and Sayre (1964). Although this two-layer model, in principle, is different from the alternative model which views the transport system as the upper layer of the bed being slowly advected downstream, it can be shown that monitoring the tracer concentration in the lower, immobile layer will yield the correct transport rate. In fact, DeVries (1973) has demonstrated that tracer behavior in the upper layer of the bed, whether moving or not, can be described by the diffusion equation and, by proper choice of the parameters, can be made to yield tracer dispersal patterns nearly identical to those of Einstein's model. In view of the comprehensive literature available on diffusion and dispersion (e.g., Csanady, 1973; Fischer et al., 1979) we therefore adopt the diffusion analogy as our conceptual model for tracer behavior in the surf zone.

Based on the diffusion analogy we may identify the two mechanisms, diffusion and dispersion, which must be dominated by advection in order to obtain reliable results from a tracer experiment. Taking only the transport of tracers in the longshore direction, y, we may locally separate the instantaneous advective velocity into a mean value, V, over a few wave periods, and the temporal fluctuations caused by the action of individual breaking waves, V'. Performing the same separation of tracer concentration, c and c', the transport rate of tracers in the longshore direction is given by

$$Q_t = \int_{x_1}^{x_2} dx \int_{z_1}^{z_2} (V + V')(c + c')dz \qquad (6A\text{-}1)$$

where x is the shore-normal coordinate, $x = x_1$ being the breakline, $x = x_2$ the shoreline; while z_1 and z_2 denote the lower and upper vertical boundary of the transport system, respectively.

Locally time averaging the transport over a few wave periods, denoted by $\langle\ \rangle$, yields

$$Q_t = \int_{x_1}^{x_2} dx \int_{z_1}^{z_2} Vc\,dz + \int_{x_1}^{x_2} dx \int_{z_1}^{z_2} \langle V'c' \rangle dz \qquad (6A\text{-}2)$$

The second term on the right-hand side of (6A-2) represents the diffusive transport of tracers. In analogy with turbulent diffusion (Csanady, 1973) this transport may be expressed as a Fickian diffusion, i.e.,

$$\langle V'c' \rangle = -D_y c_{(y)} \qquad (6A\text{-}3)$$

in which D_y is the diffusion coefficient and subscripts in parentheses denote partial differentiation. From an actual tracer experiment one may obtain information about the order of magnitude of V, c, and $c_{(y)}$. To assess to which extent the effect of diffusion is negligible, mixing length arguments may be used to estimate the magnitude of D_y as the product of the near-bottom wave orbital excursion amplitude in the longshore direction, $a_0 \sin\alpha$, and the near-bottom wave orbital velocity, u_0.

In the assumed absence of diffusion (6A-2) reduces to the advective transport of tracers in the longshore direction. Separating the advective velocity as well as the tracer concentration into their cross-sectional averages, denoted by overbars, and their deviations therefrom, denoted by double primes, we obtain

$$Q_t = \overline{V}\ \overline{c}\ A + \int\limits_{x_1}^{x_2} dx \int\limits_{z_1}^{z_2} V''c''dz \tag{6A-4}$$

where $A = \int\limits_{x_1}^{x_2} dx \int\limits_{z_1}^{z_2} dz$ denotes the cross-sectional area of the transport system, and the last term represents the transport associated with nonuniform advective velocity and tracer concentration over the cross-section, i.e., dispersive transport. Whereas information about c'' relative to $\overline{c}$ may be obtained from a tracer experiment, the variability of the advective velocity, V'', is unknown. Thus it appears difficult to quantify the dispersive transport and evaluate to which extent it is small relative to the advective transport, $\overline{V}\ \overline{c}\ A$. Only if one has allowed sufficient time for the tracer to become completely mixed in the transport system, i.e., $c'' \ll \overline{c}$, may the dispersive transport be estimated, e.g., using the method discussed by Fischer et al. (1979) with an assumed variation of the advective velocity across the surf zone.

With the assumptions of negligible diffusive and dispersive transport of tracers, the principle of conservation of tracer applied to a shore-normal slice of the transport system yields the equation governing the tracer concentration

$$(\overline{c}\ A)_{(t)} + (\overline{V}\ \overline{c}\ A)_{(y)} =$$

$$\dot{r} + \left[c\ Z_0\ (x_{(t)} - \overline{U}) \right]_{x_1}^{x_2} + \left[c\ X_B\ z_{(t)} \right]_{z_1}^{z_2} \tag{6A-5}$$

where $\dot{r}$ denotes the rate of tracer injection per unit length of the transport system, which is assumed to be confined within the surf zone width, $X_B = x_2 - x_1$ and have a vertical extent $Z_0 = z_2 - z_1$. The last terms in (6A-5) represent the net gain of tracer through the shore-normal and horizontal boundaries, which for generality have been assumed functions of time to reflect the influence of tides and/or time varying incident wave conditions.

6A.3 Tracer Methodologies for the Determination of Longshore Sediment Transport Rates

In the following sub-sections various tracer methodologies for the determination of longshore sediment transport rates are examined. Each methodology is first described by briefly outlining the simplistic, physical reasoning behind its use. Following this introduction the methodology is derived formally from the adopted model equation, (6A-5), to bring out the nature of the assumptions underlying a particular methodology, thus identifying possible error sources, and to discuss possible ways in which to ascertain the validity of employing a methodology in its simplistic form.

An assumption common to all tracer methodologies, in addition to the dominance of advection over diffusion and dispersion, is the obvious one that the tracers do not change the characteristics of the transport system and that the tracers behave in the transport system as the native material whose transport rate is to be determined. For the longshore sediment transport system this translates into a requirement of a tracer grain size distribution equal to that of the transport system *only* if it is desired to treat the entire population of grain sizes as one. If one is willing to accept the additional effort involved in performing the analysis of a tracer experiment on the basis of different grain size classes, as were Duane and James (1980), any grain size distribution of the tracers may be used so long as the range of grain sizes in the transport system is covered by the tracers. In fact, using a grain size distribution of tracers different from that of the transport system offers certain possible advantages. First of all, errors associated with the detection of the grain sizes transported, diffused and dispersed the most rapidly (usually expected to be the fines) may be improved by using tracers which are abundant in these size classes. Secondly, this procedure may reveal the dependency of transport rate on grain size (as did the experiments of Duane and James, 1980). The second advantage of this choice of tracers may well be worth the additional effort since it will produce results which may be of a more general and transferable nature than the results obtained by using tracers manufactured from the local bed material where the tracer experiment is performed.

Since the Spatial Integration Method (SIM) has been used extensively in the pioneering work of Inman and co-workers (Komar and Inman, 1970) and the Corps of Engineers (Duane, 1970) this tracer methodology is examined first. The scrutiny with which this method is examined should not be construed as an attempt to "put down" this particular methodology. Rather, the in-depth examination of the SIM facilitates the subsequent examination of the Time Integration (TIM) and the Continuous Injection Methods (CIM), since many of the underlying assumptions appear to be common to all tracer methodologies.

6A.4 Spatial Integration Method (SIM)

The SIM is Lagrangian in nature in that the behavior in space and time of a cloud of tracers is monitored. At time $t = 0$ a known quantity of tracers, M, is introduced into the surf zone at $y = 0$, say. At later times, $t = t_n$, the centroid of the tracer cloud is determined from a suitable approximation to the exact expression

$$Y_{g,n} = \frac{\int_{y_1}^{y_2} y \, \bar{c} \, A \, dy}{\int_{y_1}^{y_2} \bar{c} \, A \, dy} \tag{6A-6}$$

where the limits of integration in the longshore direction, y_1 and y_2, are chosen such that all tracers are accounted for, i.e., $\bar{c} = 0$ for $y < y_1$ and $y > y_2$.

The centroid velocity, taken to represent the average particle velocity in the longshore direction, V, is obtained from

$$\bar{V} = Y_{g(t)} = \frac{Y_{g,n} - Y_{g,m}}{t_n - t_m} \tag{6A-7}$$

and the transport rate is found from

$$Q_s = \bar{V} A = \bar{V} X_B \bar{Z}_0 \tag{6A-8}$$

where $\bar{Z}_0$ is the average vertical dimension of the transport system (interpreted physically as the thickness of the moving layer or the depth of disturbance).

To derive (6A-7) from the adopted model equation, (6A-5) is multiplied by y and integrated from y_1 to y_2, where $\bar{c}$ is assumed negligibly small. The result of this manipulation is

$$\left[\int_{y_1}^{y_2} y \, \bar{c} \, A \, dy \right]_{(t)} - \int_{y_1}^{y_2} \bar{V} \, \bar{c} \, A \, dy =$$

$$\left[\int_{y_1}^{y_2} y \, c \, Z_0 \, (x_{(t)} - \bar{U}) \, dy \right]_{x_1}^{x_2} + \left[\int_{y_1}^{y_2} y \, c \, X_B Z_{(t)} dy \right]_{z_1}^{z_2} \tag{6A-9}$$

It is readily seen that this equation may be written in a form similar to (6A-7) if and only if $\bar{V}$ is independent of y, i.e., the transport system must be uniform in the longshore direction. Provided this assumption is satisfied (6A-9) may be written

$$\bar{V} = Y_{g(t)} + Y_g \frac{M_{(t)}}{M} + \frac{1}{M} \left[\int_{y_1}^{y_2} y \, c \, Z_0 \, (x_{(t)} - \bar{U}) \, dy \right]_{x_2}^{x_1}$$

$$+ \frac{1}{M} \left[\int_{y_1}^{y_2} y \, c \, X_B Z_{(t)} \right]_{z_2}^{z_1} \tag{6A-10}$$

where $M_{(t)}$ is the rate of increase of tracers in the transport system.

The net rate at which tracers are lost from the transport system, $-M_{(t)}$, is clearly given by the integrals in (6A-10) without the y-multiplier. Uniformity of the transport system in the longshore direction assures us that Z_0, $\bar{U}$ and $x_{(t)}$ are independent of y, and (6A-10) reduces to (6A-7) if c, the tracer concentration leaving or entering the transport system depends on y in the same manner as $\bar{c}$, or if $c = 0$ at all boundaries of the transport system. The former condition is met for tracers leaving the transport system within which tracers are completely mixed, but not for tracers entering the transport system. Imagine, for example, tracers to be placed in a shorenormal trench ($y = 0$) on the dry beach during low tide. As the tide rises tracers will enter the transport system, i.e., $M_{(t)} > 0$, while the contribution of entering tracers to the integrals in (6A-10) is zero. The latter condition that $c = 0$ at the boundaries indicates that complete mixing within the transport system has not been achieved thus making the basic assumption, underlying all tracer methodologies, of negligible dispersive transport, (6A-4), suspect.

From the preceding discussion of the influence of tracers leaving or entering the transport system it is evident that the use of the SIM requires (1) tracers to be introduced directly into the system at $t = 0$; (2) the centroid velocity to be determined from sampling within the transport system, i.e., not on the dry bed following high tide; and (3) that no tracers enter the transport system during the duration of an experiment. The last of these requirements translates, for the longshore sediment transport system, into a requirement of steady state conditions despite the fact that the SIM in principle does not require a steady state transport system.

The SIM offers the advantage that it automatically leads to a determination of the amount of tracers in the transport system M_n at time t_n so that conservation of tracers may be checked. In addition the SIM appears to have the advantage of giving the correct centroid velocity (6A-7) so long as concentrations of tracers are obtained from core-samples taken to a depth greater than the depth of the transport system. This advantage is, however, only apparent since the transport rate, the ultimate objective of the tracer experiment, is obtained from (6A-8) which will yield a value of Q_s directly proportional to the depth of the transport system, $\overline{Z}_0$. The SIM has the disadvantage of weighting small concentrations heavily since the tracer concentration is expected to be small at large distances from the injection point. Sampling errors associated with the determination of small tracer concentrations, the lack of complete mixing in the transport system far from the injection site with the associated possibility of dispersion dominating advection, the possibility of completely missing a size-fraction which was advected outside the sampling area, are some of the legitimate concerns one may have about the applicability of the SIM. The automatic ability to perform a tracer mass balance when using the SIM takes care of the last concern mentioned above. The ability to perform consecutive determinations of $Y_{g,n}$ and therefore determine several values of $\overline{V}$ does provide means of assessing the degree to which diffusion, dispersion, unsteadiness of the transport system, and sampling errors have influenced the result. Thus, if several estimates of $\overline{V}$ are obtained and show a relatively small variability one may be reasonably confident in the accuracy of the value of $\overline{V}$ obtained from the SIM. This leaves the all important quantity in the determination of the longshore sediment transport rate as the average vertical extent (average depth) of the transport system across the surf zone.

6A.5 Time Integration Method (TIM)

The TIM is basically the Eulerian counterpart to the Lagrangian SIM in that a known quantity of tracers, M, is released in the transport system at $y = 0$ and $t = 0$. Some distance downstream at $y = y_0$, the variation with time of tracer concentration is monitored. A simple argument of conservation of tracers then yields

$$M = \int_0^\infty \bar{V} \bar{c} A \, dt \tag{6A-11}$$

and assuming $\bar{V}$ and A to be constant with time at $y = y_0$ the rate of transport is

$$\bar{V} A = Q_s = \frac{M}{\int_0^\infty \bar{c} \, dt} \tag{6A-12}$$

The expression equivalent to (6A-12) may be formally derived from (6A-5) by first integrating from $y_1 < 0$ where $\bar{c} = 0$ at all times to y_0 and then integrating the resulting equation over time from $t = 0$ to ∞. In this manner one obtains

$$\int_0^\infty \left[\int_{y_1}^{y_0} \bar{c} A \, dy \right]_{(t)} dt + \int_0^\infty \left[\bar{V} \bar{c} A \right]_{y=y_0} dt =$$

$$-M + \int_0^\infty \left[\bar{V} \bar{c} A \right]_{y=y_0} dt = -\Delta M \tag{6A-13}$$

where ΔM is the quantity of tracers lost from the transport system between y_1 and the location of concentration measurements, $y = y_0$.

Equation (6A-13) is identical to (6A-12) only if $\bar{V} A$ is assumed to be constant for the duration of the experiment, i.e., steady conditions at $y = y_0$, and if no tracers are lost from the transport system. Provided these conditions are met the use of the TIM does not require the transport system to be steady and uniform between injection site and the location of measurements. The requirement of advection dominating diffusion and dispersion is for the TIM applicable at $y = y_0$ which translates into a requirement of the concentration measurements being performed sufficiently far downstream for the tracers to have become completely mixed in the transport system.

In an Eulerian method as the TIM one may actually account, at least approximately, for the effect of tracers lost from the transport system by performing a spatial integration of tracers left upstream of the measurement location at the end of an experiment, i.e., at $t = t_0$ when $\bar{c} = 0$ at $y = y_0$. Provided stationary conditions prevailed at $y = y_0$ during the experiment (6A-

13) yields

$$Q_s = \overline{V} A = \frac{M - \Delta M}{\int_0^{t_0} (\overline{c})_{y=y_0} \, dt} \tag{6A-14}$$

On the surface the TIM may seem superior to the SIM since (6A-12) or (6A-14) lead directly to the determination of the transport rate, Q_s. It must, however, be recalled that whereas the SIM determination of $\overline{V}$ was insensitive to the depth of the transport system the application of (6A-12) or (6A-14) depends in a crucial fashion on the absolute value of the concentration. Thus, if the concentration of tracers is determined from 10 cm deep core-samples this would not affect the determination of the advection velocity $\overline{V}$ for the SIM so long as the depth of the transport system, Z_0, was less than 10 cm, whereas the assumption of the depth of the transport system, for which $\overline{c}$ in (6A-12) or (6A-14) is to be determined, directly influences the predicted transport rate obtained in the TIM. The TIM and the SIM therefore depend in exactly the same manner on our ability to determine the actual depth of the transport system and they differ only in the assumptions underlying each of the two methodologies.

Just as was the case for the SIM, the TIM offers possibilities for assessing the validity of the assumptions behind the methodology. For the TIM this would be provided by performing tracer concentration measurements at several sections downstream of the injection point, i.e., at $y = y_0 + \Delta y_n$. Constancy of the predicted longshore sediment transport rate should provide confidence in the prediction.

In passing it is noted that the so-called Cloud Velocity Method (CVM) can be derived formally from (6A-5) by first integrating from $y_1 < 0$ to y_0 and then multiplying the resulting equation by t prior to integration from $t = 0$ to $t = \infty$. Without loss of tracers and introducing the additional assumption of a steady state transport system between $y = 0$ (the injection) and $y = y_0$ (the measurement) this results in the equation

$$\overline{V} = y_0 \frac{\int_0^\infty (\overline{c})_{y=y_0} \, dt}{\int_0^\infty t (\overline{c})_{y=y_0} \, dt} \tag{6A-15}$$

This methodology has many of the same features as the SIM. Since it is essentially the same as the TIM its reliance on the additional assumption of stationarity of the transport system for $y_1 < y < y_0$ and its high weighting of low concentrations (sampling errors) at large times makes this methodology less desirable than the TIM; in particular, since the measurements required for the CVM and TIM are identical.

6A.6 Continuous Injection Method (CIM)

This methodology is essentially the same as the TIM except the CIM continuously injects tracers into the transport system at a known rate $\dot{r}$ at $y = 0$. A simple application of the principle of conservation of tracers, assuming steady state to have been reached, yields that the rate at which tracers are introduced into the transport system equals the rate at which tracers leave the transport system, i.e.,

$$Q_s = \frac{\dot{r}}{\bar{c}_{eq}} \qquad\qquad (6A\text{-}16)$$

where $\bar{c}_{eq}$ is the steady state (equilibrium) concentration measured at a station $y = y_0$ downstream from the injection line.

Equation (6A-16) may be derived from the adopted model equation by assuming steady state, $\partial/\partial t = 0$, and integrating (6A-5) from $y = y_1 < 0$ where $\bar{c} = 0$ at all times to $y = y_0$, the location of concentration measurements. Denoting the rate at which tracers are lost from the transport system between y_1 and y_0 by $\Delta \dot{r}$ the resulting equation becomes

$$\bar{V} A \, \bar{c}_{eq} = \dot{r} - \Delta \dot{r} \qquad\qquad (6A\text{-}17)$$

The CIM has all the same underlying assumptions as does the TIM with the additional assumption of steady state transport for $y_1 < y < y_0$. The sensitivity to the way in which $\bar{c}_{eq}$ is determined, i.e., the assumed depth of the transport system, is the same as that of the TIM. The same comment may be made about the manner in which $\Delta \dot{r}$, the rate of loss of tracers, may be determined by spatial sampling and the ability to ascertain the accuracy of the results by performing measurements at several stations removed from the injection line by different distances, $y_0 + \Delta y_n$.

The CIM does appear to offer a distinct advantage over the SIM and CVM in that the concentration increases with time thus rendering this methodology much less sensitive to sampling errors associated with low concentrations.

6A.7 Summary and Conclusions

The preceding detailed examination of tracer methodologies has revealed the striking similarity of the assumptions underlying the use of tracers for the determination of longshore sediment transport rates. In summary these assumptions are:

(a1) The tracers must behave as the native material whose transport rate is to be determined,

(a2) Advection must dominate diffusion,

(a3) Advection must dominate dispersion,

(a4)No tracers can leave or enter the transport system,

(a5)The transport system must be steady and uniform.

The first of these assumptions is an obvious one. A way to ascertain its validity in a given tracer experiment is far less obvious. Aside from the similarity between the behavior of the native material and the tracers, one has to consider the manner in which the tracers are introduced into the transport system.

The validity of the second assumption is hard to prove, although one may attempt to do this by invoking the analogy between the longshore sediment transport system and the turbulent diffusion in pipe flows.

The third assumption may be restated as a requirement of complete mixing of the tracers in the transport system, i.e., if complete mixing has been achieved dispersion should be negligible or its effect at least be quantifiable. The statement of Komar (1969) that the depth to which tracers were found near the injection line was reasonably well defined while decreasing with distance from the injection line casts some doubt on the validity of this assumption.

The fourth and fifth assumptions are in direct conflict with the third assumption. While one assumption, (a3), requires a sufficient distance or time for complete mixing of tracers to take place, the assumptions (a4) and (a5) may be interpreted to mean that a tracer experiment must be carried out before unsteadiness (tides, changing incident wave conditions) becomes significant. Thus, one assumption imposes a lower limit while other assumptions impose an upper limit on the duration of tracer experiments.

Whether or not the longshore sediment transport system invalidates the assumptions behind the use of tracer methodologies for the determination of longshore sediment transport rates (and the extent to which it so does) is in this writer's opinion not yet resolved. The use of tracer theory does, however, appear (Greer and Madsen, 1978) to be the sole "sound" methodology available for the determination of longshore sediment transport rates on a short time scale.

6A.8 References

Csanady, G. T., 1973, Turbulent diffusion in the environment, *Reidel*, Dordecht, Holland, 248 p.

DeVries, M., 1973, Applicability of fluorescent tracers, *Proceedings*, Use of Tracers in Sediment Transport, International Atomic Agency, Wien.

Duane, D. B., 1970, Tracing sand movement in the littoral zone, U.S. Army Corps of Engineers, CERC, MP4-70.

Duane, D. B. and W. R. James, 1980, Littoral transport in the surf zone elucidated by Eulerian sedimant tracer experiment, *Journal of Sedimentary Petrology*, 50(3): 929-942.

Einstein, H. A., 1936, Der Geschiebetrieb als Wahrscheinlichkeitsproblem, Thesis, E.T.H., Zurich.

Fischer, H. B., E. J. List, R. C.Y. Koh, J. Imberger and N. H. Brooks, 1979, *Mixing in Inland and Coastal Waters*, Academic Press, 483 p.

Greer, M. N. and O. S. Madsen, 1978, Longshore sediment transport data: a review, *Proceedings*, Sixteenth Coastal Engineering Conference, August 27-September 3, 1978, Hamburg, Germany, American Society of Civil Engineers, 2: 1563-1576.

Hubbel, D. W. and W. W. Sayre, 1964, Sand transport studies with radioactive tracers, *Journal of Hydraulics Division*, ASCE, 90(HY3): 39.

Komar, P. D., 1969, The longshore transport of sand on beaches, Ph.D. thesis, University of California, San Diego, 142 pp.

Komar, P. D. and D. L. Inman, 1970, Longshore sand transport on beaches, *Journal of Geophysical Research*, 75: 5914-27.

Chapter 6

TRANSPORT DETERMINATION BY TRACERS

B. Application of Tracer Theory to NSTS Experiments

Thomas E. White and Douglas L. Inman
Center for Coastal Studies
Scripps Institution of Oceanography
University of California
La Jolla, California 92093

6B.1 Methods and Controls on the Use of Tracer Sand

The successful use of tracer sand to determine sediment transport rates relies on the two basic premises that the tracer behaves exactly in the same way as the native sand and that the motion of the tracer centroid can be determined. The centroid velocity provides an estimate of the transport velocity, which when multiplied by the thickness of the moving sand, yields the transport rate.

6B.1.1 Determination of Tracer Quality

Verification of the assumption that the tracer behaves as the native sand involves both testing the properties of the tracer and determining that its initial conditions at the time of injection match those of the native sand. Generally, the tests for physical properties of the tracer are size analyses, hydraulic tests, and tests for dye permanence.

To determine whether transport estimates at Torrey Pines and Santa Barbara were biased due to a change in grain-size distribution during dyeing, the size distributions of the native sand for each experiment and of the tracer sand were measured by standard $\Phi/2$ interval sieve analyses. The moments of the size distributions, as described by Inman (1952), are presented in Inman *et al.* (1980) for the beach at Torrey Pines and in Table 6B-1 for Santa Barbara. Two methods of computing sample moments were used: moments with geometric class intervals, utilizing information from all size intervals, and graphical analogies to moment measures (in parentheses), which utilize percentile points on the cumulative frequency curve. Sample locations are expressed in both absolute and surf zone normalized coordinates, and median and mean sizes are in both Φ units and microns. The three principal moments of median size, mean

Table 6B-1. Sand Grain Size-Distributions at Santa Barbara
(graphic moments are in parentheses)

SAMPLE	MEDIAN Φ	MEDIAN μ	MEAN Φ	MEAN μ	SORTING σ	SKEWNESS	2ND SKEWNESS	KURTOSIS
Natural Sand Collected for Dyeing (9 NOV 79)	2.13	229	2.12 (2.08)	231 (237)	.40 (.41)	.01 (−.13)	(.11)	3.31 (4.02)
5 FEB 80 Spatial Grab Sample								
1. x = 60 m	1.99	251	1.98	253	.44	−.12		3.94
x/X₆ = .30			(1.98)	(253)	(.44)	(−.02)	(−.06)	(4.92)
2. x = 70 m	2.23	213	2.19	219	.51	−.55		3.48
x/X₆ = .50			(2.20)	(217)	(.53)	(−.06)	(−.25)	(3.60)
3. x = 80 m	1.76	296	1.56	339	1.02	−1.73		7.39
x/X₆ = .70			(1.57)	(337)	(.90)	(−.21)	(−.74)	(5.82)
Red Dyed Sand (Used on 5 and 21 FEB 80)	2.05	242	2.03 (2.02)	246 (247)	.44 (.43)	−.27 (−.08)	(−.10)	4.19 (4.98)
21 FEB 80 Spatial Grab Sample								
1. x = 26 m	2.09	235	1.97	256	.76	−10.53		90.40
x/X₆ = .19			(2.02)	(247)	(.48)	(−.15)	(−.33)	(6.06)
2. x = 46 m	2.19	220	2.16	224	.47	−.60		5.10
x/X₆ = .37			(2.16)	(225)	(.48)	(−.06)	(−.12)	(3.78)
3. x = 66 m	2.18	221	2.15	226	.48	−.74		6.89
x/X₆ = .55			(2.15)	(226)	(.48)	(−.06)	(−.12)	(3.84)
6 FEB 80 Spatial Grab Sample								
1. x = 60 m	2.20	218	2.11	232	.63	−1.43		8.17
x/X₆ = .29			(2.13)	(228)	(.60)	(−.11)	(−.34)	(3.54)
2. x = 70 m	2.34	198	2.32	201	.48	−.34		2.82
x/X₆ = .57			(2.32)	(201)	(.51)	(−.04)	(−.17)	(2.82)
3. x = 80 m	2.00	250	1.87	273	.90	−3.40		19.54
x/X₆ = .86			(1.94)	(260)	(.73)	(−.08)	(−.35)	(4.02)
Green Dyed Sand (Used on 6 and 22 FEB 80)	2.10	233	2.08 (2.05)	237 (242)	.42 (.42)	−.09 (−.13)	(−.04)	3.83 (4.74)
22 FEB 80 Spatial Grab Sample								
1. x = 32 m	2.10	233	2.03	245	.63	−5.55		44.72
x/X₆ = .13			(2.05)	(242)	(.47)	(−.13)	(−.25)	(5.52)
2. x = 62 m	2.06	240	2.04	243	.53	−2.04		20.82
x/X₆ = .44			(2.04)	(243)	(.46)	(−.04)	(−.03)	(5.10)
3. x = 92 m	1.98	254	1.97	255	.54	−.91		9.10
x/X₆ = .76			(1.99)	(251)	(.49)	(.03)	(−.01)	(4.92)

size, and dispersion show unexpectedly good agreement between tracer and natural sand. The values of the higher moments are extremely sensitive measures of size distribution and are seen to be more strongly a function of computational method (sample *vs.* graphical moments) and of sample location than of whether the sand was dyed or undyed.

The tracer sand used at Torrey Pines and Santa Barbara was obtained from the beach and later dyed. This insured that the mineralogical content of the tracer was the same as the native sand. Furthermore, the hydraulic properties are observed to be the same for both dyed and undyed sand, if a wetting agent is used. Settling tubes were used to demonstrate the hydraulic similitude of the sand (Inman, 1953; Ingle, 1966, Figure 17). The sediment transport studies of Ingle (1966), Inman *et al.* (1968), Komar and Inman (1970), Inman *et al.* (1980), and the present NSTS experiments all employed the same sand dyeing process.

It is concluded from the size analyses that the size distributions of native and tracer sand did not differ significantly. This is important since size can affect sediment transport velocity. For example, (Murray, 1967) found that injected sand with larger coarse fractions than the native sand had a net offshore velocity (toward the breaker) for the coarse fractions. Provided that the tracer did not move outside of the sampling grid, this would not necessarily seriously affect the estimate of longshore transport velocity. Komar (1977) found that very coarse particles (1200 microns) moved much faster longshore (more than 100%) than sand of 600 micron median diameter. But he determined that the velocity difference between different size particles in the 100-225 micron range (roughly the size of Torrey Pines and Santa Barbara sand) was less than 10%. Chang and Wang (1978) computed variance in the estimate of the mean longshore tracer velocity due to mismatch with the natural sand. They performed four injections of both 300 and 400 micron median tracer at locations of both 300 and 400 micron median native sand, situated on the same beach, and found grain-size difference between injected and native sand to account for only 2.5% of the variance of the longshore transport velocity estimates.

In the experiments performed previously at El Moreno, Mexico, tracer sand from the Torrey Pines site (about 180 micron median size) was injected at El Moreno (about 600 micron median size). By observing the presence of tracer on the beach at low tide, it was concluded that the Torrey Pines tracer had travelled on the order of a kilometer in just a few hours. This indicated a transport velocity orders of magnitude larger than the usual bedload velocity, but comparable to suspended transport velocities. Such fine tracer would obviously not have been of use in a quantitative tracer experiment.

It may be generally concluded that a mismatch of tracer sand and native sand on the order of ten percent of the modal diameter will not greatly affect the quality of the experiment. On the other hand, use of tracer which is considerably coarser than native sand may result in the tracer moving outside of the sampling grid in the offshore direction. Use of tracer which is much too fine can result in most of the tracer travelling in suspension, which violates the assumption made in tracer experiments that most of the tracer is travelling as bedload and can be sampled with bedload samplers.

6B.1.2 Tracer Injection

Tracer theory used in determining longshore transport in the surf zone assumes that the tracer is incorporated as part of the moving bed. Initial conditions are assumed to be a line of tracer extending across the surf zone perpendicular to the beach orientation. The tracer is introduced as bedload by containing the tracer in plastic bags until it is placed on the sand bed.

Since the cross-shore series of point injections are assumed to form a line of tracer, the tracer should mix in the cross-shore direction over a distance equal to the injection spacing, before the tracer reaches the first longshore sampling station. Waves, whose principal velocity is directed normal to the shoreline, can be expected to induce cross-shore dispersal of tracer very quickly. In a surf zone with small currents and 36 cm wave heights, Murray (1967, Figure 2) found cross-shore mixing to be proportional to about 1.5 times the downstream distance, 15 minutes after injection.

Accordingly, we computed the quantity 1.5 x (distance between injection and temporal sampling lines)/(cross-shore injecting spacing) for the four Santa Barbara experiments. This quantity was 9.0, 4.5, 9.0 and 2.2 for 5, 6, 21, and 22 February. All of these numbers exceed one, implying complete cross-shore mixing 15 minutes after injection. Only the very first set of temporal samples was obtained earlier than this.

6B.1.3 Determining Tracer Concentration

Preparation and injection of the tracer has been discussed. An additional control on the quality of tracer data incurs from an examination of the laboratory methods for determining tracer concentration of the bed samples from the surf zone.

Any tracer counting method, whether manual or by machine (photoelectric, radiation measurement, etc.), makes assumptions as to how the measurement relates to mass concentration of tracer. Photoelectric measurements assume tracer mass concentration to be proportional to tracer surface-area concentration, which is what is measured. This assumption cannot in general be considered a good assumption. Spheres for instance depend on diameter cubed for mass and squared for area. A tedious and subjective measure of grain sphericities would be necessary to measure the error in this assumption.

We determined tracer concentrations in our experiments by visually counting the number of dyed grains per gram in a sample (Waldorf et al., 1981 and Chapter 5A of this volume). This is not a measure of surface-area concentration since each grain is either counted or not counted, according to dye presence, regardless of size. This method will give more weight to fine size fractions, which have larger numbers of grains per unit mass. However, a visual cutoff excluding grains less than about 3-1/2 Φ (88 microns) was observed in counting, so error due to size was minimal. The counts of the different operators who performed the counting were compared over a series of widely varying tracer concentrations to determine operator bias. The standard deviation of the counts averaged over all operators was found to be 2.3% of the mean for the red tracer (with a range over all operators of 1.8 to 3.1%) and 6.0% of the mean for the green tracer (with a range of 4.7 to 8.7%). (These values are listed in Table 5A-1.)

It is important that the sand tagging not be affected by time or abrasion. Control samples of tracer were kept during the entire process, from dyeing to end of counting. These control samples did not change tracer concentration with time. Other samples were counted and then placed in a tumbler, to simulate surf zone abrasion. At various times the samples were removed and recounted. The concentrations were only weakly dependent on tumbling time up to 24 hours (Figure 6B-1).

6B.2 Monitoring Tracer Motion with a Spatial Grid

Two independent methods referred to as spatial and temporal, were used to monitor the tracer grain distribution following injection (Figure 6B-2). The first method employed traditional grab samples to obtain a rapid, synoptic, spatial view of the tracer distribution. The second is described in a later section.

Spatial sampling of the sand bed is the more traditional method of monitoring the time-averaged motion of the tracer centroid [Crickmore, 1967 (river transport); Inman, Komar, and Bowen, 1968; Komar and Inman, 1970; Chang and Wang, 1978; Kraus, Farinato, and Horikawa, 1981].

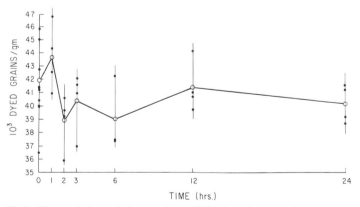

Figure 6B-1. The variation of the number of dyed grains as a function of time in a tumbler. Each solid dot represents one count of a subsample. The open dots connected by the line represent the average of all subsamples.

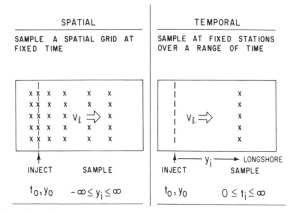

Figure 6B-2. Spatial and temporal grids for monitoring sand-tracer motion (after Inman *et al.*, 1980).

In the NSTS experiments, fluorescently dyed tracer sand was injected on a line across the surf zone. A line source was approximated by a series of closely spaced (4-10% of the surf zone width) point injections. Synoptic spatial sampling was performed on a grid extending across the surf zone and extending longshore on both sides of injection. Grab samplers, designed to sample the upper 2 cm of the bed and yield one concentration per sample, were used to obtain spatial tracer distributions. Core samplers, designed to sample the vertical distribution of tracer (extending 10-15 cm into the bed), were used in the temporal sampling and on one longshore line in the spatial sampling. The design and application of the instruments and laboratory procedures for obtaining tracer concentrations are described in Waldorf *et al.* (1981) and in Chapter 5A of this monograph. An example of the resulting picture of tracer concentration in the surf zone obtained from grab sample counts is given in Figure 6B-3.

6B.2.1 Longshore Sand Velocity

The longshore transport velocity of the average sand grain is determined by measuring the distance travelled by the sand-tracer centroid in a known period of time. Individual grains may travel in different transport modes (i.e., sheet flow, saltation, etc.) and at different velocities according to size fraction (Murray, 1967; Komar, 1977). Nevertheless, the mean transport velocity must be that given by the tracer centroid, provided that a representative tracer recovery is obtained, as discussed later.

Once the two-dimensional distribution is sampled (in longshore distance, y, and time since injection, t), the digital equation for estimating the centroid location of a two-dimensional density distribution is then applied

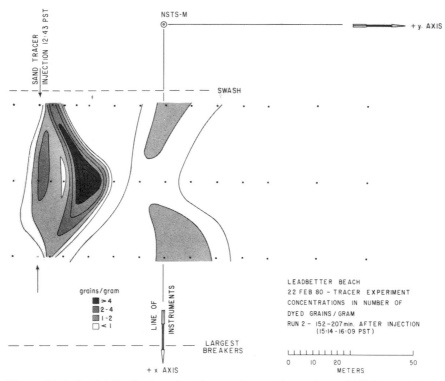

Figure 6B-3. Spatial distribution of sand-tracer from grab samples on 22 February 1980.

$$V(x) = \frac{\sum\limits_{y,t} N(x,y,t) \cdot \frac{y}{t}}{\sum\limits_{y,t} N(x,y,t)} \tag{6B-1}$$

where V is the centroid velocity, N is the sample tracer concentration (vertically integrated in the case of core samples), x is the offshore distance, and the sum is over all y for the spatial grid. This velocity is an average over the time period between tracer injection and end of sampling, and in the case of spatial sampling, over the longshore grid, but applies at one point in x. If an average velocity for the entire surf zone width is desired, it can be obtained from a weighting of the local velocities by the width of their representative sections of the surf zone

$$\overline{V} = \frac{\sum\limits_{x} V(x) \cdot \Delta x(x)}{\sum\limits_{x} \Delta x(x)} \tag{6B-2}$$

The proper method for computing the velocity of the centroid is formalized in
(6B-1). This method differs from previous computations (Komar, 1969; Inman
et al., 1980; Kraus, Farinato, and Horikawa, 1981) in which the summation in
(6B-1) was also simultaneously performed over *x*. The latter method is valid
only if $V(x)$ is a constant and the cross-shore regions represented by the
samples, $\Delta x(x)$, are also not functions of *x*.

The question of cross-shore tracer motion biasing longshore velocity
estimates was also examined as it is potentially a problem for both methods of
estimating longshore velocity. However, only two of the eight Santa Barbara
spatial samplings were found to exhibit significant (>0.1 cm/sec) cross-shore
motion of tracer, and these were related to the presence of beach cusps and the
location of rip currents.

6B.2.2 Transport Rates

The rate of sand transport is not simply the velocity of the sand, but the mass
flux of a layer extending from swash to breakers and with a thickness equal to
the time-averaged thickness of the moving bed. This bedload thickness is
obtained from core samples, most of which were part of the temporal grid. The
concept of bedload thickness will be examined later.

The local longshore bed and near-bed transport rate is computed by
multiplying the bedload velocity, $V(x)$, by the bedload thickness, $Z_o(x)$,

$$i_b(x) = g(\rho_s - \rho) N_o \cdot V(x) \cdot Z_o(x) \tag{6B-3}$$

where $i_b(x)$ is the local immersed-weight longshore bed and near-bed transport
rate, ρ_s the sand density, ρ the fluid density, and N_o the at-rest solids
concentration. The total longshore transport for the surf zone is obtained by
summing (6B-3) across the surf zone

$$I_b = \sum_x i_b(x) \cdot \Delta x(x) \tag{6B-4}$$

where both the local transport, i_b, and the cross-shore region represented by it,
Δx, are functions of cross-shore distance *x*. Note that (6B-3) and (6B-4) allow
variation in V, Z_o, and Δx across the surf zone and do not assume that one or
more of these variables is constant, as was done in previous studies (Komar,
1969; Komar and Inman, 1970; Inman *et al.*, 1980).

6B.2.3 Tracer Recovery

In order to estimate transport velocities by following the true tracer centroid,
the motion of most of the mass of tracer must be measured. This is tested by
estimating the amount of injected tracer that has been recovered and comparing
this with the amount injected. The procedure is quite straightforward for a
spatial grid. The concentration of tracer in a particular sample is representative
of the concentration in the surrounding region. The regional concentrations are
then summed over the entire grid. A problem arises, however, when samplers

are used which may not penetrate to the depth of tracer burial in the bed. Grab samplers yield only a near-surface concentration (approximately the top 2 cm), and the depth to which this concentration must be extended to yield the same value as the total, vertically integrated concentration is unknown. This representative thickness for the tracer layer, Z_o, may be estimated by various methods. The bedload thickness estimates from the temporal cores may be used. They are averaged over long periods of time and quickly attained a steady-state value with low variance.

In Chapter 13 these various estimates of Z_o are used in the spatial summation equation for tracer mass recovered

$$M = \frac{N_o \rho_s}{F} \sum_x \sum_y N(x,y) \cdot \Delta x(x) \cdot \Delta y(y) \cdot Z_o(x) \tag{6B-5}$$

where M is the mass of injected material recovered, N the sample tracer concentration in grains per unit mass, F the dyed sand concentration of the injected material in grains per unit mass, Δx the representative region in the cross-shore direction, Δy the representative region in the longshore direction, and other symbols as defined previously.

6B.2.4 Stationarity and Longshore Uniformity

A spatial sampling grid has relatively few requirements for uniform tracer behavior in time, t, or space (longshore distance, y). We generally observed a lack of correlation of bedload thickness with t or y. Even if there were significant temporal variations in the tracer motion, this would not necessarily invalidate the transport estimates from the spatial grid. The time-averaged transport measured with the spatial grid may be tested in models using time-averaged forcing, provided that tracer recovery is sufficiently high and that the wave or current data are averaged over the period between injection and sampling.

Since transport depends both on thickness and velocity, significant small spatial scale variations in tracer concentration must not be allowed to bias the transport velocity estimates. This nonuniformity problem was determined not to have introduced bias into the transport estimates for three of the four Santa Barbara experiments because of the high tracer recovery rates. However, if very low tracer recovery rates are obtained in a tracer experiment, and tracer concentrations do not vary smoothly longshore, the principal causative mechanism is usually thought to be rip currents (Inman, Tait, and Nordstrom, 1971). During NSTS experiments, dye was placed in the water in order to determine the intensity and location of such currents (Figure 6B-4).

6B.3 Monitoring Tracer Motion with a Temporal Grid

An estimate of time-averaged bedload transport may also be determined from repeated sampling along a line spanning the surf zone. Sampling is

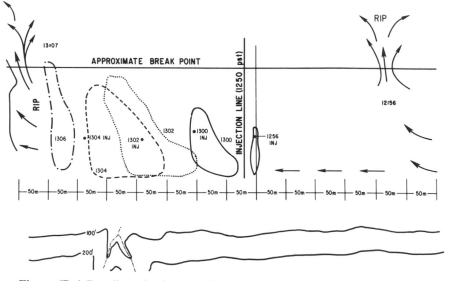

Figure 6B-4. Dye dispersion in the surf zone at Torrey Pines on 6 December 1978.

performed with coring devices along a fixed cross-shore range at successive intervals of time (Figure 6B-2). This method is referred to as temporal sampling and is analogous to sampling the dye plume in a river as it flows under a bridge (Inman *et al.*, 1980). Just as tracer concentrations vary in the longshore direction of a spatial grid in a dispersion-like manner (Figure 6B-3), similar curves are encountered in time on a temporal grid. The centroid motion is followed in time instead of in space.

6B.3.1 Transport Rates

Transport velocity, $V(x)$, is determined by the same equation (6B-1) as for the spatial grids, except that for temporal grids the summation is over time, t, instead of longshore distance, y. Local transport is determined by the product of transport velocity and bedload thickness (6B-3) and then integrated across the surf zone to obtain total transport (6B-4).

A temporal sampling grid has somewhat different requirements for uniformity than a spatial grid. Since the sampling is at one point in y (longshore distance), longshore variability in tracer distribution will not affect the transport estimate, which may be considered unbiased for that longshore location. Even though the estimate for transport is true for that location, the question arises as to whether that transport is typical of the entire beach. However, if the temporal sampling line is not located on some anomalous topographical feature (i.e., a cusp horn or valley) or in an area of anomalous current (i.e., a rip current), then there is no *a priori* reason to suspect that the longshore transport is any different than for the rest of the beach.

Temporal sampling grids must also be examined for biased estimates due to transport variation in time. Greer and Madsen (1978) conclude,

The stationarity requirement therefore does not apply to the time scale of the wave motion. Rather it should be interpreted to mean that wave conditions do not change during the experiment.

This conclusion can be reached in exact analogy to the argument made for the spatial grids, in which it was concluded that tracer recovery data and the use of a large number of samples implied that any error in spatial grid transport estimates was random, not biased. Likewise, high recovery of temporal grid tracer implies that transport pulsations result in random, not biased, error of transport estimates. The question of whether the general wave conditions changed during the experiment can be determined by observing successive wave spectra.

6B.3.2 Bedload Thickness

Equation (6B-3) requires estimates of the time-averaged thickness of the moving bed, determined after the bed has come to rest (since N_o in (6B-3) is an at-rest solids density). Estimates of this thickness have progressed from simply observing the depth of penetration of tracer (King, 1951; Inman and Chamberlain, 1959; Komar, 1969) to objective semi-empirical estimators. Crickmore (1967) first applied an objective estimator for this thickness. His estimator gave realistic results only for vertical concentration profiles in which there is no increase of concentration with depth in the bed. We will define a Crickmore profile as modifying the concentrations in those horizontal slices of bed core samples, such that a given slice would have a concentration no smaller than the layer immediately below it. The following estimator of bedload thickness (Crickmore, 1967) may then be applied

$$Z_o = \frac{\sum\limits_z N \cdot \Delta Z}{N_{\max}} \qquad (6B-6)$$

The summation is in the vertical, Δz is the vertical thickness of the horizontal slice, and $N_{\max}$ is the maximum tracer concentration in the core. Although Crickmore applied this method to transport in rivers, Gaughan (1978) later used it in surf zone studies and found the desired result of relative uniformity of Z_o in time and space. The standard deviation was equal to 42% of the mean in his fall/winter studies and 106% of the mean in his spring/summer studies. Equation (6B-6) was applied in the four Santa Barbara and two Torrey Pines experiments and confirmed Crickmore's observation that this equation yielded realistic results only if applied to the Crickmore profile of concentration instead of to the original profile. (When applied to the original profile, (6B-6) often yielded values for Z_o far less than the observed thickness of the preponderance of tracer.) Equation (6B-6) was also modified by substituting the average concentration in the core for N_{max} in the denominator, but was found to often yield values for Z_o greater than the depth of penetration of any tracer.

It is found that estimates based on the depth of maximum penetration of the tracer (Komar, 1969; Komar and Inman, 1970) overestimate Z_o. When a core tube is pressed into a sand bed, some tracer can be carried down the sides of the tube and later be counted at a greater depth than its native depth. Since this penetration occurs only at the core's outer edge (e.g., Inman and Chamberlain, 1959, Figure 5), this problem has been nearly eliminated by removing the outer layer of core samples before determining tracer concentration. Sampling experiments have confirmed that a sufficiently thick outer layer has been removed from the Torrey Pines and Santa Barbara cores. However, it is possible that a few dyed sand grains could still be present at greater than native depths. Once the number of these penetrating grains has been reduced to a small percent of the total number of dyed grains in the core, their effect will be minimized by applying any of the Z_o estimators, except the maximum-penetration estimator. If the maximum-penetration estimator of Z_o were used, these grains would completely determine Z_o. Therefore an estimator was applied which equated Z_o to the maximum penetration of a concentration of 1.0 dyed grains per gram of sand (Inman et al., 1980), in an attempt to eliminate this problem. In the analysis of the NSTS data, we compared the behavior of a number of different estimators, including the maximum-penetration estimator (Chapter 13, Tables 13-2 through 13-4).

Kraus, Farinato, and Horikawa (1981) applied an estimator which set Z_o equal to a depth of penetration of a certain percentage of the total amount of tracer found in the core. They plotted average Z_o from several cores versus the percent cutoff used to estimate it. The curve was found to depart from linearity between 60 and 90% cutoffs. The estimator they selected was the 80% cutoff. In order to observe and compare the behavior of this type of estimator with other methods, we also computed 80 and 90% tracer cutoff estimates of Z_o as listed in Tables 13-2 through 13-4.

Another objective estimate of Z_o was used by Inman et al. (1980). This estimator was based on the concept that a completely uniform-with-depth distribution of tracer, which abruptly decays to zero at a certain depth, could be judged to have a bedload thickness equal to that decay depth. This estimator yields unbiased results for a completely uniform vertical tracer distribution. This estimator is expressed as

$$Z_o = 2\frac{\sum_z N \cdot z}{\sum_z N} \tag{6B-7}$$

where the sum is taken vertically over the entire core, and z is the depth of the midpoint of each core slice. However, (6B-7) exhibits extremely aberrant behavior in the case of a buried profile. (As used here, buried means that the concentration profile decays to zero with depth and then increases again.) For example, consider the buried concentration profile $N(z) = N_1 \cdot \delta(z - d)$, where δ is the Kronecker delta function and N_1 is any nonzero concentration. This $N(z)$ function has the value N_1 at $z = d$ and zero everywhere else. Equation (6B-7) applied to such a profile yields $Z_o = 2d$, an obviously unrealistic answer. In order to avoid this problem, the concentration profile was changed to a Crickmore profile, and then (6B-7) was applied. Of course, with this method (6B-7) will yield the same answer for both uniform and buried profiles.

In general, we found that four of the transport thickness estimators described here gave comparable results. Equations (6B-6) and (6B-7) applied to the Crickmore profile, the 1.0 grain/gram cutoff estimator, and the 80% cutoff estimator generally yielded transport thickness values within 50% of each other. The other four estimators gave appreciably larger estimates.

6B.3.3 Tracer Recovery

Estimates of tracer recovery in the cross-shore temporal sampling line may be made from arguments of mass flux. The total flux of tracer through this line during the sampling period is

$$M = \frac{1}{F} \; \frac{4}{\pi D^2} \sum_t \sum_x N(x,t) \cdot \frac{y}{t} \cdot \Delta x(x) \cdot \Delta t(t) \tag{6B-8}$$

where F is the dyed sand concentration of the injected material in grains per unit mass, D is the diameter of the core sample, and N is the total tracer concentration of the sample in units of grains. Thus $4N/\pi D^2$ is concentration per unit area. The ratio y/t is tracer velocity (where y is distance from injection and t is time since injection), and Δx and Δt are the representative regions in the cross-shore direction and in time. Tracer recoveries for the NSTS experiments are listed in Table 13-2.

6B.4 References

Chang, T. T. and Y. W. Wang, 1978, Field verification of sediment transport model, *Proceedings*, 26th Annual Hydraulics Division Specialty Conference: Verification of Mathematical and Physical Models in Hydraulic Engineering, American Society of Civil Engineers, New York: 737-744.

Crickmore, M. J., 1967, Measurement of sand transport in rivers with special reference to tracer methods, *Sedimentology*, 8: 175-228.

Gaughan, M. K., 1978, Depth of disturbance of sand in surf zones, *Proceedings*, Sixteenth Coastal Engineering Conference, August 27-September 3, 1978, Hamburg, Germany, American Society of Civil Engineers, New York: 1513-1530.

Greer, M. N. and O. S. Madsen, 1978, Longshore sediment transport data: a review, *Proceedings*, Sixteenth Coastal Engineering Conference, August 27-September 3, 1978, Hamburg, Germany, American Society of Civil Engineers, New York: 1563-1576.

Ingle, J. C., Jr., 1966, The movement of beach sand, *Developments in Sedimentology*, 5, 221 pp.

Inman, D. L., 1952, Measures for describing the size distribution of sediments, *Journal of Sedimentary Petrology*, 22: 125-145.

_____. 1953, Areal and seasonal variations in beach and nearshore sediments at La Jolla, California, *Beach Erosion Board*, Corps of Engineers, Technical Memo 39, 134 pp.

Inman, D. L. and T. K. Chamberlain, 1959, Tracing beach sand movement with irradiated quartz, *Journal of Geophysical Research*, 64: 41-47.

Inman, D. L., P. D. Komar, and A. J. Bowen, 1968, Longshore transport of sand, *Proceedings*, Eleventh Conference on Coastal Engineering, London, England, American Society of Civil Engineers, New York: 298-306.

Inman, D. L., R. J. Tait, and C. E. Nordstrom, 1971, Mixing in the surf zone, *Journal of Geophysical Research*, 76: 3493-3514.

Inman, D. L., J. A. Zampol, T. E. White, D. M. Hanes, B. W. Waldorf, and K. A. Kastens, 1980, Field measurements of sand motion in the surf zone, *Proceedings*, Seventeenth Coastal Engineering Conference, March 23-28, 1980, Sydney, Australia, American Society of Civil Engineers, New York: 1215-1234.

King, C. A. M., 1951, Depth of disturbance of sand on sea beaches by waves, *Journal of Sedimentary Petrology*, 21: 131-140.

Komar, P. D., 1969, The longshore transport of sand on beaches, Ph.D. thesis, University of California, San Diego, 142 pp.

_____. 1977, Beach sand transport: distribution and total drift. *Journal of the Waterway, Port, Coastal and Ocean Division*, American Society of Civil Engineers, 103(WW2): 225-239.

Komar, P. D. and D. L. Inman, 1970, Longshore sand transport on beaches. *Journal of Geophysical Research*, 75: 5914-5927.

Kraus, N. C., R. S. Farinato, and K. Horikawa, 1981, Field experiments on longshore sand transport in the surf zone, *Coastal Engineering in Japan*, 24: 171-194.

Murray, S. P., 1967, Control of grain-dispersion by particle size and wave state, *Journal of Geology*: 612-634.

Waldorf, B. W., J. A. Zampol, R. E. Flick, and D. L. Inman, 1981, Sediment transport studies in the nearshore environment, *Report on Data from the Nearshore Sediment Transport Study Experiment*, C. G. Gable, ed., Institute of Marine Resources, Ref. No. 80-5: 102-120.

Chapter 7

DATA RECORDING

A. NSTS Data System

Robert L. Lowe
Scripps Institution of Oceanography
Center for Coastal Studies

The Scripps data acquisition system (SAS), as described in Lowe *et al.* (1973), was used in the first two field experiments. Some important modifications were made to adapt that system to the specific requirements of the Santa Barbara experiment. The basic system consists of six pulse code modulated (PCM) encoders connected via very high frequency (VHF) radio-telemetry links or coaxial cable to a receiving/recording station where the signals were recorded with a time code on a tape deck. Because of an increase in data channels at Santa Barbara, four of the six encoders were expanded from fifteen data channels to thirty-one. This gave the total system a capacity of 154 channels of data.

One of the fifteen channel encoders was dedicated to acquiring the data from the two slope arrays (Task 4F) located offshore and cabled back to the beach. It sampled each channel at thirty-two samples per second. This encoder was mounted in an environmental box on shore and transmitted to the receiving/recording station via its own VHF radio link. The other five encoders were housed in a mobile instrumentation van located on the beach. This arrangement did away with the need for offshore encoders and transmitters as at Torrey Pines, and thus increased reliability while decreasing the need for offshore maintenance. Individual sensor cables were fed into a hole in the van and then connected to data channel inputs and power supply outputs. Four of these encoders were for general sensor use and were of the thirty-one channel type. The data channels were sampled at thirty-two samples per second controlled by a common clock to prevent time differences between encoders. The fifth encoder in the van was run in one of three different modes. In normal mode it ran as the other four did with a common clock at sixty-four samples per second but with only fifteen channels of data. In high speed mode it ran by its own internal clock at one hundred twenty-eight samples per second for use with the Acoustic Bedload Sensor, described in Chapter 5B, which required a higher sample rate. In the third case, the encoder could be removed from the van and powered independently and its data transmitted through its own antenna to the receiving/recording station. This was done for the suspended sediment sensors described in Chapter 5C, to provide electrical isolation between these sensors and the rest of the system.

In each encoder, the data signals are multiplexed, digitized and sent out serially in bi-phase format. The encoder inserts a unique frame synchronization word, both to identify individual encoders and to maintain the relative order of the multiplexed signals. This word is essential in the playback mode. During normal operation of this system, all five encoders sent their data streams via individual 164 meter lengths of 50 ohm coaxial cable to the receiving/recording station housed in a trailer located near the beach. Coaxial cable, rather than a radio link, was used to minimize signal dropouts due to local radio signals from the nearby marina. The overall signal quality at Santa Barbara was far superior to that at Torrey Pines.

The receiving/recording station was housed in a portable office trailer located in a parking lot approximately 150 meters from the beach van and main instrument range. The office trailer was used to simplify transportation, installation and removal logistics. This proved very useful when the trailer was threatened on 20 February by rising water and had to be moved quickly to safety.

The receiving/recording station consisted of the necessary VHF receivers and a patch panel to accept the signal cables. This was fed directly into the tape deck, a Honeywell Model 7600. Time code in IRIG-B format was also recorded. A PCM bit synchronizer, frame synchronizer, an eight-channel data selector digital to analog converter and an eight-track strip chart recorder were installed for immediate viewing of data. After locking into a particular encoder's signal, using its unique frame synchronizing pattern, the channel selector could extract up to eight data channels, and convert the digital signals to their original analog form (Figure 7A-1). The strip chart recorder could then be used to verify sensor operation in real time. All active sensors were examined daily and many instrument failures were identified. Data tapes were also shipped daily via courier service to the Shore Processes Lab at Scripps. Analog tapes were played back and quick-look programs were run on the Interdata 70 computer to check data quality and sensor performance. Malfunctioning instruments were thus identified and replaced.

After the experiment, preliminary analysis of these analog tapes was performed to reduce the total number of data points to a manageable level, to remove extraneous spikes from the data caused by transmission failures, and to process the Santa Barbara surf spider current meter data. Figure 7A-2 shows the data flow from the sensors to the final data tapes. All data except those from the acoustic bedload sediment measuring devices were recorded at 32 samples/sec and then block averaged to 4 samples/sec while transferring the data from the

Figure 7A-1. SAS data receiving system.

7-track analog tape to computer compatible multiplexed tapes. At the next stage, the multiplexed tapes are combined to produce raw demultiplexed tapes which contain all sensors for a given time interval in a single data file. At this stage, sensors treatments are different. The suspended sediment sensors described in Chapter 5 were simply block averaged to 2 samples/sec. The surf spider current meters, described in Chapter 4A, had their non-immersed and very near surface data points set equal to zero.

Surf spider wave staffs were linearized by application of nonlinear calibrations. These were then block averaged to 5 samples/sec. The remaining sensors, which constitute the bulk of the dynamics experiment, had spikes removed and then were low pass filtered and decimated to 2 samples/sec. The spike removal consisted of smoothing all data points which, considering adjacent data points, showed abnormally large sea surface slopes or fluid accelerations (for pressure sensors and current meters, respectively). Abnormally large slopes were defined in terms of the number of standard deviations away from the rms value of slope. Care was taken to not confuse

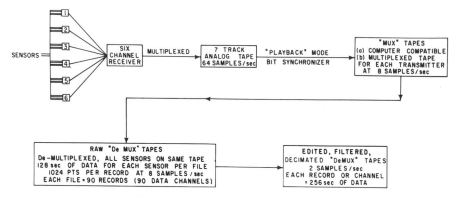

Figure 7A-2. Data processing flow chart.

steep bore faces with bad data points. The number of data points altered is generally very small. The 19-weight digital filter employed has zero phase shift, small side lobes, and a half amplitude transfer function at 1 Hz. Frequencies below 0.7 Hz are unaffected by filtering.

7A.1 References

Lowe, R. T., D. L. Inman, and B. M. Brush, 1973, Simultaneous data system for instrumenting the shelf, *Proceedings*, Thirteenth Coastal Engineering Conference, July 10-14, 1972, Vancouver, B.C., Canada, American Society of Civil Engineers, New York, 1: 95-112.

Chapter 7

DATA RECORDING

B. Wave Network System

Meredith H. Sessions and David Castel
Scripps Institution of Oceanography
Institute of Marine Resources

The sediment trapping experiments, with durations of six months to more than a year, required a completely different data recording system than the one employed for the one month intensive experiments. It must be capable of reliably logging data samples several times each day without requiring any resident personnel. In 1976, the Nearshore Research Group at SIO had completed the development of a system for monitoring wave climatology at distant locations (Figure 7B-1). This wave network, which is described in its initial form in Seymour and Sessions (1976), is funded jointly by the U. S. Army Corps of Engineers and the California Department of Boating and Waterways.

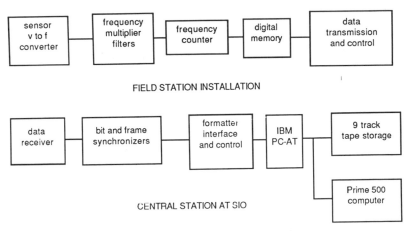

Figure 7B-1. Block diagram of wave monitoring system.

133

The system presently supports about 100 measuring instruments at approximately 25 locations from Hawaii to North Carolina. The slope arrays at Santa Barbara and at Rudee Inlet were added as regular stations in the network prior to the initiation of each of the two trap experiments.

Electrical signals sensed by the pressure transducers on the slope array are converted to a variable frequency proportional to the applied pressure. These signals in the range of one to three Kilohertz are transmitted via a specially designed 12 conductor armored underwater cable to a site on shore.

The shore station consists of a watertight housing containing the data conversion, data storage, control system and power and telephone interfaces. The frequency signals from as many as eight channels are simultaneously counted for a one second period into a digital counter. This digital 15 bit number at each channel is loaded into a digital buffer memory of usually 1024 words in length. The memory size is variable depending on sample length requirements up to 2048 words per channel. When the memory is filled it then continues to add new words at one per second expelling the oldest words at the top of the memory as each new word is added, resulting in the memory always containing the most recent 1024 words of data.

When a data sample from a shore station is desired the computer controlled central station initiates a telephone call to the shore station over normal dial-up telephone lines. The shore station responds by answering the telephone call request, and locking the most current 1024 words in the memory of each channel. All the data stored in the shore station are sent in a special 1200 Baud synchronous format to a digital data receiver connected to the control computer. A typical four channel slope array with 1024 samples per channel transmits all data in slightly over one minute and then disconnects itself from the telephone line. The 15 bit digital data words which represent approximately 10 meters water depth with a resolution of 1 mm are transferred raw to the control computer, a NOVA 1200, where station number, data, time, channel number and other header information are added prior to writing the information to magnetic tape files for raw data storage. Raw data are also simultaneously sent to a larger minicomputer for rapid analysis and display via disk files.

This system can support over 100 individual shore stations to be called in computer programmed sequence. Each station can consist of up to eight channels with a total maximum memory capacity of 8192 words of data per station. Individual channels may however contain up to 2048 samples. Each channel can be set at the site to sample data from one sample per second to one sample every eight seconds.

Data received by the central stations acquisition computer are combined with a leading header section which includes pertinent station information such as station identification, chronological information and system status flags. For archiving purposes header and data are sequentially merged and stored on the receiving computer magnetic tape.

The massive daily influx of data into the CDIP program coupled with the need for real-time analysis and timely data reporting dictate the need for an automated data quality assurance scheme. To meet these requirements an editing program was designed to examine the data following their reception into the processing computer.

The editor is programmed to objectively recognize certain classes of anomalies, correct some of the more obvious ones and reject others as bad data.

a) Spikes: Spikes are considered the most frequent cause for data rejection. They are most often caused by electronic noise of one source or another and in a sense are the easiest fault to detect. The editor calculates the standard deviation of the series and passes for a spike any value which is more than four times the standard deviation away from the mean. Spikes are replaced with the previous value in the time series; however, the occurrence of more than one percent of the number of data points as spikes will cause the time series to be rejected.

b) Flat spots: A series for which on more than N separate occasions M successive points were found to have an identical value is considered to be unacceptable.

c) Mean shift exceedence: Bad data are sometimes characterized by a significant mean shift between successive groupings of data points. The editor subdivides the series into N sub-series, calculates the mean for each subset and does an inner comparison between the means. A difference in the means greater than a predetermined threshold will cause the series to be rejected. The threshold level and length of sub-series depends on the type and station location.

d) Absence of zero crossing: A time series which does not cross the zero mean level for a specified number of points is considered unacceptable. This condition must be adjusted (dynamically) for a low energy, low period time series since an inflection point in the tidal range may mask an otherwise legitimate zero crossing occurrence.

e) Maximum and minimum wave height exceedence: An exceedence test is held to verify that recorded wave heights do not exceed an established expected maximum wave height for the station. As well, the editor verifies that the recorded wave height is greater than some minimum threshold.

f) Filtering: Where necessary, the editor filters the time series and removes the tidal component. This is done so that the energy in the tidal band does not leak into higher frequencies and mask the lower level energy bands found in the infragravity portion of the spectrum.

g) Inner comparisons: In the case of the four gage directional slope array, comparisons are performed between the individual gage variances. Differences greater than preset thresholds will cause the odd gage to be excluded from the directional analysis process. More than one deviating gage will reject the record.

Upon completion of the editing stage the data are Fourier transformed, where necessary modified by the extinction coefficients, and stored on temporary disk files. Grouped spectral and statistical information are permanently stored in a central computer data bank and become available to users with access rights and capabilities. Additionally, the data in their grouped spectral form, along with relevant statistical data are reported in real time to the National Weather Service for analysis and transmission over their maritime broadcasting frequencies. On a monthly basis data for the previous month are published in tabulated and graphical form and made available to selected libraries and interested users.

7B.1 References

Seymour, R. J. and M. H. Sessions, 1976, Regional network for coastal engineering data, *Proceedings*, Fifteenth Coastal Engineering Conference, July 11-17, 1976, Honolulu, Hawaii, American Society of Civil Engineers, New York, 1(5): 60-71.

Chapter 8

WIND WAVE TRANSFORMATION

Edward B. Thornton
Naval Postgraduate School

R. T. Guza
Scripps Institution of Oceanography
Center for Coastal Studies

8.1 Introduction

Wave-induced velocities are the primary driving force for littoral sand transport. For this reason, a major component of the NSTS program was to measure wave associated velocity and elevation fluctuations. A description of the shoaling wave transformation is a necessary ingredient in the development of any sediment transport model.

Torrey Pines proved to be an excellent field laboratory to study wave transformation, and much of this chapter deals with describing the results of that experiment. The beach profiles show no well-developed bar structure and are remarkably free from longshore topographic inhomogeneities. Measurements described here are from sensors located along the on-offshore transect from 10 m depth to the inner surf zone. A cross section of the instrument transect inside 3 m depth is shown in Figure 8-1. The wavelength has been drawn to scale (vertical scale distorted 1:20) showing that the dominant 14 second period swell is spatially well resolved. A primary difference between these and past experiments is the relatively close spacing of measurement locations compared to the dominant wavelength, and the relatively simple, almost plane, beach profile.

This chapter summarizes a number of papers written concerning the NSTS program, plus some additional findings. §8.2 shows that, in the wind wave frequency band, linear wave theory spectral transfer functions can be used locally (i.e., at a single horizontal location) to transform either pressure or current velocity to sea surface elevations (Guza and Thornton, 1980). §8.3 demonstrates that a linear, inviscid shoaling model (which uses measured spectra in ≈ 10 m depth as input data) predicts the total wind wave variance at depths between 3 m and 10 m, with errors of less than 20%. However,

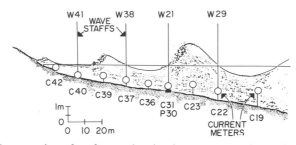

Figure 8-1. Cross-section of surf zone showing instrument spacing and elevations relative to measured waves on 20 November 1978 at Torrey Pines Beach, California. (Thornton and Guza, 1983)

nonlinear energy transfers lead to much larger errors at individual frequency bands. The energy conserving shoaling model is not applicable in the surf zone region, because of wave breaking (Guza and Thornton, 1980). Data presented in §8.4 shows that wind wave heights in the surf zone are depth limited (Thornton and Guza, 1982). §8.5 presents observations of wave phase speed. At the deepest stations, frequency dispersion as predicted by linear theory is observed. In shallower water, the observed phase speeds exhibit weak amplitude dispersion, and are nearly frequency non-dispersive (Thornton and Guza, 1982). The next two sections deal with observed statistical properties of nearshore wave fields. §8.6 examines various velocity moments. Very substantial fractions of the velocity variance are shown to be at surf beat periods ($T > 20$ sec). Surf beat motions must be filtered out of the raw data when considering properties of the wind wave field; for example, the inner surf zone saturation discussed in §8.4. §8.6 also compares the predictions of Gaussian and sinusoidal wave models with observations of various high order velocity moments. At Torrey Pines at least, the assumption of a Gaussian process is reasonable for many of the statistics, but is not valid for others, particularly the odd moments such as skewness (Guza and Thornton, 1985). §8.7 shows that the Rayleigh distribution provides a good description of the observed probability distribution of wave heights, not only offshore but everywhere including the surf zone (Thornton and Guza, 1983). Finally, in §8.8, elements from many of the preceding sections are used in the development of a transformation model for random wave heights, valid both inside and outside the surf zone (Thornton and Guza, 1983). This model is an important component of the longshore current model discussed in Chapter 16.

8.2 Local Transformation Using Linear Spectral Theory

Wave properties on a cross-shore transect at Torrey Pines Beach were measured using a few wave staffs, six pressure sensors and about ten current meters. Because of the very limited number of direct measurements of any single dynamic quantity, it was desirable to use measurements of one quantity to draw inferences about the others. For example, we later apply linear theory to pressure (P) and velocity (u) measurements to infer surface elevation (η) statistics. This use of linear wave theory was justified by intercomparison of pressure, velocity and sea surface elevation measurements at the same horizontal location (Guza and Thornton, 1980).

A number of studies have shown reasonable results using linear theory transformations well outside the surf zone [Bowden and White (1966); Simpson (1969) 4 to 6 m depths; Thornton and Krapohl (1974) 19 m depth; Cavaleri et al. (1978) 16 m depth; to name a few]. But it is not obvious that linear theory can adequately relate P, u, and η for waves that are so clearly nonlinear. The Korteweg-deVries equations show that $O(a/h)$ errors arise in using linear theory to relate P, u, and η, and the size of this term can be significant, about 0.2 in the surf zone. However, the present data show the local P, u, η agreement using linear theory to be reasonable everywhere.

Linear wave theory describes the local sea surface elevation $\eta(t)$ as the superposition of an infinite number of independent sinusoids

$$\eta(t) = \sum_{n=1}^{\infty} a_n \cos\left[\vec{k}_n \cdot \vec{x} + \omega_n t + \varepsilon_n\right] = \sum_{n=1}^{\infty} \eta_n, \qquad (8.2\text{-}1)$$

where a_n is the amplitude, $\vec{x}$ is the horizontal coordinate vector, t is the time, ε_n is the phase angle, $\vec{k}_n$ is a horizontal vector wave number and ω_n is the frequency related in linear theory to k_n by

$$\omega_n^2 = g\,|k_n|\tanh|k_n|h \qquad (8.2\text{-}2)$$

where g is the acceleration of gravity and h is the total local water depth. The equations for the horizontal velocity and pressure are

$$u(t) = \sum_{n=1}^{\infty} \frac{a_n \omega_n \cosh|k_n|(h+z)}{\sinh|k_n|h}\cos\left[\vec{k}_n \cdot \vec{x} + \omega_n t + \varepsilon_n\right]$$

$$= \sum_{n=1}^{\infty}\left[\frac{\omega_n \cosh|k_n|(h+z)}{\sinh|k_n|h}\right]\eta_n \qquad (8.2\text{-}3)$$

and

$$P(t) = -\rho g z + \sum_{n=1}^{\infty}\left[\frac{\cosh|k_n|(z+h)}{\cosh|k_n|h}\right]\rho g \eta_n \qquad (8.2\text{-}4)$$

respectively, where z is the depth of interest measured positively upward from mean water level.

Velocity and pressure spectra measured at depth are related to sea surface elevation, using linear theory, through the spectral transfer functions (8.2-3) and (8.2-4). The spectral densities of the two orthogonal components of horizontal velocity measured by the current meter were summed, yielding the horizontal velocity spectral density. Figure 8-2 compares surface elevation spectra predicted from horizontal velocity (E_u) with that calculated from pressure (E_p) or directly measured sea surface elevation (E_η). The ratios of total energy, $E_u/E_{p,\eta}$, are 0.91, well outside the surf zone $(h = 563$ cm), 0.7 near the mean breakpoint $(h = 176$ cm), and 1.08 in the inner surf zone $(h = 111$ cm), where E is summed over the frequency range 0.05-0.3 Hz. These correspond to errors in surface elevation standard deviation of 6, 17, and 5%. The larger error just outside the visually observed average breakpoint may be due to the very peaky shape of these waves just prior to breaking.

In general, the agreement is good between 0.05 and 0.3 Hz, and much of the observed difference can be attributed to calibration inaccuracies. For frequencies higher than about 0.3 Hz, instrument noise is being amplified through the large correction to sea surface elevation. Thus, there is a maximum frequency at which sea surface can be accurately inferred from near bottom measurements of other quantities.

It is shown in §8.7 that the significant wave height (H_s) can be accurately obtained from

$$H_s = 4\sigma_\eta \tag{8.2-5}$$

where σ_η^2 is the sea surface elevation variance obtained by summing the area under the (inferred) surface elevation spectrum. The ratio between the significant wave height using linear theory on the measured velocity (H_s^u) and that obtained from either depth-compensated pressure (H_s^p) or directly measured elevation (H_s^η) are plotted as a function of depth in Figure 8-3. Each data point represents a 34-min data run, with variances summed between 0.05 and 0.3 Hz. Sensor pairs not near the breakpoint usually show a discrepancy less than 10% both inside and outside the surf zone. Pairs near the breakpoint have as much as 20% disparity. The comparisons on a frequency band by frequency band basis are always about as accurate as the total variance comparisons, as in Figure 8-2. The data shown here are from eight different days with rather different incident wave conditions, varying from narrow-banded (November 20, Figure 8-2) to very broad banded (November 11, Figure 8-6).

It is concluded that a measurement of the P, u, or η spectrum in shallow water allows a reasonably good prediction of spectra of the other variables at that same location, for wind wave frequencies.

8.3 Wave Shoaling Using Linear Spectral Theory

Linear shoaling theory assuming no dissipation was used at Torrey Pines to predict spectra in the shoaling regime, given the input spectrum measured in 10 m depth (Guza and Thornton, 1980). Torrey Pines proved to be an exceptionally simple beach on which to test shoaling theory because not only can it be assumed the bottom contours are straight and parallel, but the waves can be assumed to approach normal to the beach.

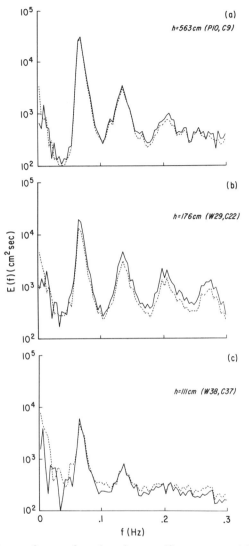

Figure 8-2. Comparisons of sea surface elevation ___ [from pressure (a), surface piercing staff (b, c)] and horizontal current ---------- at various total depths. 20 November 1978 at Torrey Pines Beach, California.

The hypothesis that the contours can be treated as straight and parallel was tested by running a version of Dobson's (1967) linear refraction program over a topographic grid using all survey lines. As a test case, waves of 0.067 Hz (corresponding to a typical swell peak in the data) and varying angles of

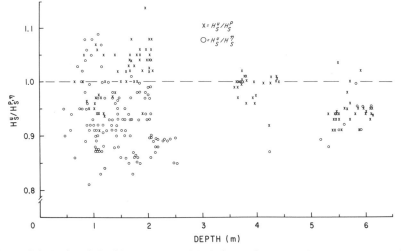

Figure 8-3. Ratio of significant wave height inferred from velocity measurements (H_s^{η}) at various water depths. Values less than 0.9 at 2 m depth are from a single sensor pair, suggesting calibration error. Torrey Pines Beach, California.

incidence in 10 m depth were refracted from 10 to 3 m depth on the sensor range line. The percent difference between the resulting amplifications of wave heights and theoretical values calculated by assuming normally incident waves on plane-parallel contours are shown (Figure 8-4) to be less than 5% for any directional band within the 15 degrees angular spread. The analysis was limited to ±15 degrees since waves in 10 m depth at Torrey Pines Beach usually do not have significant energy at angles larger than 15 degrees because of refraction further offshore and sheltering by offshore islands.

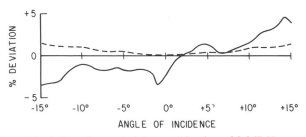

Figure 8-4. Percent deviation of wave height amplification of 0.067 Hz waves, and varying angles of incidence, shoaled from 10 m to 3 m depth over, _____ real topography, and ---------- plane parallel contours, from the amplification of normally incident waves on plane parallel contours. (Guza and Thornton, 1980)

Figure 8-4 also shows that on plane-parallel contours, 0.067 Hz waves with a 15 degree (or less) angle of incidence in 10 m differ by less than 1.2% in amplification at 3 m compared to normally incident waves. If the measured contours were perfectly plane-parallel, the solid and dashed curves in Figure 8-4 would overlap. Thus waves from the northern quadrant (positive angles) traverse essentially parallel contours, while those from the south exhibit a weak convergence. The deviation will be considerably smaller for smooth directional distributions and higher frequencies. Therefore in the following comparisons of energy spectra, the effects of directional distributions of energy are neglected; all waves are assumed to impinge normally onto parallel contours and shoaled over the measured bathymetry on the instrument transect.

On a more complex topography as at Leadbetter Beach, it is necessary to measure the directional spectrum offshore and individually refract each frequency-directional component to the desired location and integrate across all directions to calculate the shoaled energy spectrum.

Elevation energy spectra calculated from the deepest pressure sensor ($\approx$10 m depth) were linearly shoaled to more shoreward locations and compared with observed spectra (obtained from surface corrected pressure and velocity records). Figure 8-5 is typical of narrow-banded (in frequency) incident spectra, and Figure 8-6 of broadbanded cases. Harmonic amplification due to nonlinearities is evident in Figure 8-5. The linearly predicted spectrum underestimates the energy at harmonic frequencies and overestimates the energy at the primary frequency. The total variance is more accurately predicted than the energy content in a particular frequency band (Table 8-1).

This is in contrast to the previously discussed comparisons of locally measured P, u, and η, where the amount of disagreement is essentially constant with frequency (Figure 8-3). For broadbanded incident waves (Figure 8-6) the harmonics of the spectral peak are submerged among energetic high-frequency incident waves, and the agreement is reasonably good across the entire range of frequencies. The ability of linear shoaling theory to predict the total variance is summarized in Figure 8-7 for a variety of days selected for different total energy levels. Current meters, pressure sensors, and wave staffs were all expressed as equivalent (according to linear theory) sea surface elevation spectra, which results in several points at the same depth on the same day. The data, plotted as H_s, are in general agreement with linear theory for depths greater than about 3 m. Inside of 3 m depth dissipation cannot be ignored, particularly once the waves start to break.

Nonlinear spectral shoaling models based on variants of Boussinesq's equations have recently been developed (Freilich and Guza, 1984). The models are cast as a set of coupled evolution equations for the amplitudes and phases of the temporal Fourier modes of the wave field. Triad interactions across the entire wind-wave frequency band (0.05-0.25 Hz) provide the mechanism for cross spectral energy transfers and modal phase modifications as the wave

Table 8-1. Predicted and Observed Variances
(depth = 395 cm, 20 November 1978)

	Harmonic Variance, cm^2		
Harmonic	Predicted	Observed	Error, %
f	401	331	+21
2f	23	93	-75
3f	17	49	-65
4f	9	22	-59
Total Variance	502	563	- 7.6

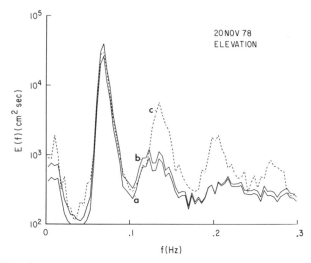

Figure 8-5. Elevation spectra. (a) measured in 1019 cm total depth, P4 (b) predicted at h = 395 cm, P16, 32 degrees of freedom. (Guza and Thornton, 1980)

propagates shoreward through the shoaling region (10-3 m depth). A field experiment at Torrey Pines Beach, during the summer of 1980, provided data on shoaling wave parameters over a wide range of conditions. All the sensors were outside the surf zone. Energy spectral comparisons, as well as spectra of coherence and relative phase between model predictions and data, indicate that the nonlinear models accurately predict Fourier coefficients of the wave field through the shoaling region for all data sets. Typical observed and predicted power spectra are shown in Figures 8-8 and 8-9. Wave spectra measured in 10 m depth provided initial conditions for the models. With narrow-banded swell conditions, the nonlinear model correctly predicts the growth of 2nd and 3rd harmonics (Figure 8-8). With a broad incident wave spectrum, the nonlinear prediction agrees with the observed growth of the entire high

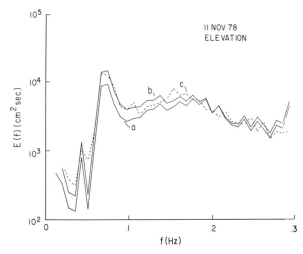

Figure 8-6. Elevation spectra. (a) measured at h = 1012 cm. (b) predicted at h = 353 cm (c) observed at h = 353 cm, 64 degrees of freedom. (Guza and Thornton, 1980)

frequency ($f > 0.15$ Hz) portion of the spectrum (Figure 8-9). Although generally inferior to the nonlinear models, linear, finite-depth theory accurately predicts Fourier coefficients in regions of physical and frequency space where nonlinear evolution of the power spectrum is not observed, thus verifying the validity of the linear, finite-depth dispersion relation in limited portions of physical and frequency space in the shoaling region.

In shallower water, wave breaking totally invalidates the dissipationless model. The inadequacies of linear spectral shoaling lead us to develop parametric models for the shoaling and breaking of random wave height distributions in which the nonlinearities and breaking effects are lumped within an adjustable parameter (§8.8).

8.4 Wave Height in the Inner Surf Zone

Waves in the sea-swell band of frequencies ($0.05 < f < 0.5$) of the inner surf zone were examined by Thornton and Guza (1982). The depth dependence of the observed rms wave heights at Torrey Pines can be seen in Figure 8-10. Thirty-four minute averages of H_{rms} were estimated from the surface elevation variances inferred from the horizontal velocity spectra. Depths were obtained from the nearest wave staff or pressure sensor. Set-up is taken into account by using direct measures of the depth. Starting offshore and following a particular curve in Figure 8-10, the wave height initially increases and then curves over until in shallow water all waves tend to decay in approximately the same manner. Wind waves that are locally depth controlled are described as *saturated*, and this region is defined as the *inner surf zone*. The waves plotted in

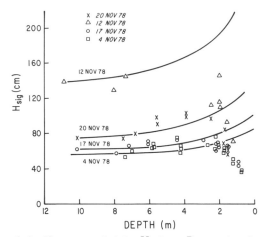

Figure 8-7. Measured significant waveheight, H_s from P, η and u for various days and depths compared to theoretical predictions (solid lines) using 10 m depth measured spectra and linear shoaling theory. (Guza and Thornton, 1980)

Figure 8-10 demonstrate the wind-wave height in the inner surf zone is essentially independent of the wave height outside the surf zone.

The temporal variability of wave height for 17.1 minute averages is shown in Figure 8-11. Offshore (C15) and about the breakerline (C22), H_{rms} shows considerable temporal variability due to the statistical variability of the incoming wave variance. The furthest offshore location is C15 with inshore stations increasing in number. The variability of H_{rms} decreases inside the surf zone as the waves are no longer governed by outside conditions, but become locally depth controlled. The four inshore measurement locations form quasi-parallel lines. The slow decreasing trend in H_{rms} of the waves inside the surf zone, while the heights of the waves outside do not show a decreasing trend, is due to the mean depth of the water decreasing during ebb tide, again emphasizing the depth dependence of wave height inside the surf zone.

H_{rms} values calculated using both velocity sensors and wave staffs for the frequency range $0.05 < f < 0.5$ are plotted against mean depth for all days at Torrey Pines in Figure 8-12. H_{rms} values calculated using current meters and measured using wave staffs give essentially the same results. An approximate envelope for wave heights in the inner surf zone suggests a linear relationship

$$H_{rms} = \gamma h \approx 0.42\, h \qquad (8.4\text{-}1)$$

The envelope curve (8.4-1) represents a maximum H_{rms} relationship with depth. The strong depth dependence in the inner surf zone is because a high proportion of waves are breaking. Wave heights in the outer surf zone region where only

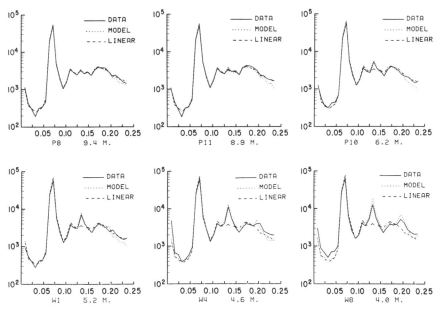

Figure 8-8. Comparison of observed and predicted power spectra on 11 September 1980. Torrey Pines Beach. *Linear* is linear spectral shoaling theory and *model* is a nonlinear Boussinesq theory. The mean depth of each observation is given below each spectrum. (Freilich and Guza, 1984)

some of the waves have broken, i.e., unsaturated, are not represented by a simple depth dependent relationship such as in (8.4-1).

Wind-wave heights in the inner surf zone at Leadbetter Beach were similarly examined (Figure 8-13). Average H_{rms} for 68 minutes were calculated using the zero-up-cross method to define wave heights from sea surface elevations inferred from either pressure sensors or current meters. Although the mean value of γ is nearly identical to the 0.42 obtained over a wide range of wave conditions at the Torrey Pines experiment (Figure 8-13), there is considerably more scatter in the Leadbetter Beach data. This may be partially due to the differences in breaker type at the two beaches. On the low slope ($\tan\beta \approx 0.01$) Torrey Pines Beach, the waves broke as spilling or mixed plunging/spilling. On the steeper Leadbetter Beach ($\tan\beta \approx 0.04$), the waves were sometimes clearly plunging (2-4 February) and sometimes mixed. Laboratory measurements suggest greater γ for plunging breakers (Battjes, 1974). Additionally, well formed cusps occurred during the first week (1-5 February) at Leadbetter Beach, which definitely disturbed the nearshore flow. The results from the two experiments suggest that the wave heights of the inner surf zone are primarily

9 SEPT 80   HI RESOLUTION

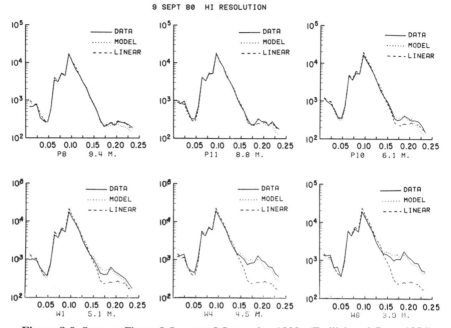

Figure 8-9. Same as Figure 8-8 except 9 September 1980. (Freilich and Guza, 1984)

dependent on the local depth as given by (8.4-1), and secondary influence due to bottom slope, wave steepness, or breaker type, at least for the range of conditions encountered during the two NSTS experiments.

8.5 Celerity

Wave phase speed (celerity) of individual waves has been measured in a number of studies. Svendsen *et al.* (1978) and Hedges and Kirkgoz (1981) measured individual waves in laboratory studies of monochromatic breaking waves on various plane beaches (slopes 0.02-0.08) and found better agreement with bore theory than linear wave theory. Field studies by Inman *et al.* (1971) and Suhayda and Pettigrew (1977) measured the phase speed of individual waves in the surf zone and found the phase speeds slightly greater than the linear prediction of ($\sqrt{gh}$) and less than the theoretical value for a solitary wave or inviscid bore theory.

But the measurement of phase speed by observing individual waves is only applicable in a non-dispersive (or very weakly dispersive) system, or with monochromatic waves. Calculation of the phase difference between sensors, as a function of frequency, allows calculation of the phase speed as a function of frequency, i.e., the celerity spectrum. Celerity spectra have been measured in the laboratory for deep water wind generated waves (Ramamonjiarisoa *et al.*,

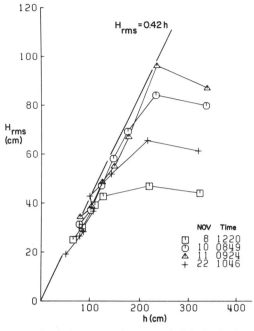

Figure 8-10. H_{rms} versus depth demonstrating wave height in the inner surf zone is locally depth controlled. Torrey Pines Beach. (Thornton and Guza, 1982)

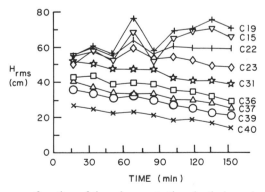

Figure 8-11. H_{rms} as a function of time demonstrating depth dependence of variance in inner surf zone, 10 November 1978. Current meter number is indicated. Torrey Pines Beach. (Thornton and Guza, 1982)

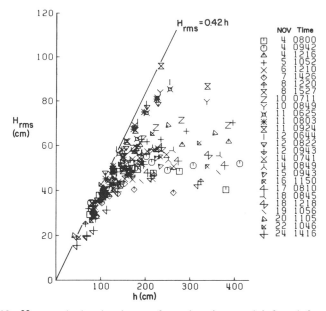

Figure 8-12. H_{rms} calculated using surface elevations and inferred from velocities versus depth at Torrey Pines Beach. (Thornton and Guza, 1982)

1977; and Mitsuyasu *et al.*, 1979) and in the surf zone (Thornton *et al.*, 1976; and Busching, 1978). These measurements show discrepancies with linear theory in that the observed wave celerities are the same at higher frequencies (where linear theory indicated the celerity should decrease) as at the peak of the spectrum.

The discrepancy between linear theory and constant measured celerities stimulated a number of papers to explain the differences. The various explanations offered include: directional wave spreading (Huang and Tung, 1977), nonlinear coupling (Crawford *et al.*, 1981), surface drift currents (Plant and Wright, 1979), and changes in celerity when shorter waves ride on longer waves (Phillips, 1981). All offer plausible explanations and individually or collectively could explain the observed differences. The celerity measurements described here (Thornton and Guza, 1982) for Torrey Pines also show similar discrepancies with linear theory. Because of the generally narrow directional spectra measured at Torrey Pines due to offshore island sheltering (Chapter 2), the directional spreading effect is not considered significant. Likewise, the surface drift current contribution is considered not important because of the general lack of significant winds during the experiment.

For a linear wave system, all wave components travel at their own speed as prescribed by the linear theory dispersion relationship for phase speed

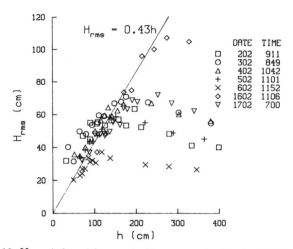

Figure 8-13. H_{rms} inferred from velocities versus depth at Leadbetter Beach.

$$C = \left[\frac{g}{k} \tanh kh \right]^{1/2} \qquad (8.5\text{-}1)$$

For weakly nonlinear steady uniform wave trains (e.g., Stokes waves), all the harmonics travel at the speed of the fundamental frequency, i.e., the celerity spectrum is constant. In general, a system of finite amplitude waves is composed of both free waves and forced (bound) waves. The behavior of the system will depend on the relative importance of each type of wave.

Celerity spectra were measured by using pairs of wave gages or current meters separated in a line perpendicular to the beach. Consider a spectral wave component propagating toward shore and choosing for convenience the measurement points $(x_1, y_1) = (0, 0)$ and a shoreward location on a line perpendicular to the beach at $(x_2, y_2) = (\Delta x, 0)$, given by

$$\eta_1(\omega) = a_1(\omega)\cos(-\omega t)$$

$$\eta_2(\omega) = a_2(\omega)\cos(k_x \Delta x - \omega t) \qquad (8.5\text{-}2)$$

where $k_x \approx k\cos\overline{\alpha}$, is the x-component of wave number $\overrightarrow{k}$, and $\overline{\alpha}$ is the mean angle of wave incidence at frequency ω.

The phase difference between the two measurement points is given by

$$\phi(\omega) = k_x \Delta x = \omega \frac{\Delta x}{C_x} \qquad (8.5\text{-}3)$$

where $C_x = \omega/k_x$ is the x-component of celerity which is calculated using

$$C_x = \omega \frac{\Delta x}{\phi(\omega)} \qquad (8.5\text{-}4)$$

For small angles of wave approach ($\bar{\alpha} < 10°$), $C \approx C_x$, with less than 2% error.

Measured celerity spectra (squares denote spectral estimates) at various depths proceeding from offshore (top of figure) to onshore (down the figure) for 20 November 1978 are shown in Figure 8-14. Up to 13 instruments along the main range line representing 12 adjacent instrument pairs were used to make the celerity calculations. Only 5 pairs are shown here. The sensors with a P designation are pressure sensors and C are current meters.

The curving solid lines in Figure 8-14 are the celerity calculated using the linear wave theory relationship (8.5-1). The theoretical celerity predicts frequency dispersion in that the wave speed decreases with increasing frequency. The theoretical frequency dispersiveness becomes less as the depth decreases. For frequencies of about 0.3 Hz and higher, however, there is still significant theoretical frequency dispersion even at the shallowest measurement depth of 1.0 m. The measured celerity spectra and the theoretical spectra generally compare well at the deepest measurements (Figure 8-14) demonstrating frequency dispersion. In shallower water, there is a marked departure between measurements and linear theory for all data sets. The measured celerity becomes essentially constant with frequency as the waves approach and transect the surf zone. A horizontal line of best fit has been drawn through the calculated shallow water celerity values to indicate an average celerity. Except for offshore, the celerity spectra are essentially constant.

Again, November 20th is an example of extreme narrow-bandedness with rich harmonic growth as the waves propagate shoreward (Figures 8-2, 8-5). The ratio of energy in the first harmonic to the fundamental in 10 m depth is 0.08 but in 1.8 m depth (location of C22 in Figure 8-1) the ratio increased to 0.39. The strength of the harmonics is an indication of the nonlinearity. Based on the strength of the nonlinearities and narrowness of the directional spectra, it is hypothesized the constancy of the celerity spectra is due in large part to the nonlinearity of the waves. The waves on November 20 appear to be a good example of a nearly one-dimensional narrow-band, modulated wave train described by Crawford et al. (1981) for which the celerity of the harmonics are bound to the fundamental.

Higher order, nonlinear wave theory shows that waves may be amplitude dispersive, i.e., the wave celerity is amplitude dependent. The amplitude dependence could account for some of the differences between the measured and linear theory celerities. For example, both simple bore and shallow water Cnoidal (i.e., solitary) wave theories predict

$$C_S \approx \sqrt{gh} \left[1 + \frac{1}{2} \frac{H}{h} \right] \qquad (8.5\text{-}5)$$

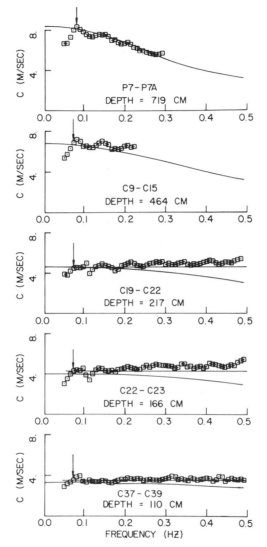

Figure 8-14. Celerity spectra for moderate, very narrow-band waves on 20 November 1978. Torrey Pines Beach. (Thornton and Guza, 1982)

Comparing bore and shallow water linear wave theories

$$\frac{C_S - C_L}{C_L} = \frac{1}{2}\frac{H}{h} \tag{8.5-6}$$

In order to compare all celerity calculations, a single value was chosen to represent the spectra. The celerity corresponding to the coherence peak (denoted in Figure 8-14 by an arrow over the celerity spectrum) was found to be representative of the mean value which is reasonable for the constant celerity cases. The percent difference between *mean* measured and linear theory celerity as a function of the rms wave height, H_{rms}, to depth ratio is shown in Figure 8-15. All mean measured celerity values, with the exception of 2, are within +20% and −10% of linear theory even at breaking and inside the surf zone. Recalling from the measurements (Figure 8-12) that inside the surf zone, $H_{rms} \leq 0.44\,h$; substituting this relationship into (8.5-6) (assuming H_{rms} is a correct parameterization) says that bore theory predicts the celerity to be a maximum of about 20% greater than linear theory. This is indeed the upper bound on the observed differences. The vertical line in Figure 8-15 is the saturation envelope relationship which forms an upper bound on the abscissa. Although there is considerable scatter, there does appear to be a definite increase in the deviation from linear phase speed with increasing H_{rms}/h. The dependence of celerity on wave height is indicative of amplitude dispersion.

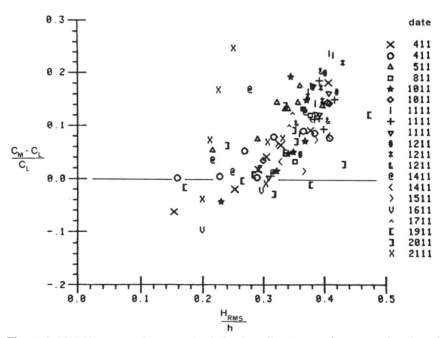

Figure 8-15. Differences of measured celerity from linear wave theory as a function of wave height suggesting weak amplitude dispersion, Torrey Pines Beach. (Thornton and Guza, 1982)

In conclusion, the measured celerities show the waves inside 4 m are weakly amplitude dispersive, and generally frequency nondispersive (even for high frequencies which are not, in a linear context, shallow water waves).

8.6 Velocity Statistical Moments

Several of the sediment transport models to be discussed in later chapters depend on the statistical moments of the wave induced velocity field. For example, in the special case of no flows other than cross-shore oscillatory motions, the longshore transport rate is zero, and the cross-shore bedload transport rate (i_x) is (Bailard, 1981)

$$\langle i_x \rangle = \rho c_f \ u_{rms}^3 \ \varepsilon_b \ \tan\phi^{-1} \left[\frac{\langle u^3 \rangle}{u_{rms}^3} - \frac{\tan\beta}{\tan\phi} \frac{\langle |u|^\beta \rangle}{u_{rms}^3} \right] \qquad (8.6\text{-}1)$$

where ρ, c_f, ε_b, $\tan\phi$ are constants, $\tan\beta$ is the beach slope and $\langle \ \rangle$ denotes time averaging.

The value of the moments appearing in the transport equations obviously depend upon what sort of wave field is assumed for the nearshore. Many models have determined u_{rms} by assuming linear shoaling seaward of the breakpoint, and wave height saturation inside the breakpoint. §8.3 and 8.4 have shown these to be reasonably good approximations for the wind wave frequency bands (Figures 8-7, 8-10). We emphasize the fact that those results have intentionally excluded important contributions from surf beat frequencies (defined here somewhat arbitrarily as f < 0.05 Hz). There is, however, no reason why surf beat motions cannot contribute to sediment motion.

Figure 8-16 shows cross-shore and longshore velocity variances as functions of depth, partitioned into low and high frequency components. The high frequency $(f > 0.05\,\mathrm{Hz})$ portion of the cross-shore velocity variance first increases during shoaling, and then decreases as breaking becomes an important energy dissipation mechanism. But the low frequency (surf beat) band energy increases monotonically shoreward with a maximum in the swash. The decreasing wind wave energy is almost balanced by the increase in surf beat energy resulting in an almost constant cross-shore velocity variance in depths ranging from 40 to 300 cm. The standard modelling hypothesis that velocity fluctuations are vanishingly small at the shoreline is not accurate because of energetic surf beat. Surf beat is discussed in more detail in Chapter 9.

If the local value of u_{rms} is given, then certain assumptions about the wave field allow the calculation of other important moments. Wave-induced sediment transport models in the past have generally assumed monochromatic, sinusoidal wave fields (e.g., Bowen, 1980; Bailard, 1981). We now show that more accurate prediction of velocity moments follows from the assumption of a Gaussian wave field. Only a few simple moments will be considered here; for a discussion of the complicated moments appearing in the full sediment transport

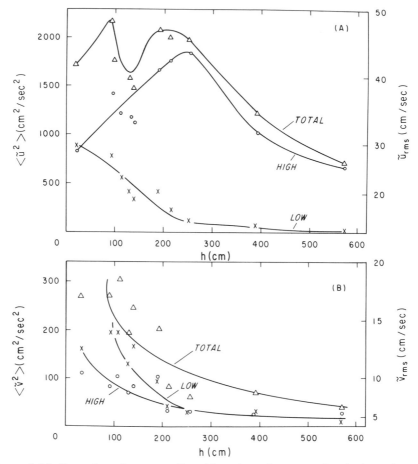

Figure 8-16. Cross-shore (upper panel) and longshore (lower panel) velocity variances versus depth on 20 November 1978. Low passed ($f < .05$ Hz), high passed ($.05 < f < .05$ Hz), and total ($0 < f < .5$ Hz) variances are shown. (Guza and Thornton, 1985)

equations see Guza and Thornton (1985). The first four non-zero moments about the mean are, for a simple monochromatic wave train,

$$
\begin{array}{rcll}
\langle u^2 \rangle & = & .5 u_m^2 & (a) \\
\langle |u^3| \rangle & = & 1.20 \langle u^2 \rangle^{3/2} & (b) \\
\langle u^4 \rangle & = & 1.5 \langle u^2 \rangle^2 & (c) \\
\langle |u^5| \rangle & = & 1.92 \langle u^2 \rangle^{5/2} & (d)
\end{array}
\tag{8.6-2}
$$

where u_m is the velocity amplitude of the sinusoid wave component.

Instead of assuming sinusoidal motions, moments can be calculated on the alternative assumption of a Gaussian distribution of velocities, as occurs in a linear random sea,

$$P(u) = \frac{1}{\sigma_u \sqrt{2\pi}} e^{\frac{-(u' - \bar{u})^2}{2\sigma_u^2}} \tag{8.6-3}$$

where u' denotes the total velocity, u the velocity deviations about the mean, $\bar{u}$, $P(u)$ is the probability density function of u, and σ_u is the standard deviation. The average (expected) value of any function of u ($f(u)$) is found from

$$\langle f(u) \rangle = \int_{-\infty}^{\infty} f(u) P(u) du \tag{8.6-4}$$

so

$$\langle u^2 \rangle = \sigma_u^2$$

Using (8.6-4) to calculate the other moments in (8.6-2) yields

$$\begin{array}{llll}
\langle |u^3| \rangle & = & 1.59 \langle u^2 \rangle^{3/2} & (a) \\
\langle u^4 \rangle & = & 3.0 \ \langle u^2 \rangle^2 & (b) \\
\langle |u^5| \rangle & = & 6.38 \langle u^2 \rangle^{5/2} & (c)
\end{array} \tag{8.6-5}$$

It is clear from (8.6-2) and (8.6-5) that a random sea with a Gaussian distribution and an *equivalent monochromatic wave* (i.e., with the same variance) have substantially different higher even moments. The higher the moment, the greater the differences. The infrequently occurring large velocities which occur with the Gaussian distribution (but not with monochromatic waves) contribute very substantially to the higher moments.

The moments (about the mean) of (8.6-2) and (8.6-5) were calculated for two different days. November 20 is a day with a narrow swell peak and pronounced harmonics (Figure 8-2) while November 17 has substantial energy at both swell (0.075 Hz) and locally generated wind wave (0.19 Hz) frequencies. Significant wave heights in 10 m depth were 76 cm on 20 November, and 64 cm on 17 November. The third moment of the absolute value of the normalized, cross-shore velocity (8.6-2b, 8.6-5a) is shown in Figure 8-17. Each plotted point represents the value of this moment averaged over an entire run, 120 minutes for the 17th and 170 minutes for the 20th of November.

The average value of all sensors is 1.60 on 17 November and 1.69 on 20 November (Table 8-2), slightly higher than the Gaussian value of 1.59, but significantly larger than the sinusoidal value of 1.2. Similar averaged daily values for the other moments of (8.6-5) are given in the first six columns of Table 8-2. Both cross-shore and longshore velocity moments agree much better with a Gaussian model than with a sinusoidal process, at least for the lowest order non-zero moments.

However, substantial deviations from Gaussian behavior are observed. For example, the third moment about the mean, or skewness, is zero for a Gaussian process while the observed cross-shore velocity skewness is always negative

Table 8-2. Mean Statistics for Oscillating Velocity Components

MOMENT	$\dfrac{\langle |u|^3 \rangle}{\langle u^2 \rangle^{3/2}}$	$\dfrac{\langle |v|^3 \rangle}{\langle v^2 \rangle^{3/2}}$	$\dfrac{\langle u^4 \rangle}{\langle u^2 \rangle^2}$	$\dfrac{\langle v^4 \rangle}{\langle v^2 \rangle^2}$	$\dfrac{\langle |u|^5 \rangle}{\langle u^2 \rangle^{5/2}}$	$\dfrac{\langle |v|^5 \rangle}{\langle v^2 \rangle^{5/2}}$	$\dfrac{\langle u^3 \rangle}{\langle u^2 \rangle^{3/2}}$	$\dfrac{\langle v^3 \rangle}{\langle v^2 \rangle^{3/2}}$	$\dfrac{\langle u^5 \rangle}{\langle u^2 \rangle^{5/2}}$	$\dfrac{\langle v^5 \rangle}{\langle v^2 \rangle^{5/2}}$
17 November	1.60	1.68	2.86	3.41	7.77	8.06	− .55	− .036	− 4.95	− .05
20 November	1.69	1.67	3.50	3.44	8.58	8.56	− .50	+ .006	− 5.39	− .52
Gaussian	1.59	1.59	3.0	3.0	6.38	6.38	0.	0.	0.	0.
Sinusoid	1.2	1.2	1.5	1.5	1.92	1.92	0.	0.	0.	0.

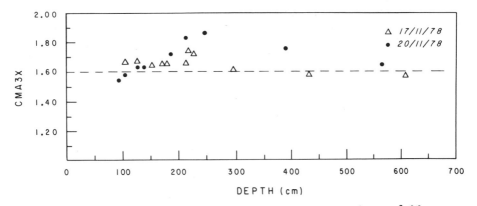

Figure 8-17. $CMA\,3X$ versus depth, where $CMA\,3X = \langle\,|\,\tilde{u}\,|^3\,\rangle / \langle\,|\,\tilde{u}^2\,|^{3/2}\,\rangle$. Torrey Pines Beach. (Guza and Thornton, 1985)

(Figure 8-18) which (in the present coordinate frame) indicates that the strongest flows are directed shorewards.

Skewness is almost zero offshore where the wave field is nearly Gaussian; it becomes increasingly negative shoreward and is most negative around 2 m depth, near the mean breakerline. The skewness then tends to zero inside the surf zone and processes in the inner surf zone are again nearly Gaussian. The trend of the skewness, normalized by the total variance, is indicative of the strength of nonlinearities. Table 8-2 shows that, averaged over all sensors, the longshore third (skewness) and fifth moments are much closer to the Gaussian value of zero than the cross-shore velocities. This occurs primarily because the highly skewed incident wind waves approached almost normal to shore in such shallow water.

In summary, modelling the nearshore velocity field as a linear, Gaussian, random process results in reasonable accurate predictions of many of the moments, but poor predictions of others, most obviously the odd moments of cross-shore velocity. No moments are accurately given by a monochromatic, unidirectional wave model.

8.7 Wave Height Distribution Observations

The approximation that wave heights are Rayleigh distributed is commonly made when considering statistical properties of wave heights. The Rayleigh wave height distribution was shown by Longuet-Higgins (1952) to apply to deep water waves on the assumption that the sea waves are a narrow-banded, linear Gaussian process. The Rayleigh wave height probability density function pdf is

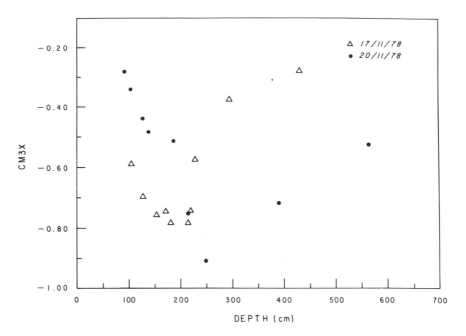

Figure 8-18. $CM\,3X$ versus depth, where $CM\,3X = \langle \tilde{u}^3 \rangle / \langle \tilde{u}^2 \rangle^{3/2}$ is the velocity skewness. Torrey Pines Beach. (Guza and Thornton, 1985)

$$p(H) = \frac{2H}{H_{\text{rms}}^2}\, e^{-\left[\frac{H}{H_{\text{rms}}}\right]^2} \tag{8.7-1}$$

which is entirely specified by H_{rms}. The Rayleigh distribution is strictly only valid for linear waves (Longuet-Higgins, 1975). The distribution derived with a linear assumption would not be expected to hold for waves approaching maximum height, i.e., close to breaking, as in the surf region or for broader-banded *sea* conditions with whitecaps.

The theoretical Rayleigh distribution has been found by several authors (Chakrabarti and Cooley, 1977; Forristall, 1978) to over-predict the number of large waves in the tail compared with observations. Various explanations including nonlinearity (Forristall, 1978; Tayfun, 1980; Longuet-Higgins, 1980) whitecapping (Tayfun, 1981) and finite bandwidth (Longuet-Higgins, 1980) have been examined as causes for the deviation from a Rayleigh distribution. Tayfun's (1981) breaking effect and Longuet-Higgins' (1980) finite bandwidth mechanism both have some success in explaining field data. These studies seek to explain deviations from a Rayleigh distribution, but the relevant point here is that wave heights appear to be nearly Rayleigh under a much wider range of conditions than the strict assumptions of a narrow-band Gaussian (linear) process would imply. The results of the field experiments at Torrey Pines show that wave height data even within the surf zone are reasonably well described by the Rayleigh distribution (Thornton and Guza, 1983).

Six days at Torrey Pines were selected for analysis covering a wide range of conditions. Wave heights were determined from 68 minutes surface elevation records using the zero-up-crossing method in which the wave height is defined as the difference of maximum and minimum occurring between two consecutive up-crossings. Surface elevations were either measured directly using wave staffs, or inferred using pressure or current meter records and linear theory to convolve the records (Thornton and Guza, 1983). Empirical pdf's of wave heights derived from velocity and pressure measurements are compared with the Rayleigh pdf for selected depths on 20 November (Figure 8-19). The Rayleigh pdf appears to reasonably describe the measured wave heights everywhere. The largest discrepancies of the measured waves with the Rayleigh pdf are deficits at the lowest and highest waves. But the bulk of the distribution is reasonably well described and therefore, the central moments such as $\bar{H}$ and H_{rms} should be well described using the Rayleigh pdf for model comparisons.

A more exact standard for comparisons is to use directly measured wave heights from wave staffs. For wave staffs on all six days, measured H_{rms} are first compared with the commonly used approximation for H_{rms} in (8.7-1)

$$H_{rms} = \sqrt{8}\sigma_\eta \qquad (8.7-2)$$

where σ_η^2 is the surface elevation variance. The calculated correlation coefficient between the two variables is 0.995 and the mean error is +0.3%. Measured $H_{1/3}$ (average of the highest one-third waves), $H_{1/10}$ are next compared with their respective Rayleigh-derived statistics in Figures 8-20 and 8-21 where for the Rayleigh distribution

$$H_{1/3} = 1.42 H_{rms} \qquad (8.7-3)$$

$$H_{1/10} = 1.80 H_{rms} \qquad (8.7-4)$$

A 45 degree solid line is drawn to show a perfect correlation. The linear regression curve forced through zero is drawn as a dashed line, but cannot be differentiated from the 45 degree line in Figure 8-20. The maximum wave height for the Rayleigh pdf was also calculated using (Cartwright and Longuet-Higgins, 1956)

$$H_{max} = \left[(1n\ N)^{1/2} + 0.2886(1n\ N)^{-1/2} \right] H_{rms} \qquad (8.7-5)$$

where N is the total number of waves in the distribution. The correlation coefficients are 0.997, 0.988, and 0.924, for $H_{1/3}$, $H_{1/10}$ and H_{max}, respectively. The linear regression curves forced through zero give an average percent error of -0.2%, -1.8%, and -6.8%, for $H_{1/3}$, $H_{1/10}$, and H_{max}, respectively. The

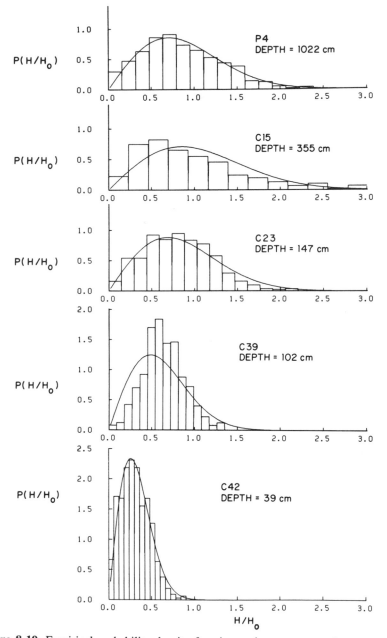

Figure 8-19. Empirical probability density functions using pressure and current meters plotted against Rayleigh pdf for 20 November 1978, Torrey Pines Beach. H_o is H_{rms} in $\approx$10 m depth = 50 cm. (Thornton and Guza, 1983)

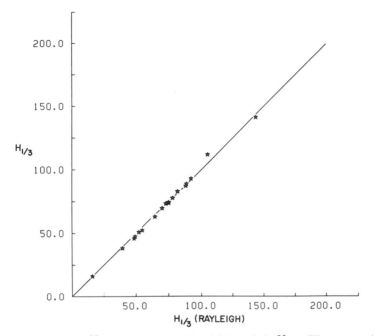

Figure 8-20. Measured $H_{1/3}$ plotted against Rayleigh statistic $H_{1/3}$. (Thornton and Guza, 1983)

increasing spread of the points going from $H_{1/3}$ to $H_{1/10}$ to H_{max} may be because fewer points are used to calculate the statistics, H_{max} being the extreme which is represented by a single point. The heights may also be non-Rayleigh in the extreme tail. The results show that the central moments of H_{rms}, $H_{1/3}$, and even $H_{1/10}$ are well predicted using the Rayleigh distribution. Therefore it is concluded that the Rayleigh distribution can be used to give a reasonable description of wave heights, even in the surf zone, at least for the spilling breakers measured at Torrey Pines Beach.

8.8 Wave Height Distribution Transformation

The transformation of wave height distributions from offshore to the beach was described by Thornton and Guza (1983) in an extension of work by Battjes and Janssen (1978) solving the energy flux balance equation assuming straight and parallel contours

$$\frac{dEC_{gx}}{dx} = \langle \varepsilon_b \rangle \tag{8.8-1}$$

where E is the energy density, C_{gx} is the shoreward component of group velocity, and $\langle \varepsilon_b \rangle$ is the ensemble averaged dissipation due to wave breaking.

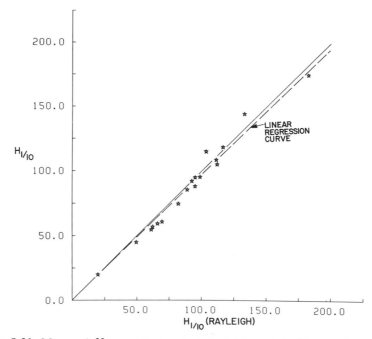

Figure 8-21. Measured $H_{1/10}$ plotted against Rayleigh statistic $H_{1/10}$. (Thornton and Guza, 1983)

Dissipation due to bottom friction was also considered in the analysis, but was shown to be negligible compared with dissipation due to wave breaking and is not considered here.

It was shown in §8.5 that the waves shoreward of 4 m depth can be frequency non-dispersive across the sea-swell frequency range. The phase speeds are the same as the linear phase speed at the spectral peak, approximately given by $\sqrt{gh}$. These measurements suggest using a lowest order model for the energy density and group velocity given by the linear wave theory relationships

$$E = \frac{1}{8} \rho g \, H_{\text{rms}}^2 = \frac{1}{8} \rho g \int_0^\infty H^2 p(H) dH \qquad (8.8\text{-}2)$$

and

$$C_{gx} = \frac{C}{2} \left[1 + \frac{2 k h}{\sinh 2 k h} \right] \cos \overline{\alpha} \qquad (8.8\text{-}3)$$

where $\overline{\alpha}$ is the mean wave direction and k is the wave number associated with average frequency $\overline{f}$ corresponding to the peak of the spectrum.

It is assumed in the analysis that the breaking wave dissipation can be modeled after a bore. The details of the turbulence dynamics in the bore (spilling breaker) are avoided by applying conservation of mass and momentum at regions of uniform flow upstream and downstream of the bore. The average rate of energy dissipation per unit area is calculated (Stoker, 1957)

$$\varepsilon_{bore} = \frac{1}{4} \rho g \frac{(h_2 - h_1)^3}{h_1 h_2} Q \simeq \frac{1}{4} \rho g \frac{(BH)^3}{h^2} \frac{Ch}{L} \qquad (8.8\text{-}4)$$

where the wave height H is measured as the maximum to minimum of the bore, Q is the volume discharge per unit area across the bore described by simple linear bore theory (Hwang and Divoky, 1970), and B is a breaker coefficient of $0(1)$. The coefficient B accounts for the differences in various breaker types and is considered as a function of the proportion of the foam region on the face of the breaker (Battjes and Janssen, 1978). The coefficient B is the only unspecified parameter in the model and is determined from the data.

The formulation starts by describing the offshore wave heights in terms of the Rayleigh distribution, $p(H)$, with the implied assumptions of narrow-bandedness such that the waves can be described by a single mean frequency, $\bar{f}$, and direction, $\bar{\alpha}$. The initially Rayleigh height distributions in 10 m depth were shown in §8.7 to be modified by shoaling and breaking into new distributions which are again nearly Rayleigh, but with some energy loss. Therefore, the distribution for all waves is assumed Rayleigh everywhere.

The bore dissipation function (8.8-4) is only applied to the breaking waves, and it is therefore necessary to identify which waves are breaking. The distribution of breaking waves describe which waves of the Rayleigh distribution are breaking, and can be expressed as a weighting of the Rayleigh pdf

$$p_b(H) = W(H) p(H) \qquad (8.8\text{-}5)$$

where the weighting function $W(H) \leq 1$, since $p_b(H)$ must be a subset of $p(H)$. The area under the distribution is equal to the percent of breaking waves, A_b, such that

$$A_b = \int_0^\infty p_b(H) \, dH \qquad (8.8\text{-}6)$$

Defining $p_b(H)$ in this manner means it is not a pdf, but is defined to keep track of the fraction of breaking waves. In deep water, $A_b \to 0$, as $h \to \infty$; and in the surf zone $A_b \to 1$, as $h \to 0$; i.e., all waves are breaking. To obtain a simple analytic result, Thornton and Guza (1983) hypothesized that the probability of a wave breaking depends only on H_{rms} and h, so that

$$W(H) = A_b = \left(\frac{H_{rms}}{\gamma h} \right)^4 \qquad (8.8\text{-}7)$$

where $\gamma = 0.44$ is used from (8.4-1). $p_b(H)$ described using (8.8-7) was shown by Thornton and Guza (1983) to reasonably predict the fraction of waves that break from offshore to saturation conditions when compared with actual field measurements. More complicated forms for $W(H)$, which make the larger waves (at a given depth) more likely to break, do not improve the wave height calculations compared with the simpler model discussed here.

The average rate of energy dissipation is calculated by multiplying the dissipation for a single broken wave of height H by the probability of wave breaking at each height, as given by $p_b(H)$. For the ensemble, using (8.8-4),

$$\langle \varepsilon_b \rangle = \frac{B^3}{4} \rho g \frac{\overline{f}}{h} \int_0^\infty H^3 p_b(H) dH \qquad (8.8-8)$$

Substituting $p_b(H)$ yields

$$\langle \varepsilon_b \rangle = \frac{3\sqrt{\pi}}{16} \rho g \frac{B^3 \overline{f}}{\gamma^4 h^5} H_{rms}^7 \qquad (8.8-9)$$

Substituting the linear theory formulations for energy density (8.8-2) and group velocity (8.8-3) to specify the energy flux in (8.8-1) and describing the breaking wave dissipation using (8.8-9), the energy flux balance is given by

$$\frac{d}{dx}\left[\frac{1}{8} \rho g \, H_{rms}^2 \, C_g \right] = \frac{3\sqrt{\pi}}{16} \frac{\rho g \, \overline{f} \, B^3}{\gamma^4 h^5} H_{rms}^7 \qquad (8.8-10)$$

For an arbitrary bottom profile, H_{rms} is determined by numerically integrating (8.8-10) starting with an initial wave height in deep water.

By limiting the analysis to normally incident waves in shallow water and to a plane sloping beach, Thornton and Guza (1983) obtained an analytical solution giving

$$H_{rms} = a^{1/5} h^{9/10} \left[1 - h^{23/4} \left[\frac{1}{h_o^{23/4}} - \frac{a}{y_o^{5/2}} \right] \right]^{-1/5} ; \qquad (8.8-11)$$

$$0 \le h \le h_o$$

where

$$y_o = H_o^2 h_o^{1/2}$$

and

$$a = \frac{23}{15} \left[\frac{g}{\pi} \right]^{1/2} \frac{\gamma^4 \tan\beta}{B^3 \overline{f}}$$

where the subscript o refers to the input conditions at the outer shallow water depth limit. The asymptotic case as the depth gets very shallow is

$$H_{rms} = a^{1/5} h^{9/10}; \text{ as } h \to 0 \qquad (8.8\text{-}12)$$

which says the wave height in the inner surf zone is related to the depth and independent of the initial conditions in deeper water. This result is similar to the observations at Torrey Pines Beach (Figures 8-8, 8-10, 8-11) showing that waves of all deep water height eventually become saturated in the inner surf zone with the heights given by (8.4-1).

8.9 Model Comparisons with Torrey Pines Beach Data

Using the measured bottom profiles, H_{rms} values are calculated by numerically integrating (8.8-10) and compared with the six days of measured H_{rms} values representing a relatively broad range of wave conditions. The bottom contours at Torrey Pines were shown to be well approximated as straight and parallel and the waves normally incident, simplifying the analysis since $C_{gx} = C_g$ in (8.8-1). A best fit coefficient B was determined by iteration. The sums of the square error of the model H_{rms} compared with the measured H_{rms} were calculated for all sensors for all six days using various values of B. An optimal value was found for $B = 1.72$ with a standard error of 8.3 percent.

The calculated H_{rms} values using (8.8-10) are shown in Figure 8-22 to be capable of predicting the wave height increase due to shoaling and subsequent decrease due to wave breaking. The waves peak up to a maximum, a point defined here as the mean breaker-line, and then decrease. In the inner surf zone, all the waves become locally depth controlled and changes in the depth are reflected in changes in wave height. The wiggles in the results in the inner surf zone for 20 November (Figure 8-22) are due to variations in the bottom profile.

All the measured H_{rms} versus predicted values are shown in Figure 8-23. The calculated results do not appear to depend on the magnitude of H_{rms}, i.e., as the wave height increases, the differences from the 45 degree line do not increase. The percentage error between predicted and measured values are all within ±20% of the measured values, and usually much less. The error standard deviation (standard error) is 0.083.

H_{rms} measurement errors are due to sensor errors and errors inherent in transforming the velocity and pressure measurements to surface elevations with linear wave theory. The current meter and pressure sensor calibration errors were on the order of ±5%. The combined errors associated with linear theory and sensor error are less than ±20% (Figure 8-3), which is not substantially different from errors between predictions and measurements (Figure 8-23).

The wave transformation results at Torrey Pines and Leadbetter Beaches demonstrate the usefulness of linear theory and identify nonlinear effects. Harmonic amplification, coupled harmonics resulting in frequency nondispersiveness, and amplitude dispersion were identified as nonlinear wave phenomena. Linear wave theory was found useful in the local spectral transformation of kinematic quantities, and also in describing variance shoaling,

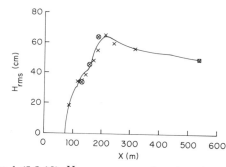

Figure 8-22. Calculated (8.8.10) H_{rms} versus x plotted against measured values for 20 November 1978. Torrey Pines Beach. Offshore distance is measured relative to NSTS datum. Shoreline is located where H_{rms} goes to zero. The measured H_{rms} values were determined from sea surface elevations measured using wave staffs (circles), and by applying linear theory to the velocity (x's) and pressure (squares) measurements to infer sea surface elevations. (Thornton and Guza, 1983)

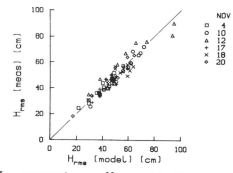

Figure 8-23. H_{rms} measured versus H_{rms} model. (Thornton and Guza, 1983)

and the phase speed at the peak frequency. In addition, Gaussian statistics including the Rayleigh distribution description of wave heights, appear to give at least a first order description of the random wave processes. It is pointed out that the waves during the Torrey Pines experiment and during much of the Leadbetter Beach experiments were low steepness waves ($H_{rms}/L_o <\sim 0.02$). The low steepness is due to generally long period, long wavelength, incident swell. A simple first order wave transformation model based on the energy flux balance is shown to reasonably compare with the measurements. The wave transformation model incorporates linear shoaling, linear bore theory to describe dissipation due to wave breaking, and wave saturation in the inner surf zone. This wave transformation description will be incorporated into longshore current models described in Chapter 15.

8.10 References

Bailard, J. A., 1981, An energetics total load sediment transport model for a plane sloping beach, *Journal of Geophysical Research*, 86: 10938-10954.

Battjes, J. A., 1974, Surf similarity, *Proceedings*, Fourteenth Coastal Engineering Conference, June 24-28, 1974, Copenhagen, Denmark, American Society of Civil Engineers, New York: 466-480.

Battjes, J. A. and J. P. F. M. Janssen, 1978, Energy loss and set-up due to breaking of random waves, *Proceedings*, Sixteenth Coastal Engineering Conference, August 27-September 3, 1978, Hamburg, Germany, American Society of Civil Engineers, New York: 569.

Bowden, K. F. and R. A. White, 1966, Measurements of the orbital velocities of sea waves and their use in determining the directional spectrum, *Geophysical Journal Royal Astronomical Society*, 12: 33-54.

Bowen, A. J., 1980, Simple models of nearshore sedimentation; beach profiles and longshore bars; in the coastline of Canada, *Geological Survey of Canada*: 21-30.

Busching, F., 1978, Anomalous dispersion of Fourier components of surface gravity waves in the nearshore area, *Proceedings*, Sixteenth Coastal Engineering Conference, August 27-September 3, 1978, Hamburg, Germany, American Society of Civil Engineers, New York: 247-267.

Cartwright, D. E. and M. S. Longuet-Higgins, 1956, The statistical distribution of the maxima of a random process, *Proceedings of the Royal Society of London*, Series A, 237: 212-232.

Cavaleri, L., J. A. Ewing and N. D. Smith, 1978, Measurement of the pressure and velocity field below surface waves, in *Turbulent Fluxes Through the Sea Surface Wave Dynamics and Predictions*, NATO Conference, Series V., Plenum, New York: 257-272.

Chakrabarti, S. K. and R. P. Cooley, 1977, Statistical distributions of periods and heights of ocean waves, *Journal of Geophysical Research*, 82: 1363-1368.

Crawford, D. R., B. M. Lake, P. G. Saffman and H. C. Yuen, 1981, Effects of nonlinearity and spectral bandwidth on the dispersion relation and component phase speeds of surface gravity waves, *Journal of Fluid Mechanics*, 112: 1-32.

Dobson, R. S., 1967, Some applications of a digital computer to hydraulic engineering problems, *Technical Report 80*, Stanford University, Stanford, California.

Forristall, G. Z., 1978, On the statistical distribution of wave heights in a storm, *Journal of Geophysical Research*, 83: 2353-2358.

Freilich, M. and R. T. Guza, 1984, Nonlinear effects on shoaling surface gravity waves, *Philosophic Transcript of the Royal Society of London*, A-311: 1-41.

Guza, R. T. and E. B. Thornton, 1980, Local and shoaled comparisons of sea surface elevations, pressures, and velocities, *Journal of Geophysical Research*, 85: 1524-1530.

_____. 1985, Velocity moments in the nearshore, *Journal of Waterways, Port, Coastal and Ocean Engineering*, 111(2): 235-256.

Hedges, T. S. and M. S. Kirkgoz, 1981, An experimental study of the transformation zone of plunging breakers, *Coastal Engineering*, 4: 319-333.

Huang, N. E. and C. C. Tung, 1977, The influence of the directional energy distribution on the nonlinear dispersion relation in a random gravity wave field, *Journal of Physical Oceanography*, 7: 403-414.

Hwang, Li-San and D. Divoky, 1970, Breaking wave set-up and decay on gentle slopes, *Proceedings*, Twelfth Coastal Engineering Conference, September 13-18, 1970, Washington, D.C., American Society of Civil Engineers, New York: 377-389.

Inman, D. L., R. J. Tait and C. E. Nordstrom, 1971, Mixing in the surf zone, *Journal of Geophysical Research*, 76(15): 3493-3514.

Longuet-Higgins, M. S., 1952, On the statistical distribution of the heights of sea waves, *Journal of Marine Research*, 11(3): 245-266.

_____. 1975, On the joint distribution of the periods and amplitudes of sea waves, *Journal of Geophysical Research*, 80: 2688-2694.

_____. 1980, On the distribution of the heights of sea waves: some effects of nonlinearity and finite bandwidth, *Journal of Geophysical Research*, 85: 1519-1523.

Mitsuyasu, H., Y. Kuo and A. Masuda, 1979, On the dispersion relation of random gravity waves, part 2: an experiment, *Journal of Fluid Mechanics*, 92(4): 731-749.

Phillips, O. M., 1981, The dispersion of short wavelets in the presence of a dominant long wave, *Journal of Fluid Mechanics*, 107: 465-485.

Plant, W. J. and J. W. Wright, 1979, Spectral decomposition of short gravity wave systems, *Journal of Physical Oceanography*, 9: 621-624.

Ramamonjiarisoa, A., S. Baldy and I. Choi, 1977, Laboratory studies on wind-wave generation, amplification and evolution, *NATO Symposium on Turbulent Fluxes through the Sea Surface, Wave Dynamics and Prediction*, Marseille, France: 402-420.

Simpson, J. H., 1969, Observation of the directional characteristics of waves, *Geophysical Journal Royal Astronomic Society*, 17: 93-120.

Stoker, J. J., 1957, *Water Waves*, Interscience, New York, 567 pp.

Suhayda, I. N. and N. R. Pettigrew, 1977, Observations of wave height and wave celerity in the surf zone, *Journal of Geophysical Research*, 82(9): 1419-1424.

Svendsen, I. A., P. A. Madsen and J. Buhr Hansen, 1978, Wave characteristics in the surf zone, *Proceedings*, Sixteenth Coastal Engineering Conference, August 27-September 3, 1978, Hamburg, Germany, American Society of Civil Engineers, New York: 520-539.

Tayfun, M. A., 1980, Narrow-band nonlinear sea waves, *Journal of Geophysical Research*, 85: 1548-1552.

_____. 1981, Breaking-limited wave heights, *Journal of Waterway, Port, Coastal and Ocean Division*, Proceedings, American Society of Civil Engineers, 107(WW2): 59-70.

Thornton, E. B. and R. F. Krapohl, 1974, Water particle velocities measured under ocean waves, *Journal of Geophysical Research*, 79: 847-852.

Thornton, E. B., J. S. Galvin, F. L. Bub and D. P. Richardson, 1976, Kinematics of breaking waves, *Proceedings*, Fifteenth Coastal Engineering Conference, July 11-17, 1976, Honolulu, Hawaii, American Society of Civil Engineers, New York: 461-476.

Thornton, E. B. and R. T. Guza, 1982, Energy saturation and phase speeds measured on a natural beach, *Journal of Geophysical Research*, 84: 9499-9508.

_____. 1983, Transformation of wave height distribution, *Journal of Geophysical Research*, 88: 5925-5938.

Chapter 9

RUN-UP AND SURF BEAT

R. T. Guza
Scripps Institution of Oceanography
Center for Coastal Studies

Edward B. Thornton
Naval Postgraduate School

Measurements of velocity and elevation in the inner surf zone usually show that a substantial fraction of the total variance is at surf beat periods, roughly 30-200 sec. (Inman, 1968; Suhayda, 1971, 1974; Goda, 1975; Huntley, 1976; Wright et al., 1979; Holman, 1981; Wright et al., 1982; and others). A typical current record from an inner surf zone sensor at Torrey Pines Beach, and a low passed version of the same record are shown in Figure 9-1. Although wind wave motions are usually the most obvious component of the unfiltered cross-shore velocity field (upper panel), surf beat motions are certainly significant (lower panel). Below, we first briefly review some theoretical ideas about the nature and origin of surf beat. Then experimental evidence supporting the various theories is discussed, placing special emphasis on the NSTS Torrey Pines results.

There are two fundamentally different hypotheses about the generation mechanism and dynamics of surf beat. The first model is due to Munk (1949) and Tucker (1950), who suggested that incident wave groups (formed by the superposition of incident wind waves of differing frequency) have associated low frequency components progressing shoreward with the group. They reasoned that this forced motion is somehow released when the incident waves break, reflects off the beach, and then propagates seaward as a free wave. Longuet-Higgins and Stewart (1962) subsequently showed theoretically that variations in the radiation stress due to grouping do indeed drive a forced wave, lending support to the basically intuitive arguments of Munk and Tucker. More recent modeling focuses on the possibility that the time varying breakpoint and sct-up associated with grouped incident waves can also be an effective surf beat generator (Symonds et al., 1982). Note that the Symonds et al. (1982) model predicts long wave generation inside the surf zone, and is not a release mechanism for the Munk-Tucker forced wave. This first class of models has no

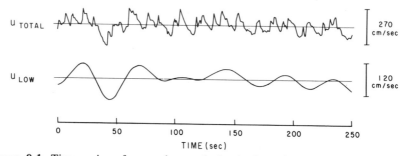

Figure 9-1. Time series of cross-shore velocity in 1 m depth, Torrey Pines Beach, 12 November 1978. U_{total} is all frequencies, U_{low} is passed, $f \leq .035$ Hz.

longshore variation in the incident wave groups. Surf beat is seen as the locally forced surf zone response to low frequency modulations in the incident wave height.

Gallagher (1971) extended the idea of forcing by incident wave groups to include the possibility of longshore variations in wave height when waves approach the shore at an oblique angle. Certain combinations of group frequency and longshore wave number can cause the resonant excitation of edge waves, the free wave modes trapped to the shoreline by refraction. Gallagher (1971) excluded the surf zone from his model.

Edge waves are gravity waves which, if excited in shallow water, cannot radiate energy to deep water. Hence, given a shallow water energy source having the proper combination of frequency and longshore wave number, edge waves can (in theory) grow to large amplitudes. Eckart's (1951) linear shallow water solutions give the edge wave dispersion relation between the angular frequency σ and the longshore wave number k as

$$\sigma^2 = gk(2n + 1)\beta \quad n = 0, 1, 2, \ldots \quad (9\text{-}1)$$

where β is the (small) beach slope. In Gallagher's (1971) theory, resonant edge wave excitation occurs if the incident wave group longshore wave number (k) and frequency (σ) satisfy (9-1). This model views surf beat as the resonant surf zone response to group forcing. Surf beat at any one time and surf zone location would be the result of past forcing over time scales of many surf beat periods.

Field studies of surf beat have been hampered both by the small number of sensors usually deployed and the limited range of incident wave conditions encountered during the generally short-term experiments. In spite of these difficulties, Suhayda (1971, 1974) and Huntley (1976) were able to demonstrate convincingly that surf beat elevations and cross-shore velocities have many of the characteristics of linear, shallow water waves that are standing in the cross-shore direction. Huntley (1976) used three biaxial current meters on a shore-perpendicular line to show also that the phase relationships and relative magnitudes of cross- and longshore surf beat velocity components were qualitatively consistent with the presence of low mode edge waves. Holman (1981) found a linear relationship between visually observed breaker heights and typical mid-surf zone orbital velocities at surf beat periods.

With a greatly increased number of sensors and a month-long deployment, the Torrey Pines experiment has quantified and extended many earlier results. Figure 4-1 shows a plan view of the experiment. The ten current meters on the 0.6 meter depth contour were used as a linear array to study the longshore structure of surf beat velocities (Huntley *et al.*, 1981). Two days, 20 and 21 November, were investigated. Cross spectra between all pairs of sensors were analyzed with the high resolution Maximum Likelihood Method (MLM) (Davis and Regier, 1977). At each frequency, this technique yields an estimate of the relative amounts of energy at various longshore wave numbers. The objective is to see if the observed distribution of energy in (σ, k) space agrees with the edge wave dispersion relation (9-1). Longshore and cross-shore velocities, which were considered separately, give markedly different results.

The longshore velocity fluctuations at surf beat periods show energy concentrated on the dispersion curves for low mode edge waves (Figure 9-2). The longest periods ($T > 80$ sec) have energy primarily in mode zero, while shorter periods ($40 < T < 80$ sec) have energy in mode 1. Roughly equal amounts of edge wave energy were propagating in up and down coast directions. The ten current meters on a line perpendicular to the shore (Figure 4-1) were used to verify that the offshore variation of longshore surf beat velocities was consistent with the edge waves found using the longshore array of sensors.

Huntley *et al.* (1981) show that, while the cross-shore currents and surface elevations are not inconsistent with the presence of the progressive edge waves deduced from the longshore currents, there are clearly some additional sources of low frequency energy which contribute particularly to the cross-shore flow. They suggest the presence of high mode standing edge waves, and/or locally forced (i.e., non-edge waves) motions. These types of flow can have cross-shore velocities much larger than the longshore component.

Analysis on other days of the Torrey Pines data set, and on the Santa Barbara experiment, is continuing. Questions of interest include:

(1) Is the surf beat longshore velocity field always composed primarily of low mode edge waves?

(2) Are low mode progressive edge waves ever a substantial fraction of the elevation and cross-shore velocity fields?

(3) Can the edge waves observed in the longshore velocities be related to the incident wind wave directional spectra as suggested by Gallagher (1971)?

(4) What are the energy sources and dynamics of the low frequency cross-shore velocities, which at least some of the time are not low mode edge waves?

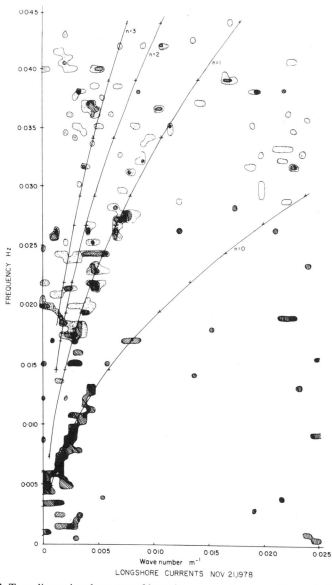

Figure 9-2. Two-dimensional spectra of longshore current energy. Shown is the average energy propagating northward and southward along the shore. Contours give the enhancement factor, defined as the ratio of the observed energy to the energy which would be observed if the energy at a given frequency were uniformly distributed across k space to an upper wave number of 0.015 m^{-1}. The solid lines give the predicted dispersion curves for edge wave modes on a plane beach of slope $\beta = 0.023$ (Huntley *et al.*, 1981).

The results and questions discussed above are interesting dynamical details, but there are further useful empirical conclusions which can be drawn without full theoretical understanding. Most sediment transport models do not include any surf beat motions in their descriptions of the surf zone velocity field. This is a serious flaw since, in the inner surf zone, surf beat velocities can be as large as the wind wave velocities (Figure 9-1). It would therefore be useful for modeling purposes to empirically relate cross-shore velocity energy levels at surf beat periods to offshore wind wave heights. Since the incident wave field is presumably the energy source for low frequency oscillations, some correlation between the two might be expected.

Because standing waves of all frequencies have antinodes at the shoreline, a run-up meter is an efficient way of monitoring surf beat energy levels. A run-up meter is essentially a very long wave staff deployed along the beach face, measuring the time history of the point where the swash intersects the dry beach. Knowledge of the beach profiles enables the measured swash motions in the plane of the beach face to be converted to vertical displacements. Figure 9-3 shows that the significant vertical run-up height at Torrey Pines is about 70 percent of the significant wave height in 10 m depth.

The solid line is

$$\overline{R_s^v}(cm) = 3.48(cm) + 0.71\overline{H_s}(cm) \tag{9-2}$$

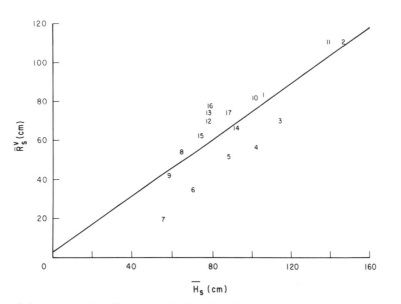

Figure 9-3. Average significant wave height in 10 m depth (H_s) and vertical run-up height ($\overline{R_s^v}$). Solid line is (9-2). Each number corresponds to a data run of several hours. (Guza and Thornton, 1982).

$\overline{R_s^v}$ is related to the variance of (vertical) swash oscillations (σ^2) by

$$\overline{R_s^v} = 4\sigma \qquad (9\text{-}3)$$

analogous to the relationship between significant wave height ($\overline{H_s}$) and the variance of sea surface fluctuations. $\overline{R_s^v}$ is the run-up equivalent of $\overline{H_s}$. Equation (9-2) essentially states that vertical elevation fluctuations are almost as large at the shoreline as in 10 m depth. Of course, the shoreline fluctuations are at much lower frequencies.

Figure 9-4 shows several spectra from data runs representative of those used in Figure 9-3. Energy levels in the wind wave band ($f > 0.05$ Hz) are relatively constant, independent of offshore wave height. This is not unexpected since, as discussed in Chapter 8, wave breaking strongly limits the height of wind waves in very shallow water. Regardless of the height of offshore waves, only very small bores reach the run-up. The surf zone and swash are saturated with energy at wind wave frequencies.

In contrast to the situation at wind wave frequencies, surf beat energy levels in the swash can differ by an order of magnitude (Figure 9-4). Surf beat frequencies are so low that very large amplitudes can be reached before wave breaking occurs. Thus, the increase in vertical swash excursions with increasing wind wave height (Figure 9-3) is due almost entirely to increases in surf beat energies (Guza and Thornton, 1982).

Given that incident wave heights and run-up surf beat levels can be empirically related through (9-2), it still remains to show that the cross-shore

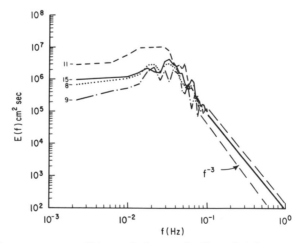

Figure 9-4. Run-up spectra: offshore wind wave significant heights are 140, 73, 65, 59 cm for data runs 11, 15, 8, 9, respectively. At high frequencies, all spectra fall within the dashed lines (Guza and Thornton, 1982).

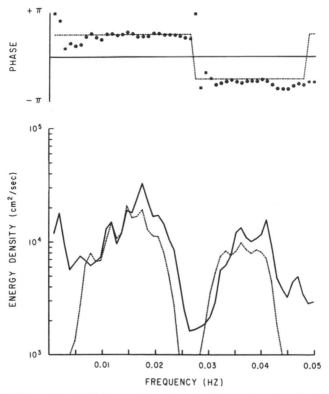

Figure 9-5. 6 November 1978, Torrey Pines Beach. Observed and predicted cross-shore velocity spectra (solid and dashed lines, lower panel), and observed and predicted phase difference between velocity sensor and run-up meter (symbols and dashed line, upper panel). * are not significant at 80% confidence. Current meter is in 1 m depth.

surf beat velocity spectrum can be related to the run-up spectrum. Following Suhayda (1974), Huntley (1976) and Holman (1981), we assume that surf beat motions in the cross-shore velocity and elevation fields are free standing waves, not strongly trapped at the coast. Given this assumption, and the measured run-up spectra, the linear shallow water equations can be numerically integrated to yield predictions of the surf beat velocity field at any location (Holman and Bowen, 1979).

Figure 9-5 shows the results of such a calculation. There is good agreement between observed and predicted velocity spectra. The spectral valley at 0.025 Hz is a node for velocities, and, as predicted, there is a $\Pi/2$ phase jump between the velocity and run-up sensors. The fact that such a simple model yields quantitative results is encouraging. However, it is unclear how these results are affected by complex topography or a steeper beach with a narrow surf zone.

Since the submission of this manuscript in 1982, additional analysis of NSTS surf beat data has been completed. Guza and Thornton (1985) show that the findings from Torrey Pines of an increase in surf beat energy level with increasing wave height (Figure 9-3), and qualitative agreement with a simple standing wave model (Figure 9-5), also hold at Santa Barbara. Oltman-Shay and Guza (1987) discuss the importance of low mode edge waves to the surf beat velocity field at both NSTS field sites. Extensive surf beat measurements at other field sites are discussed in Sallenger and Holman (1985). A preliminary intercomparison of various data sets is given by Guza et al. (1984).

9.1 References

Davis, R. E. and L. Regier, 1977, Methods for estimating directional wave spectra from multi-element arrays, *Journal of Marine Research*, 35: 453-477.

Eckart, C., 1951, Surface waves in water of variable depth, *Wave Report, SIO Reference 51-12, 100*, Scripps Institution of Oceanography, La Jolla, California, 99 pp.

Gallagher, B., 1971, Generation of surf beat by non-linear wave interactions, *Journal of Fluid Mechanics*, 49: 1-20.

Goda, Y., 1975, Irregular wave deformation in the surf zone, *Coastal Engineering Japan*, 18: 13-27.

Guza, R. T. and E. B. Thornton, 1982, Swash oscillations on a natural beach, *Journal of Geophysical Research*, 87: 483-491.

———. 1985, Observations of surf beat, *Journal of Geophysical Research*, 90: 3161-3172.

Guza, R. T., E. B. Thornton and R. A. Holman, 1984, Swash on steep and shallow beaches, *Proceedings*, Nineteenth Coastal Engineering Conference, September 3-7, 1984, Houston, Texas, American Society of Civil Engineers, New York: 708-723.

Holman, R. A., 1981, Infragravity energy in the surf zone, *Journal of Geophysical Research*, 86: 6442-6450.

Holman, R. A. and A. J. Bowen, 1979, Edge waves on complex beach profiles, *Journal of Geophysical Research*, 84: 6339-6346.

Huntley, D. A., 1976, Long period waves on a natural beach, *Journal of Geophysical Research*, 81: 6441-6449.

Huntley, D. A., R. T. Guza and E. B. Thornton, 1981, Field observations of surf beat, 1, progressive edge waves, *Journal of Geophysical Research*, 83: 1913-1920.

Inman, D. L., 1968, Mechanics of sediment transport by waves and currents, *SIO Reference Series 68-10*, Scripps Institution of Oceanography, La Jolla, California.

Longuet-Higgins, M. S. and R. W. Stewart, 1962, Radiation stress and mass transport in gravity waves, with application to "surf-beats," *Journal of Fluid Mechanics*, 13: 481-504.

Munk, W., 1949, Surf beats, *EOS Transcripts*, AGU 30: 849-854.

Oltman-Shay, J. and R. T. Guza, 1987, Infragravity edge wave observations on two California beaches, *Journal of Physical Oceanography*, 17(5): 644-663.

Sallenger, Jr., A. H. and R. A. Holman, 1985, Wave energy saturation on a natural beach of variable slope, *Journal of Geophysical Research*, 190(C6): 11939-11944.

Suhayda, J. N., 1971, Experimental study of the shoaling transformation of waves on a sloping bottom, Ph.D dissertation, Scripps Institution of Oceanography, La Jolla.

_____. 1974, Standing waves on beaches, *Journal of Geophysical Research*, 79: 3065-3071.

Symonds, G., D. A. Huntley and A. J. Bowen, 1982, Two-dimensional surf beat: long wave generation by a time-varying breakpoint, *Journal of Geophysical Research*, 87: 492-498.

Tucker, M. J., 1950, Surf beats: sea waves of 1 to 5 min period, *Proceedings, Royal Society*, Series A, 202: 565-573.

Wright, L. D., J. Chappell, B. G. Thom, M. P. Bradshaw and P. Cowell, 1979, Morphodynamics of reflective and dissipative beach and inshore systems. Southeastern Australia, *Marine Geology*, 32: 105-140.

Wright, L. D., R. T. Guza and A. D. Short, 1982, Dynamics of a high-energy dissipative surf zone, *Marine Geology*, 45: 41-62.

Chapter 10

NEARSHORE CIRCULATION

A. Conservation Equations for Unsteady Flow

Edward B. Thornton
Naval Postgraduate School

R. T. Guza
Scripps Institution of Oceanography
Center for Coastal Studies

The description of nearshore currents is of primary importance to the development of littoral transport models. The bases are given here for the current models to be described in Chapter 15. A convenient starting point for the study of unsteady flow phenomenon is a statement of the general conservation equations of mass, momentum and energy, following, for example, Mei (1983). In the first part of this chapter, the conservation equations are examined and discussed. A basic element of the NSTS dynamics task was to develop longshore current formulae for the case of statistically stationary waves (statistics do not change with time) propagating over straight and parallel contours. Making these assumptions, the momentum balances are simplified and compared with some of the NSTS field data in the second part of this chapter.

The analysis will not be concerned with the detailed internal flow structure of the fluid; hence, the derivation is simplified by integrating the mass, momentum and energy equations over the vertical depth and averaging over time. The time averaging is short compared with the total time (one can consider averaging over several wave periods), and does not preclude long term unsteadiness in the mean motion.

The conservation equations are applied here to wave motion, but they are equally applicable to turbulent motion. For algebraic simplicity, Mei (1983) defines a mean velocity

$$U_i = \frac{1}{D} \overline{\int_h^\eta u_i \, dz} \qquad i = 1, 2 \tag{10A-1}$$

where η is the free surface displacement and the total mean depth D is defined

$$D = \overline{\eta} + h \tag{10A-2}$$

where $\overline{\eta}$ is the mean derivation from the still water depth, h. Because the conservation equations are depth integrated, the vertical velocity is not explicitly included in the final equations and the indices ($i = 1,2$) refer only to horizontal components (x,y); the corresponding velocity components are denoted (u,v), (Figure 10A-1). It is noted the mean velocity includes contributions by the mean and averaged unsteady flow (e.g., mass transport velocity of surface gravity waves). The deviations about the mean include wave induced $\tilde{u}_i$, and turbulent u_i' velocities, such that the total velocity is given by

$$u_i = U_i + \tilde{u}_i + u_i' \quad i = 1,2 \tag{10A-3}$$

It follows from (10A-1) and (10A-3) that

$$\int_{-h}^{\overline{\eta}} (\tilde{u}_i + u_i')\, dz \equiv 0 \tag{10A-4}$$

An appealing feature of the conservation equations which follows from substitution of (10A-3) into the mass and momentum conservation equations is that the terms involving the mean and fluctuating quantities can be separated. This facilitates the understanding of the effect of the unsteady wave motion on the total flow phenomenon, and gives physical insight into the complicated mechanisms taking place in the nearshore.

10A.1 Conservation of Phase

In addition to the conserved quantities of mass, momentum, and energy normally considered, a fourth condition is available to aid the analysis -- conservation of phase, or more simply, conservation of wave crests [e.g., Whitham (1962)]. For simplicity, the conservation of phase is derived here for a single plane wave. The expressions derived are also applicable to more general linear wave motion which can be formed by the superposition of individual Fourier components. The expression for the water surface profile for harmonic motion is given by

$$\eta(x_i,t) = a(x_i)\cos(\int k_i dx_i - \omega t) \quad i = 1,2 \tag{10A-5}$$

where $a(x_i)$ is the amplitude, and the quantity $(\int k_i dx_i - \omega t)$ is called the phase function χ. The wavenumber k_i and radial frequency ω are defined in terms of the phase function

$$\vec{k} = \vec{\nabla}\chi$$

$$\omega = \frac{-\partial\chi}{\partial t} \tag{10A-6}$$

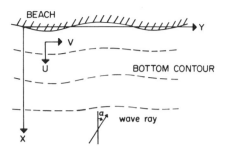

Figure 10A-1. Schematic of nearshore zone.

An important property of the wavenumber can be seen immediately by the vector identity

$$\vec{\nabla} \times \vec{\nabla}\chi = \vec{\nabla} \times \vec{k} \equiv 0 \tag{10A-7}$$

so that the wavenumber is irrotational. From (10A-6) and (10A-7), the kinematical conservation equation for the wavenumber can be written

$$\frac{\partial k_i}{\partial t} + \frac{\partial \omega}{\partial x_i} = 0 \quad i = 1, 2 \tag{10A-8}$$

If ω is expressed as a function of k_i and x_i, from local arguments (such as assuming a uniform simple harmonic wave train locally), and further allowing a mean current U_i, then,

$$\omega = \sigma(|k|, x_i) + k_i U_i \tag{10A-9}$$

σ is the local wave frequency and is the apparent frequency to an observer moving with the current. These equations will be used to describe wave refraction in the dynamic models (see Chapter 15).

10A.2 Conservation of Mass

The general conservation of mass equation is obtained by integrating over depth and averaging in time the continuity equation where it is assumed the density, ρ, is a constant

$$\frac{\partial}{\partial t} \bar{\eta} + \frac{\partial}{\partial x_i} U_i D = 0 \quad i = 1, 2 \tag{10A-10}$$

It is noted that time variability of mean sea level, $\bar{\eta}$, is allowed over long times, which could represent tidal or other long wave (compared with sea-swell) phenomenon.

10A.3 Conservation of Momentum

Integrating the full momentum equations over depth and averaging in time yields conservation equations for mean horizontal momentum per unit area

$$\rho D \left[\frac{\partial}{\partial t} U_i + U_j \frac{\partial U_i}{\partial x_j} \right] + \frac{\partial S_{ij}}{\partial x_j} =$$

$$-\rho g D \frac{\partial \bar{\eta}}{\partial x_i} + \tau_i^s - \tau_i^b \quad i,j = 1,2 \tag{10A-11}$$

where τ_i^s and τ_i^b are the mean surface and bottom stresses, and $-\rho g D \partial \bar{\eta}/\partial x_i$ is a net horizontal force per unit area due to the slope of the mean sea level which acts to balance changes in the momentum flux and boundary stresses. In arriving at (10A-11), it is assumed that molecular viscous stresses are small compared with the turbulent stresses and that the mean free surface and bottom slopes are small; both of these are reasonable assumptions in practical applications. The third term on the LHS of (10A-11) is the mean momentum flux due to the fluctuating motion, where

$$S_{ij} = \tilde{S}_{ij} + S_{ij}' \quad i,j = 1,2 \tag{10A-12}$$

The contribution to the mean momentum flux by wave motion is commonly referred to as the radiation stress (Longuet-Higgins and Stewart, 1962) and is given by

$$\tilde{S}_{ij} = \rho \overline{\int_{-h}^{\eta} \tilde{u}_i \tilde{u}_j dz} + \{ \overline{\int_{-h}^{\eta} p \, dz} - \frac{1}{2} \rho g D^2 \} \delta_{ij} \tag{10A-13}$$

$$i,j = 1,2$$

where p is the total pressure and δ_{ij} is the Kronecker delta. Explicit expressions for $\tilde{S}_{ij}$ are obtained by substituting linear wave theory descriptions for velocity and pressure to give (Longuet-Higgins and Stewart, 1964)

$$\tilde{S}_{11} = \tilde{S}_{xx} = \frac{E}{2} \left[2 \frac{C_g}{C} \cos^2 \alpha + \left[2 \frac{C_g}{C} - 1 \right] \right]$$

$$\tilde{S}_{22} = \tilde{S}_{yy} = \frac{E}{2} \left[2 \frac{C_g}{C} \sin^2 \alpha + \left[2 \frac{C_g}{C} - 1 \right] \right] \tag{10A-14}$$

$$\tilde{S}_{12} = \tilde{S}_{21} = \tilde{S}_{xy} = \tilde{S}_{yx} = E \frac{C_g}{C} \sin \alpha \cos \alpha$$

where E is the energy density, C_g is the group velocity, C is the phase speed and α is the angle of wave incidence (see Figure 10A-1). This description of the

momentum flux due to waves will be used extensively in describing the
nearshore wave dynamics in this and later chapters.

The significant turbulent contributions to S_{ij} represent the integrated
Reynolds stresses

$$S_{ij}' = \rho \int_{-h}^{\eta} \overline{u_i' u_j'}\, dz \qquad\qquad\qquad (10A\text{-}15)$$

A more complete description that includes higher order terms is given by Mei
(1983).

Since a description of the integrated Reynolds stress is not known, it has
been variously parameterized using eddy viscosity

$$S_{ij}' = -\rho \mu D\, \frac{\partial U_i}{\partial x_j} \qquad i,j = 1,2 \qquad\qquad (10A\text{-}16)$$

where μ is the kinematic eddy viscosity coefficient. The kinematic eddy
viscosity is introduced to simulate the Reynolds stresses acting to transfer
momentum laterally. Bowen (1969) was the first to apply (10A-16) and simply
assumed μ constant. The difficulty of using μ is that it is a function of the flow
field and, thus, varies from case to case. Subsequent authors have employed
Prandtl's mixing length hypothesis

$$\mu = - |q\,l| \qquad\qquad\qquad (10A\text{-}17)$$

where q and l are characteristic scales of velocity and length. Thornton (1970)
and Jonsson et al. (1974), consider the mixing length to be characterized by the
water particle excursion due to wave motion and the velocity intensity
characterized by the wave-induced velocity. Although these could be
appropriate scales, the wave-induced excursion and velocity are in quadrature
and in reality do not contribute to turbulent exchange. Longuet-Higgins (1970)
suggested that the mixing length should increase with distance from shore and
the velocity intensity be characterized by the wave speed, such that

$$\mu = N\, |x| \sqrt{gh} \qquad\qquad\qquad (10A\text{-}18)$$

where N is an adjustable coefficient in the range $0 \leq N \leq 0.016$. Ostendorf and
Madsen (1979) used a similar formulation, but assumed the characteristic
velocity to be the maximum velocity at the bed. Bowen and Inman (1972)
compared various eddy viscosity formulations with values obtained from dye
dispersion studies within the surf zone and from model fitting of measured
longshore current distributions; they concluded that the Longuet-Higgins (1970)
formulation gave the most reasonable fit with data. Although (10A-18) appears
to be a reasonable form, a recognized deficiency is that eddy viscosity increases
indefinitely offshore, which would not be expected outside the surf zone if the
turbulent momentum exchange is associated with wave breaking (Battjes, 1975).

Battjes (1975) gives a physical argument for (10A-18). He emphasizes that the maintenance of turbulence is not from the shear of the mean flow, but generated by the breaking waves; as a consequence, the length scale is limited by the vertical depth. Battjes further hypothesizes that the mean rate of wave energy dissipation equals the rate of turbulent energy production locally to obtain an appropriate velocity scale. For a plane beach, he derives the same form as (10A-18), but

$$N = (5/16 \, \gamma^2)^{1/3} (tan \, \beta)^{4/3} M \qquad (10A-19)$$

where $\gamma = H/D$ and M is a constant. μ now depends on the beach slope, $tan\beta$. The correct formulation (indeed, if one exists) has yet to be determined. A deficiency to date has been the lack of good field data to test the solutions and make determinations of the model coefficients.

The mean surface and bottom stresses are assumed to be quadratic functions of the wind and bottom fluid velocity fields. Since the winds in the data to be analyzed did not play a significant role dynamically, only the bottom stresses are addressed. The mean bottom shear stress is usually expressed as

$$\vec{\tau}^b = \rho c_f \, \overline{|\vec{u}| \, \vec{u}} \qquad (10A-20)$$

where the total velocity component is comprised of both mean and fluctuating components and c_f is the bed shear stress coefficient.

For the case of small angle of wave incidence and weak currents $U_i/|\bar{u}| \ll 1$), the bed shear stress of (10A-20) simplifies to (Longuet-Higgins 1970a, b; Thornton, 1970)

$$\vec{\tau}^b = \rho c_f \, |\bar{u}| \, \vec{U} \qquad (10A-21)$$

which has the effect of linearizing the momentum equations in U_i.

Liu and Dalrymple (1978) investigated the case of large angle of wave incidence and strong currents, and showed that (10A-21) is valid when

$$\sin \alpha \ll 4c_f \, \frac{(1+3\gamma^2/8)}{5\pi \tan \beta} \sim 0.27 \, \frac{c_f}{\tan \beta} \qquad (10A-22)$$

where $\tan \beta$ is the bottom slope and a value of $\gamma = 0.44$ has been used. Assuming the nominal value for $c_f = 0.01$ and beach slopes of 0.01 at Torrey Pines and 0.04 at Santa Barbara, (10A-22) implies that $\vec{\tau}^b$ given by (10A-21) applies at Torrey Pines ($\alpha < 16$ degrees) but only sometimes at Santa Barbara ($\alpha < 4$ degrees).

In the general case, but limiting the discussion to the condition of straight and parallel contours, the modulus of the total velocity is given by

$$|\vec{u}| = (V^2 + \tilde{u}^2 + 2V \, \tilde{u} \sin \alpha)^{1/2} \qquad (10A-23)$$

where V is the longshore velocity, and the cross-shore velocity $U = 0$ for the situation under consideration. The longshore velocity is composed of the mean current plus the wave component

$$v = V + \tilde{u} \sin \alpha \qquad \text{(10A-24)}$$

The general longshore bed shear stress formulation requires substituting (10A-23, 24) into (10A-20) and averaging over the wave period at each location giving

$$\tau_y^b = \frac{1}{(\omega T)} \int_T \rho c_f (V^2 + \tilde{u}^2 + 2V\tilde{u} \sin \alpha)^{1/2}(V + \tilde{u} \sin \alpha) d\,\omega t \qquad \text{(10A-25)}$$

This results in a difficult nonlinear solution since the dependent variable V cannot be brought outside the averaging integral, requiring V to be solved iteratively. In addition, if the mean current interaction with the wave field is significant, the mean velocity effects should be included in wave refraction

$$V + \frac{C}{\sin \alpha} = \frac{C_o}{\sin \alpha_o} \qquad \text{(10A-26)}$$

Various formulations for τ^b will be tested against data in Chapter 15.

A difficulty in applying a quadratic bed shear stress formulation is the lack of information on c_f. The friction coefficient under waves is dependent on bottom roughness and flow intensity near the bed. A number of experiments have been conducted in the field and laboratory to measure c_f with a large range of reported values. Shemdin et al. (1977) summarized a number of the field experiments, and included new data, for experiments in which the bottom friction appeared to be the dominant dissipation mechanism (mean grain size 0.1 – 0.4 mm). Additional measurements are reported by Iwagaki and Kakinuma (1967). All the past field experiments measured wave attenuation in depths of 10 – 45 m, well outside the surf zone, between two locations separated by large distances (compared to a wave length). Dissipation was calculated using either a monochromatic wave approach (Bretschneider and Reid, 1954), or a spectral technique incorporating quasi-linear dissipation developed by Hasselmann and Collins (1968). The reported values varied widely ($c_f = 0.006$–0.3), depending on experiment and location. Shemdin et al. (1977) attributed the large variability of c_f to greatly increased bottom roughness when ripples are present.

Several formulae relate c_f to bottom roughness through an equivalent roughness which is ambiguously related to sand size and ripple height (Jonsson, 1966; Riedel et al., 1972). More recent formula explicitly state the roughness in terms of well defined measures of sand size and ripple height (Vitale, 1979; Nielsen, 1981; and Grant and Madsen, 1982). But inside the surf zone, sheet flow generally exists, causing ripples to disappear and the bottom to become flat; very few measurements have been made under these conditions. Direct measures of c_f in the surf zone were made by Thornton and Guza (1978) for the Torrey Pines experiment, and for this reason are described in detail below.

10A.4 Conservation of Energy

The energy balance equation is derived in the same manner as the mass and momentum conservation equations by partitioning the fluctuating and steady motion, integrating over depth and averaging in time. The energy balance for the fluctuating motion as derived by Phillips (1977), but applying the definition for mass transport (10A-2), is given by

$$\frac{\partial E}{\partial t} + \frac{\partial}{\partial x_i} [U_i E + F_i] + S_{ij} \frac{\partial U_j}{\partial x_i} = -\varepsilon \qquad (10A\text{-}27)$$

$$i = 1,2$$

where F_i is the mean energy flux by the fluctuating motion alone and ε represents the rate of energy dissipation per unit area. In the models to be described in Chapter 15, the energy balance is applied outside the surf zone where the mean currents are assumed negligible. In addition, it is assumed the waves are statistically stationary, in which case (10A-27) simplifies to

$$\frac{\partial F_i}{\partial x_i} = \frac{\partial EC_{gi}}{\partial x_i} = -\varepsilon, \quad i = 1,2 \qquad (10A\text{-}28)$$

This equation is used to solve for the wave field input into the current models.

10A.5 Stationary Waves Over Straight and Parallel Contours

The mass and momentum flux balance equations are now used to examine some preliminary experimental results. The simplified case of stationary waves propagating over straight and parallel contours is considered, consistent with the assumptions made for the longshore current and sediment transport models presented in Chapters 15-17. Wave stationarity means the time dependent terms $(\partial/\partial t)$ in the conservation equations are zero. Excluding edge waves, straight and parallel contours implies that the wave induced alongshore gradients $(\partial/\partial y)$ are also zero. These assumptions lead to great simplification in the nearshore dynamical equations.

10A.5.1 Cross-Shore Momentum Balance

As waves shoal, there is a general decrease in wave length and increase in wave height until a wave steepness is reached at which the waves break. The velocities and resulting onshore-directed momentum flux exhibit similar changes to the wave height, increasing to a maximum at near breaking and then decreasing inside the surf zone. It has been shown theoretically for monochromatic waves (Dorrenstein, 1961; Longuet-Higgins and Stewart, 1962, 1963, 1964) that a small depression (set-down) occurs seaward of the break point

$$\bar{\eta} = -a^2 \frac{k}{2 \sinh 2kh} \tag{10A-29}$$

But once the waves start to break there is a decrease in momentum which is balanced by a larger superelevation or set-up. Bowen *et al.* (1968) performed a detailed laboratory study and verified that the set-up inside the surf zone is proportional to the beach slope, and obtained an explicit form for the constant of proportionality, K,

$$\frac{\partial \bar{\eta}}{\partial x} = -K \frac{\partial h}{\partial x} \quad K = (1 + 2.67\gamma^{-2})^{-1} \tag{10A-30}$$

Both theory and laboratory measurements show that the maximum set-up, $\bar{\eta}_M$, occurs at the shoreline. These simple monochromatic theories predict (Battjes, 1974)

$$\frac{\bar{\eta}_M}{H_b} = 0.3\gamma \tag{10A-31}$$

where H_b is the wave height at breaking.

The extension of the monochromatic set-up theory to cover incident wave spectra (Battjes, 1974; Battjes and Janssen, 1978) necessarily involves assumptions about the distribution of wave heights in the surf zone and the amount of energy lost by individual waves during the breaking process. These complications preclude explicit analytic statements about the size of $\bar{\eta}_M$ relative to H_b, which is now a statistical quantity. These models, however, are not very sensitive to variation in beach slope, and when properly normalized, are only weakly dependent on details of the offshore incident wave spectra. Theory and laboratory experiments with random waves show (Battjes, 1974) that

$$0.14 < \frac{\bar{\eta}_M}{H_{s,\infty}} < 0.21 \tag{10A-32}$$

where $H_{s,\infty}$ is the significant wave height in deep water. The range of values given in (10A-32) corresponds to an order of magnitude variation in incident wave steepness.

The mean run-up measurements at Torrey Pines (see Chapter 9) were used to determine the wave set-up (Guza and Thornton, 1981). The $\bar{\eta}_M$ is compared with the significant wave height in deep water in Figure 10A-2. $H_{s,\infty}$ was calculated by using standard linear shoaling theory to back the waves out to deep water from the deepest pressure spectrum, measured in about 10 m depth. The best fit line to the data gives

$$\bar{\eta}_M = 0.17 H_{s,\infty} \tag{10A-33}$$

which agrees essentially with (10A-32) suggested by laboratory and theoretical studies with random waves.

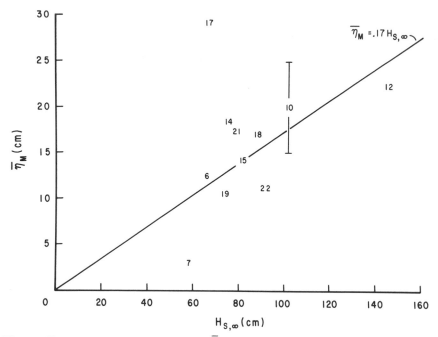

Figure 10A-2. Set-up at the shoreline ($\overline{\eta}_M$) versus offshore significant wave height (H_s, ∞). Solid line is best fit, $\overline{\eta}_M = 0.17 H_s$, ∞. Numbers correspond to day the data were recorded in November 1978. (Guza and Thornton, 1981)

The measurements of velocity and elevation at Torrey Pines are used to gain insight into the mechanism resulting in the measured set-up. It was shown in Chapter 8 that the waves at Torrey Pines could be approximated as normally incident and as having linear dynamics. The wave induced shoreward momentum flux density is then given by

$$S_{xx}(f) = E(f) \left[\frac{2kh}{\sinh 2kh} + \frac{1}{2} \right] \tag{10A-34}$$

with E(f) the wave energy spectral density. Assuming a stationary wave field, no mean cross-shore currents, and ignoring viscous stresses, gradients in the total S_{xx} must be balanced by changes in the mean sea level

$$\frac{\partial S_{xx}^T}{\partial x} + \rho g(\overline{\eta} + h) \frac{\partial \overline{\eta}}{\partial x} = 0 \tag{10A-35}$$

where

$$S_{xx}^T = \int S_{xx}(f) df \tag{10A-36}$$

All set-up theories basically hypothesize x dependencies for $E(f)$ and thereby for $S_{xx}(f)$, and integrate (10A-35) for $\bar{\eta}$. Outside the surf zone, wave energy flux is usually assumed to be conserved, which leads to a predicted shoreward increase in $S_{xx}(f)$ (10A-34) and a negative gradient, or depression, in mean level (10A-35) [the set-down of (10A-29)]. The qualitative correctness of these standard assumptions, applied to the present field data, is shown in Figure 10A-3 where S_{xx}^T is plotted as a function of depth for three representative data days. Well seaward of the surf zone, the observed values of S_{xx}^T agree well with predictions based on linear shoaling theory with the deepest observed $S_{xx}(f)$ spectrum as an input condition (solid lines). In the theoretical predictions of S_{xx}^T, the real topography at Torrey Pines can be acceptably modeled as straight and parallel contours so that shoaling effects can be treated analytically. Pressure sensors were used to calculate $S_{xx}(f)$ which requires that the local waves be close to normally incident, as in the present experiments. This is demonstrated by the similar values of S_{xx}^T obtained from a current meter and pressure sensor at the same horizontal location, shown in Figure 10A-3 as two points at the same depth (i.e., November 21, h = 380 cm, 550 cm). A typical measured $S_{xx}(f)$ in 10 m depth, and predicted and observed $S_{xx}(f)$ in 4.25 m depth, are shown in Figure 10A-4. Nonlinear shoaling effects not accounted for in the theory lead to a gross underprediction of $S_{xx}(f)$ at frequencies near the second harmonic of the spectral peak, but this is partially compensated for by an overprediction at the spectral peak. The calculated and observed S_{xx}^T differ by 15%. Because $S_{xx}(f)$ is proportional to E(f), the discussion in §8.2 and 8.3 about local and shoaled energy densities is easily extended to local and shoaled radiation stress densities. The relevant conclusions are (1) linear theory adequately relates local $S_{xx}(f)$ measurements from pressure, sea surface elevation and velocity sensors to each other, and (2) linear shoaling theory does a qualitatively good job of predicting S_{xx}^T outside the breaker zone, given an input spectrum at 10 m depth. Thus, Figure 10A-3 leads to the expectation that seaward of the breaker zone, the model predictions for set-down are based on valid assumptions about the spatial dependence of S_{xx}^T. None of the present measurements is accurate enough to measure set-down.

In the breaker zone, Figure 10A-3 shows that data from the three days approximately collapse to a single curve representing saturation as discussed in §8.4. For a random wave field in shallow water,

$$S_{xx}^T = \frac{3}{2} E^T = \frac{3}{16} \rho g H_{rms}^2 = \frac{3}{16} \rho g \gamma^2 (\bar{\eta} + h)^2 \qquad (10A\text{-}37)$$

The dashed curve in Figure 10A-3 is (10A-37) for $\gamma = 0.28$. The difference in $\gamma = 0.44$ reported in §8.4 and the γ reported here is due to the spectra being summed over a narrower band here. Given that the data approximately follow (10A-37)

$$S_{xx}^T = 0.015 \rho g (\bar{\eta} + h)^2 \qquad h < h_b' \qquad (10A\text{-}38)$$

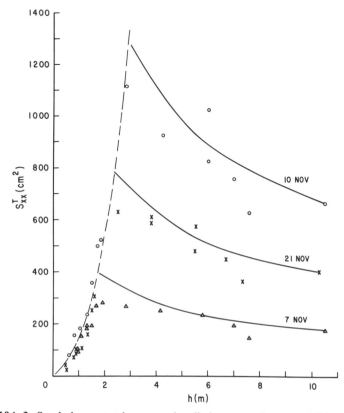

Figure 10A-3. Symbols are total measured radiation stress between 0.05 and 0.3 Hz versus depth, based on 4096 s of data. Solid lines are predictions based on linear shoaling theory; dashed line is saturation (Equation 10A-37, $\gamma = 0.28$). (Guza and Thornton, 1981)

where $h_b{}'$ is the depth of maximum S_{xx}^T, (10A-35) yields

$$\frac{\partial \bar{\eta}}{\partial x} = -0.03 \frac{\partial h}{\partial x} \qquad h < h_b{}' \tag{10A-39}$$

This is the same functional form as the classical monochromatic result (10A-30), but the constant of proportionality $K = 0.3$ is an order of magnitude greater in the monochromatic case. Integrating (10A-39) from $h_b{}'$ to the shoreline yields

$$\bar{\eta}_M = 0.03 h_b{}' \tag{10A-40}$$

where the small set-down at $h_b{}'$ is neglected. From Figure 10A-3, $h_b{}'$ varies from roughly 1.6 m on November 7 to 2.8 m on November 10 which suggests $\bar{\eta}_M$ values of 4.8 and 8.4 cm, respectively. This is substantially less than the

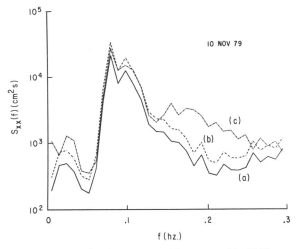

Figure 10A-4. Radiation stress density spectra: (a) measured in 1040 cm total depth; (b) predicted at $h = 425$ cm; (c) observed at $h = 425$ cm. Data record is 137 min, 64 degrees of freedom. (Guza and Thornton, 1981)

values inferred from the measurements (Figure 10A-2). A possible explanation for the discrepancy between measured set-up, and the smaller value calculated on the basis of measured values of S_{xx}^T and integration of the momentum equation, is the invalidity of the assumption of constant set-up slope across the surf zone. Laboratory measurements by Bowen *et al.* (1968) and by Van Dorn (1976) showed that set-up slope increased markedly very close to the shoreline. The measurements suggest underestimates of $\overline{\eta}_M$ by 30 – 100% using (10A-30) on a plane beach.

Variations in $\overline{\eta}$ values inferred from the measured S_{xx}^T and a crude numerical integration of (10A-35) are shown in Figure 10A-5 for typical data runs. The smooth lines are drawn to guide the eye. Offshore there is a small set-down. The approximately constant set-up slope, except very near the shoreline, reflects the fact that (10A-37), which predicts a constant slope (10A-39), provides a reasonably good fit to the observed S_{xx}^T (Figure 10A-3). The inferred set-up slope is substantially smaller than that observed in laboratory studies, primarily because the present field values of γ are less than half those typical of laboratory experiments. Set-up was measured only at the shoreline, and these data points are shown as open. The conclusion that the set-up slope markedly increases very close to the shoreline is supported by laboratory data with monochromatic incident waves (Bowen *et al.*, 1968). Data from all runs suggest $\overline{\eta}_M = 0.17 H_{s,\infty}$ (Figure 10A-2). This result could be different on beaches with a different porosity, or with a topographic (bar) structure which alters the spatial variation of S_{xx}^T.

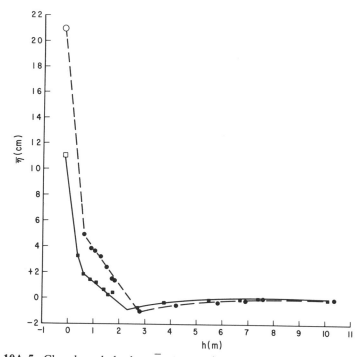

Figure 10A-5. Closed symbols show $\overline{\eta}$ obtained by numerical integration of (Equation 10A-35), using measured S_{xx}^T. Open symbols are $\overline{\eta}M$ measured by the run-up meter. Box, November 19; circle, November 10. (Guza and Thornton, 1981)

10A.5.2 Alongshore Momentum Flux Balance and Bottom Shear Stress

Alongshore momentum flux balances were calculated for the Torrey Pines data to determine an appropriate bed shear stress coefficient (Thornton and Guza, 1981). The simplest model was used in accordance with the above assumptions of stationary waves over straight and parallel contours. The alongshore momentum balance (10A-11) simplifies to the cross-shore change in the alongshore momentum flux balanced by the average alongshore bottom shear stress τ_y^b

$$\frac{\partial S_{xy}^T}{\partial x} = -\tau_y^b = -\rho c_f \overline{|\vec{u}| v} \tag{10A-41}$$

The quadratic shear stress law (10A-20) has been assumed. $S_{xy}(f)$ is the depth integrated covariance between u and v

$$S_{xy}(f) = \int_{-h}^{0} \overline{\rho u(f,z) v(f,z) dz}$$

$$= \int_{-\pi}^{\pi} E(f,\alpha) n(f) \sin \alpha \cos \alpha \, d\alpha \tag{10A-42}$$

where $n = 1/2 + 2kh/\sinh 2kh$, with f the frequency, α the angular deviation from normal incidence, and E the energy density per unit frequency and direction. The S_{xy}^T is summed over all frequencies

$$S_{xy}^T = \int_f S_{xy}(f)df \qquad\qquad (10\text{A-}43)$$

The objective here is to calculate the unknown bed shear stress coefficient c_f by measuring all other terms in (10A-41).

The measurements and analysis emphasized the difficulties in directly measuring the radiation stress, S_{xy} and its gradients. $S_{xy}(f)$ was measured in 10 m using the 5 sensor linear pressure array where the incident angle was relatively large and the array alignment well known. But it was not possible at Torrey Pines Beach to make direct measures of S_{xy} inshore of 3 m because the small angles of incidence resulted in large errors. As pointed out by Thornton and Guza (1978), the primary measurement errors in S_{xy} can be due to (1) errors in the current meter orientation, (2) using discrete measurement locations to represent a continuous function (see Chapter 4) and (3) using a narrow-band directional spectrum approximation (see Chapter 2).

Large errors in radiation stress can occur due to even small errors in resolving the direction of fluid motion. The total energy density and the quadrature-spectrum are invariant with coordinate rotation, but the cospectrum used to calculate radiation stress is very sensitive to coordinate rotation. Guza and Thornton (1978) showed that the relative error in the radiation stress at a particular frequency is given by

$$Rel\ Error = \frac{\sin \Delta}{\tan \alpha(f)} \sim \frac{\Delta}{\alpha(f)} \qquad\qquad (10\text{A-}44)$$

where α is the true incident wave angle and Δ is the angular error. A major difficulty encountered at Torrey Pines was that the relatively long period (mean wave period 13 sec) waves were highly refracted so that in shallower water the angle of wave incidence was small. For example, applying Snell's law to monochromatic plane waves with period 13 sec shows that even a relatively large deep water wave approach of 20 degrees is reduced to a small angle of about 3 degrees in 1 m depth. Hence, measurement errors must be substantially less than 3 degrees to have an acceptable measurement error in S_{xy}.

Angularity errors are caused by lack of instrument resolution, installation misalignment, and choosing the incorrect alongshore direction. Tow tank tests on the Marsh-McBirney current meter showed reasonably good directional resolution with a nominal error of ± 2 degrees at small angles of incidence. Considerable effort was made to properly orient the current meters (§4A.5). The maximum combined error due to instrumentation resolution and the orientation procedure was judged to be about 3 degrees, and this is unacceptably large.

Given perfect instrument directional response and perfect orientation, there is also the more fundamental problem of defining the longshore direction. What spatial scales should be averaged over to determine a contour orientation? Errors associated with choosing a longshore direction even on the relatively straight and parallel contours of Torrey Pines appear to be on the order of several degrees.

The combined angularity errors are at least as large as the angles of wave incidence in shallow water, resulting in S_{xy} errors of 100% or more. Therefore, direct measures of S_{xy}^T in shallow water (< 3 m) at Torrey Pines could not meaningfully be made, and an approximate method described below had to be used.

S_{xy} values for frequencies between 0.05 Hz and 0.3 Hz were measured using a current meter in 3 m depth by calculating the covariance of uv as indicated by (10A-42). The current meter was aligned properly by numerically rotating the current meter until the $S_{xy}(f)$ values for frequencies $0.05 - 0.115$ Hz matched those obtained by refracting the measured $E(f, \alpha)$ from the very accurately aligned 10 m depth array to 3 m and used to calculate $S_{xy}(f)$ with (10A-42). The current meter, after proper orientation, provided $S_{xy}(f)$ values over the frequency range $0.05 - 3$ Hz. Excellent agreement between $S_{xy}(f)$ from the refracted array inputs and the rotated current meter across a wide range of frequencies was obtained in this manner (see Figure 2; Thornton and Guza, 1981).

Once the waves start to dissipate their energies due to bottom friction or breaking, the current meter alignment correction technique is no longer appropriate. Instead, indirect measures of S_{xy} inside 3 m were obtained by making the reasonable assumption (for Torrey Pines Beach) that the bottom contours are straight and parallel. Assuming long crested spectral components specified by a single direction, α, the velocities can be represented by

$$v(f) = \vec{u}(f) \sin \hat{\alpha}(f)$$

$$u(f) = \vec{u}(f) \cos \hat{\alpha}(f) \qquad\qquad (10A\text{-}45)$$

Substituting (10A-45) into (10A-42) and multiplying the numerator and denominator by the celerity, C(f), gives:

$$S_{xy}(f) = \frac{\sin \hat{\alpha}(f)}{C(f)} \frac{C(f)}{\cos \hat{\alpha}(f)} \int_{-h}^{0} \overline{\rho u^2(f,z,t)} dz \qquad (10A\text{-}46)$$

The integral of (10A-46) can be determined using linear dynamics as demonstrated in §8.2.

The celerity was measured directly from the phase spectra between pairs of current meters (see §8.5). Since the current meters were in a line perpendicular to the beach, the x-component of celerity is measured

$$C_x(f) = \frac{2\pi f}{k_x} = \frac{C(f)}{\cos \hat{\alpha}(f)} \tag{10A-47}$$

which is the second term on the rhs of (10A-46).

For straight and parallel contours, the term $\sin \hat{\alpha}(f)/C(f)$ is a constant by Snell's law (referred to here as Snell's constant). Snell's constant contains all the local wave angle information across the surf zone and is determined from (10A-42) using the accurately measured radiation stress at 3 m.

The total radiation stress inshore is then measured using

$$S_{xy}^T = \int_f \frac{\sin \alpha_o(f)}{C_o(f)} C_x(f) \int_{-h}^o \overline{u^2(f)}\, dz\, df \tag{10A-48}$$

where $C_x(f)$ and $u(f)$ are measured locally and subscript o denotes offshore (at the 3 m contour). The unknown bed shear stress coefficient can be determined since all other terms of (10A-41) are measured.

The S_{xy} values appeared to be statistically noisy due to variability in the incident wave energy and perhaps also wave direction (Guza and Thornton, 1978). Because the incident wave angle was small, even small changes in angle resulted in relatively large changes in S_{xy}. More than a 100% variation of 17.1 minute averages of S_{xy} at 3 m depth often occurred during a 100 minute period. The variability in S_{xy} is indicated by the scatter of values for 17.1 minute averages of c_f, calculated at various distances offshore for November 4th (Figure 10A-6). The calculated c_f values show considerable scatter. Some c_f values are even negative, indicating the currents are opposing the driving force. The averages for 137 minutes are connected by a solid line. The average of all calculations is 0.006.

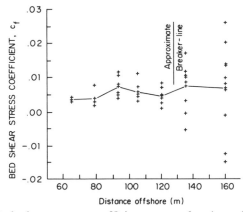

Figure 10A-6. Bed shear stress coefficients as a function of distance offshore, 10 November 1978 at Torrey Pines Beach, California. Small crosses indicate 17.1 minute averages. The line connects 137 minute averages. (Thornton and Guza, 1981)

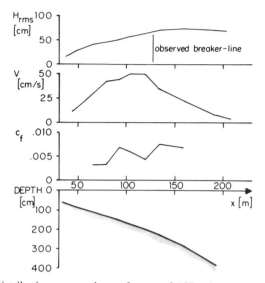

Figure 10A-7. Distribution across the surf zone of 137 minute averages of rms wave height, longshore velocity, bed shear stress coefficient and depth, 10 November 1978 at Torrey Pines Beach, California. (Thornton and Guza, 1981)

An example of the distributions across the surf zone of rms wave height, H_{rms}, mean longshore current and bed shear stress coefficient is given in Figure 10A-7. All values are 137 minute averages. The shoaling waves first increase in height and then decrease, as energy is dissipated primarily due to wave breaking. The visually observed mean breaker-line is indicated. The longshore current distribution exhibits the classical quasi-parabolic distribution with a maximum velocity of approximately 50 cm/s shoreward of the observed breaker-line. The c_f values were consistent across the surf zone for the 137 minute averages, varying between 0.003 – 0.007. Table 10A-1 gives average values of c_f obtained by averaging together c_f values (137 minute averages) from all spatial locations. The five days were chosen as having the most homogeneous currents in the alongshore direction. Other days, when rip currents were clearly present, were also analyzed and these sometimes showed negative values of c_f. The spatial variability of c_f is reflected by the standard deviation given in Table 10A-1, which often exceeds the mean. No physical reason could be found for excluding 17 November. The mean bed shear stress coefficient for the values given in Table 10A-1 (excluding the negative value) is 0.01±0.01. The results emphasize the difficulty in making a quantitative alongshore momentum flux balance. For this reason, results obtained from the Torrey Pines experiment regarding longshore currents and bed shear stress

Table 10A-1. Bed Shear Stress Coefficients

	c_f	Standard Deviation
04 Nov	.012	.034
10 Nov	.006	.002
17 Nov	(− 013)	.130
18 Nov	.030	.020
24 Nov	.002	.005
Average	.01 ± .01	

measurements should be viewed cautiously. As we shall see in Chapter 15, the longshore current results from Santa Barbara are much more encouraging.

10A.6 References

Battjes, J. A., 1974, Computations of set-up, longshore currents, run-up overtopping due to wind-generated waves, Report 74-2, Delft University of Technology, Delft, The Netherlands.

_____. 1975, Modeling of turbulence in the surf zone, *Proceedings*, Second Annual Symposium on Modeling Techniques, San Francisco, CA, September 3-5, 1975, American Society of Civil Engineers, New York: 2, 1050-1061.

Battjes, J. A. and J. P. F. M. Janssen, 1978, Energy loss and set-up due to breaking of random waves, *Proceedings*, Sixteenth Coastal Engineering Conference, August 27-September 3, 1978, Hamburg, Germany: 569.

Bowen, A. J., 1969, Rip currents 1: theoretical investigations, *Journal of Geophysical Research*, 74: 5479-5490.

Bowen, A. J. and D. L. Inman, 1972, Nearshore mixing due to waves and wave-induced currents, *Proceedings*, Symposium Physical Processes Responsible for Disposal of Pollutents in the Sea, International Council for the Exploration of the Sea, 167: 6-12.

Bowen, A. J., D. L. Inman and V. P. Simmons, 1968, Wave set-down and wave set-up, *Journal of Geophysical Research*, 73(8): 2569-2577.

Bretschneider, C. L. and R. O. Reid, 1954, Modification of wave height due to bottom friction, percolation, and refraction, Beach Erosion Board Technical Memo No. 45.

Dorrestein, R., 1961, Wave set-up on a beach, *Proceedings*, Second Technical Conference on Hurricanes: 230.

Grant, W. D. and O. S. Madsen, 1982, Movable bed roughness in unsteady oscillation flow, *Journal of Geophysical Research*, 87: 469-481.

Guza R. T. and E. B. Thornton, 1978, Variability of longshore currents, *Proceedings*, Sixteenth Coastal Engineering Conference, August 27-September 3, 1978, Hamburg, Germany, American Society of Civil Engineers, New York, 1: 756-775.

Hasselmann, K. and J. I. Collins, 1968, Spectral dissipation of finite-depth gravity waves due to turbulent bottom friction, *Journal of Marine Research*, 26(1): 1-12.

Iwagaki, Y. and T. Kakinuma, 1967, On the bottom friction factors off five Japanese coasts, *Coastal Engineering in Japan*, 10: 13-22.

Jonsson, I. G., 1966, Wave boundary layers and friction factors, *Proceedings*, Tenth Conference on Coastal Engineering, September, 1966, Tokyo, Japan, American Society of Civil Engineers, New York: 127-148.

Jonsson, I. G., O. Skovgaard and T. S. Jacobsen, 1974, Computation of longshore currents, *Proceedings*, Fourteenth Coastal Engineering Conference, June 24-28, 1974, Copenhagen, Denmark, American Society of Civil Engineers, New York: 699-714.

Liu, P. L. F. and R. A. Dalrymple, 1978, Bottom friction stresses and longshore currents due to waves with large scales of incidence, *Journal of Marine Research*, 36: 357-475.

Longuet-Higgins, M. S., 1970a, b, Longshore currents generated by obliquely incident sea waves. 1 and 2, *Journal of Geophysical Research*, 75: 6778-6789 and 6790-6801.

Longuet-Higgins, M. S. and R. W. Stewart, 1962, Radiation stress and mass transport in gravity waves with application to surf beats, *Journal of Fluid Mechanics*, 13: 481.

_____. 1963, A note on wave set-up, *Journal of Marine Research*, 21: 4.

_____. 1964, Radiation stress in water waves, a physical discussion with application, *Deep Sea Research*, 11: 529.

Mei, C. C., 1983, *The Applied Dynamics of Ocean Surface Waves*, John Wiley and Sons, New York, 740 pp.

Nielsen, P., 1981, Dynamics and geometry of wave generated ripples, *Journal of Geophysical Research*, 86: 6467-6472.

Ostendorf, D. W. and O. S. Madsen, 1979, An analysis of longshore current and associated sediment transport in the surf zone, Parsons Laboratory, Dept. Civil Engineering, Massachusetts Institute of Technology Report 241.

Phillips, O. M., 1977, *The Dynamics of Upper Ocean*, Cambridge University Press, New York, 261 pp.

Riedel, H. P., J. W. Kamphuis and A. Brebner, 1972, Measurement of bed shear stress under waves, *Proceedings*, Thirteenth Coastal Engineering Conference, July 10-14, 1972, Vancouver, B.C., American Society of Civil Engineers, New York: 587-604.

Shemdin, O. H., K. Hasselmann, S. V. Hsiao and K. Herterick, 1977, Nonlinear and linear bottom interaction effects in shallow water, *Proceedings*, NATO Symposium on Turbulence Fluxes through the Sea Surface, Wave Dynamics and Prediction, Marseille, France: 347-372.

Thornton, E. B., 1970, Variation of longshore current across the surf zone, *Proceedings*, Twelfth Coastal Engineering Conference, September 13-18, 1970, Washington, D. C., American Society of Civil Engineers, New York: 291-308.

Thornton, E. B. and R. T. Guza, 1981, Longshore currents and bed shear stress, *Proceedings*, Directional Wave Spectra Applications, American Society of Engineers, New York: 147-163.

Van Dorn, W. G., 1976, Set-up and run-up in shoaling breakers, *Proceedings*, Fifteenth Coastal Engineering Conference, July 11-17, 1976, Honolulu, Hawaii, American Society of Civil Engineers, New York: 738-751.

Vitale, P., 1979, Sand bed friction factors for oscillatory flows, *Journal of the Waterway, Port, Coastal and Ocean Division*, American Society of Civil Engineers, WW3: 229-45.

Whitham, G. B., 1962, Mass, momentum and energy flux in water waves, *Journal of Fluid Mechanics*, 12: 135-147.

Chapter 10

NEARSHORE CIRCULATION

B. Rip Currents and Wave Groups

Ernest C.-S. Tang and Robert A. Dalrymple

Ocean Engineering Group
Department of Civil Engineering
University of Delaware

10B.1 Introduction

The field data collected at Torrey Pines Beach in November 1978, were used to study the offshore wave groups, the wave-induced nearshore circulations and their generation mechanisms on a beach with relatively simple bathymetry. Unlike the theoretical models of nearshore circulation based on deterministic mechanisms, the mean currents at natural beaches of dissipative type (where the ratio of deep water wave steepness to beach slope is high and waves break cleanly) are superimposed on very energetic infragravity waves. Migrating and pulsating rip currents with long alongshore wave length scales were particularly difficult to detect, using the current meter array, despite the visual observations of rip currents. A special instrument, consisting of tripod-mounted current meter and a pressure sensor, linked to shore by telemetry, was designed to make measurements in a stationary rip current (see Chapter 4); however, it proved of little value, due primarily to the lack of stationarity of the nearshore circulation system.

This chapter reviews the various theories for rip current generation, presents data for rip currents obtained from the vector plots of all the current meters, and then, utilizing multivariate analysis, the hypothesis that the rips are driven by the offshore wave groups is examined.

The water motions in the surf zone are dominated by wind waves, the mean longshore currents, and low frequency wave motion. Superimposed on these motions is the nearshore circulation system, composed of rip currents, flowing rapidly offshore in a narrow jet, feeder currents and a slow return current from the offshore.

In the Torrcy Pines experiment, the nearshore circulation is studied for the purpose of identifying its physical features and mechanisms. Using array processing techniques, the migrating nearshore circulation system is shown to be a very low frequency motion (1000 sec or longer). Its periodic spatial structure was not easily identifiable from the field data, due to its long wave length (greater than 500 m), its nonlinear structure and the presence of other forced wave motion at this particular frequency band, causing unsatisfactory wave number resolution. However, the field observations, both the field notes and the vector plots of the current meter output confirm the presence of the rip currents.

From a review of the various theories for nearshore circulation, the mechanism thought most applicable to the Torrey Pines data set is a wave-interaction model, which postulates that the nearshore system is driven by the offshore wave climate, particularly the wave groups.

This hypothesis is examined using several multivariate methods, including empirical orthogonal eigenfunction analysis and canonical coherency analysis. These techniques are applied to the data of November 20, 1978, when rip currents, migrating to the north, were observed in the field. The same techniques are applied to the infragravity wave bands (0.005 to 0.02 Hz), which are likely to have the same forcing mechanism and which are dominated by edge wave motion, in order to compare results with the even lower frequency rip current band, which we show to contain too much energy to be due simply to free waves at this site (see §10B.5).

10B.2 Theoretical Models for Nearshore Circulation

Since the introduction of the concept of radiation stresses in waves by Longuet-Higgins and Stewart (1964), numerous models have been proposed to explain the occurrence of periodically spaced rip currents along a shoreline. While our understanding of longshore currents is comparatively adequate, the only agreement between the existing nearshore circulation models is that there must be a wave height (or wave-induced set-up) variation in the longshore direction. Dalrymple (1978) classified the rip current models as wave interaction models or structural (including the bottom) interaction models. Here we use three classes:

(1) wave-boundary interaction models,
(2) wave-wave interaction models,
(3) instability models.

The first category consists of models requiring an incident wave-bathymetric interaction to produce a periodic longshore variation in wave height and set-up. Within the surf zone, this variation leads to periodic longshore gradients in the radiation stresses and hence the formation of cellular circulation patterns. These models include:

(a) local bathymetric variation; Bowen (1969), Sonu (1972), Noda (1974), Ebersole and Dalrymple (1979) among many others;

(b) offshore bathymetric variation (leading to divergence and convergence of the wave field due to refraction); Shepard and Inman (1950);
(c) offshore structures; Liu and Mei (1976), Sasaki (1975), Mei and Angelides (1977).

This category of models may account for most of the rip currents observed in nature; however, rip currents are also observed on planar beaches, forming bottom variations. Therefore additional mechanisms are necessary.

The second category was initiated by Bowen (1969) and Bowen and Inman (1969), who showed analytically and verified in the laboratory that the interaction between a normally incident wave and either standing or progressive synchronous edge waves would lead to a spatially stable longshore variation in wave height. Dalrymple (1975) proposed a model including two or more synchronous wave trains incident on the beach with different angles, which would cause a similar effect as the Bowen model (without appealing to edge waves). Dalrymple's model has also been verified both in the laboratory (Dalrymple and Lanan, 1976) and numerically (Ebersole and Dalrymple, 1980).

The third class of model was suggested by LeBlond and Tang (1974). By perturbing the solution of the governing equations (for the mean horizontal flows and the mean water level) for uniform set-up on a planar beach created by normally incident waves, other stable modes of the nearshore circulation have been shown to exist. However, the differences in the assumed equations governing the waves, currents and the energy dissipation inside the surf zone will produce different results for the scales of the induced nearshore circulation (Iwata, 1978, Dalrymple and Lozano, 1978, Miller and Barcilon, 1978, and Hino, 1974, who includes a sedimentary response as well). All of these models are eigenvalue solutions, and the magnitude of the nearshore circulation is not determinable; however, the rip current spacing is calculable and often related to the beach slope. Only the model of Dalrymple and Lozano includes the wave-current interaction, which provides a physical mechanism for the stable nearshore current field. In their model, the outflowing rip current refracts the incoming waves towards the base of the rip current, which drives longshore feeder currents, which, in turn, maintain the rip currents. No verification of these models has been made either in the laboratory or through the use of numerical models.

Sasaki (1975) classified the rip current generation mechanisms by the surf similarity parameter, defined as the ratio of the offshore wave steepness to the beach slope. For steep, reflective beaches (the similarity parameter is small and surging or collapsing breakers occur), he suggested that the rip currents are generated by synchronous edge waves and incoming wave interactions. For beaches with milder slopes, the instability models were suggested. An infragravity wave model was proposed for beaches with mild slopes (large surf similarity parameter, spilling breakers). This classification deserves more investigation and verification.

Each of the three types of models given above have drawbacks due to the simplifications leading to the analytic solutions. Further, as discussed, they are valid only under certain conditions. For the Torrey Pines data set, which corresponds to a mild slope beach with nearly straight and parallel offshore contours, the wave-boundary interaction models were discounted. As for the wave-wave models which utilize edge waves, the recent studies of edge waves by Guza and Inman (1975), Bowen and Guza (1978), and Guza and Chapman (1979) show that edge waves at the incident wave frequency would probably not exist on mild beaches as their wave length is too short compared to the width of the dissipative surf zone. Therefore, we narrowed down the candidate models to two: interacting incident waves or instability models. Field verification of the instability models is difficult, requiring proof of strong wave-current interaction for the Dalrymple and Lozano model. The instability models also do not provide a correlation between the incoming waves and the magnitude of the rip current velocities.

Our major focus, then, was on the incoming wave-interaction models. The models of Dalrymple (1975) and Ebersole and Dalrymple (1980) were developed for steady forced circulation only, which is a special case of the infragravity wave generation model of Gallagher (1971), who proposed that beats of the incoming wave train could generate infragravity waves, particularly resonant edge waves, through nonlinear interactions. The forced motion with very low frequency may have a response similar to the steady circulation, in the form of slowly migrating circulation cells. Further the combination of a steady circulation cell and the low frequency wave motion may give an appearance of migrating and/or pulsating nearshore circulation cells. These cells may have a stronger offshore flow than inshore flow due to convective acceleration (Arthur, 1962) and become rip currents. Therefore we examine the incoming wave groups as a forcing mechanism for circulation in the surf zone.

The low frequency motion (30-300 sec) in the surf zone, including surf beat and infragravity waves, were first described by Munk (1949) and Tucker (1950). Longuet-Higgins and Stewart (1962) and Gallagher (1971) developed the theoretical background for understanding the observations. Bowen and Guza (1978) examined the generating mechanism due to incoming wave groups in a laboratory wave tank. This study confirmed that the infragravity wave motion can indeed be induced by wave groups and it also proved the existence of resonant excitation of trapped edge waves when the beat frequency ($1/T$) and the longshore wave number ($2\pi/L$) are related by the edge wave dispersion relationship (Ursell, 1952).

$$L = \frac{gT^2\sin\{(2n + 1)\beta\}}{2\pi}$$

where β is the plane beach slope angle. Holman (1979), Huntley et al. (1981)

and Symonds *et al.* (1982) indicated that most of the infragravity wave motion can be explained by edge waves. Field observations of edge waves, Huntley and Bowen (1973), Huntley (1976), Katoh (1981) and Huntley *et al.* (1981), support this conclusion. However, the wave period of the edge waves examined by these investigators was always less than 200 sec. The large scale experiment performed by Munk *et al.* (1964) at a depth of 7 m showed evidence that the edge waves with periods of 900 sec may exist.

If the nearshore circulation and the infragravity wave motion are forced by the incoming wave field, then a statistical approach is more appropriate for the analysis, since the incoming waves are random and the response is certainly random. This randomness may explain why rip currents are not always visible; they may be the extremes of a certain random process governing the nearshore circulation in and outside the surf zone.

10B.3 Deterministic Analysis of the Data

The identification of the rip currents was achieved by examining successive vector plots of mean currents (averaged over a 256 sec period) at 19 locations, ranging 520 m alongshore and 350 m offshore (Figure 10B-1), which often showed the migrating nature of the circulation pattern. The strong offshore flows associated with rip currents usually occurred at positions and times in general agreement with visual observation of rips, injected dye patterns and time lapse photographs (taken from the backshore cliff, about 100 m south of the main range line, using a Nikon camera with an automatic timer).

Examination of the vector plots for November 20 in detail showed that the point of maximum offshore flow migrated slowly to the north (which corresponds with the recorded field observations) at a speed estimated to be less than 0.3 m/sec and the longshore spacing of the rip is probably longer than 500 m since there were never two clearly discernible rip currents in the experimental area simultaneously. This conclusion is tentative, since the northern portion of the longshore array had very wide spacings between the instruments and it was not clear whether some of the weak offshore flows measured there might have been rip currents. Field observations make no mention of rip currents at the northern meters.

The measured strength of the offshore flow in the throat of the rip current had a maximum of about 50 cm/sec. However, this velocity is not likely to be the maximum associated with the rip current for several reasons:

(1) the rip currents typically have their maximum velocities in a narrow throat region near the breaker line, while the longshore instrument array was usually well inside the breaker zone. As a result only one or two instruments on the offshore profile could possibly catch the maximum velocity when a rip passed.

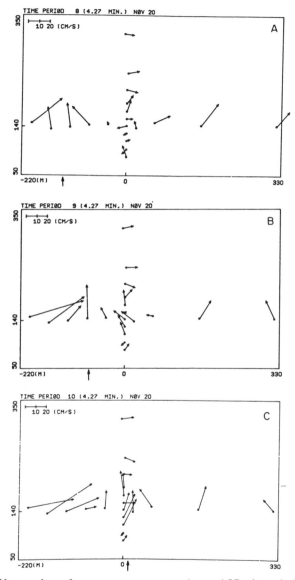

Figure 10B-1. Vector plots of mean currents averaged over 4.27 min period showing rip currents migrating to the north. The arrows on the bottom of the plots indicate probable rip head positions. a) Time period 8, November 20, starting at 11:35 AM. b) Time period 9. c) Time period 10.

(2) The currents were measured at a depth about 40-60 cm above the bottom, while the maximum of a rip current likely occurs at the surface.

(3) A 256 sec averaging period is too long for measuring the peak flow as a rip current migrates past a current meter.

The time scale for the nearshore circulation system was estimated to be (length/speed) of the order of 1500 sec. The estimated period is longer than the 7.8 min fluctuation period between the two stationary rips reported by Shepard and Inman (1950), but the generating mechanism (a divergent wave field due to an offshore canyon) is different than at Torrey Pines. Shepard and Inman noted that the beating of the rip current flows had a period much longer than the surf beat period; thus, we directed our attention to the low frequency part of the wave spectrum.

10B.4 Spectral Analysis and Statistical Methods

Spectral analysis of the longshore velocity spectra have been discussed elsewhere. For this discussion, we examine the low frequency portion of the spectra. A high resolution analysis using 2 hours and 50 minutes of data (5 blocks of data, each 2048 sec long are averaged to give 10 degrees of freedom at each current meter) shows the predominance of the low frequency motion (Figure 10B-2a); the energy in the five lowest frequency bands (0.00049-0.0024 Hz) contains 70 percent of the total energy in the infragravity wave range (periods larger than 30 sec) and there is little energy at higher frequencies. The rms amplitude for the first five bands alone is 20 cm/sec and the mean velocity averaged over the ten longshore current meters is 9 cm/sec directed northward with individual means varying from 20 cm/sec to 1 cm/sec along the array, suggesting some spatial dependency of the mean flow. [The EOF analysis discussed later shows that this low frequency motion has a longshore wave length of 0 (1000 m), suggesting that these are not tidal flows (Figure 10B-1)].

In Figure 10B-2b, the spectrum of the cross-shore flow is shown to be clearly different than the longshore flow. The first five low frequency bands contain a significant amount of energy (with rms amplitudes of 12.3 cm/sec); however, they contain only 16 percent of the total infragravity energy. There are three discernible high energy bands (<0.002, 0.006-0.015, and 0.024-0.034 Hz) with another peak at the incoming wave frequency band (0.068 Hz, not shown). The order of magnitude of the lowest frequency peak is the same as the second one, but the latter has a much wider bandwidth. The mean cross-shore flow is offshore at 12 cm/sec and it varies from 20 cm/sec to 4 cm/sec along the array. The likely cause of this offshore mean flow is the return flow induced by the onshore mass transport (Hansen and Svendsen, 1984).

It is not clear why the cross-shore flow energy is lower (about 1/3) than the longshore flow at the lowest frequency bands. The cross-shore flow seems to be dominated by a strong offshore mean flow and small oscillations with low frequencies, while the longshore flow has a large amplitude oscillation.

The strong, very low frequency longshore currents inside the surf zone cannot be pure edge wave motion since it would require a large surface elevation and stronger cross-shore currents than were measured offshore. It seems that the very low frequency motion observed in this study is due to forced wave motion and that the motion with higher (infragravity) frequencies is due to edge waves.

The wave groups in the offshore wave record were obtained with a square law detector (Rice, 1945), which is one of two convenient ways to obtain the envelope of the waves. (The second method is the Hilbert transform, which has been used by Melville, 1983). Squaring a time signal, applying a low-pass filter, and then taking the square root of the result yields the envelope of a narrow-banded wave train. Figure 10B-3 shows an example of this technique applied to the signal of one of the offshore pressure transducers. Figure 10B-2c shows the spectrum of the wave groups averaged over 4 offshore sensors. The lowest frequency bands clearly are more predominant than the higher frequencies, leading to the conclusion that the low frequency wave groups provide nearshore forcing at the lowest frequencies.

For the longshore current meter array, array processing techniques can provide a wave number-frequency spectrum, which indicates the energies present at a given wave number and frequency. The resolution of this technique depends on the number of spatial lags, or unique distances, between the various meters. Huntley *et al.* (1981) have used this method to examine the infragravity waves present in the Torrey Pines experiment.

Standard Fourier methods do not produce a clear peak in the wave number spectrum for the nearshore circulation system, as the strong narrow rip current must be described by higher harmonics in wave number space, spreading out the wave number peaks. The same problem occurs in the frequency domain for migrating nearshore circulation cells. A more objective technique, that does not *a priori* specify the shape of the wave form, is more useful in this case; for example, the empirical orthogonal eigenfunction (EOF) method. Another approach is to examine the statistical properties of the low frequency motion and to find its correlation with the supposed forcing mechanism -- the incoming wave groups. If the response of the nearshore circulation is self-correlated and has a high correlation with the forcing, then the low frequency motion is meaningful and induced by the wave groups.

The longshore wave number-frequency (W.-F.) spectrum is obtained by a Maximum Likelihood Method (MLM), which was originally developed by Capon *et al.* (1967) for the analysis of seismic data containing random Gaussian noise, and later presented in Capon (1969). For details of the method, see McDonough (1982) and Davis and Regier (1977). The advantage of the MLM is that it has a higher resolving power than Fourier methods for narrow-banded data; however, for broadbanded signals, the MLM estimate is unreliable (Cox, 1973).

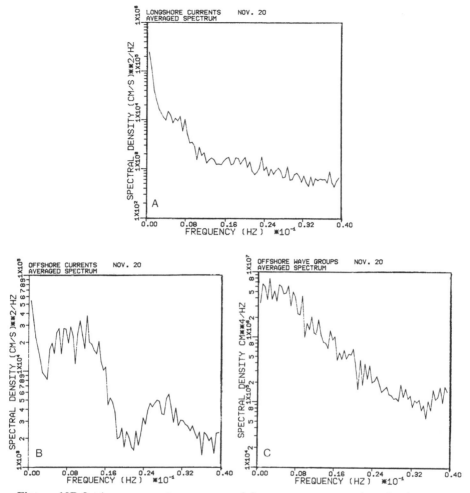

Figure 10B-2. Average spectrum measured by sensors at approximately the same offshore position. a) Longshore currents, 10 sensors on the main longshore line. b) On-offshore currents, 9 sensors on the main longshore line, excluding C26. c) Offshore wave groups, 4 sensors on the deep water longshore line, excluding P5.

The MLM processor is a nonlinear processor -- it is data adaptive, so the statistical properties of the result are hard to assess. It is difficult to find a criterion to determine whether the coherence in the wave number-frequency domain between two sets of supposedly correlated data sets is significant or not. To examine the relationship between the forcing and the response, we still need to resort to the frequency domain coherency analysis.

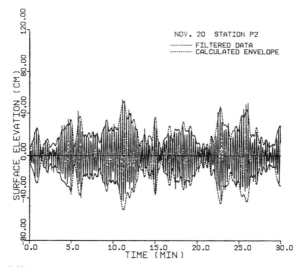

Figure 10B-3. Offshore wave groups detected by the square law detector, sensor P2. The data are band-filtered before processing (0.04 Hz to 0.3 Hz).

The often used (bivariate) coherence analysis examines the relationship between data obtained with a pair of instruments. It is sometimes misleading: for example, it is known that the coherence between data measured at two sensors becomes small as the distance between them becomes large. This is not a result of noise or lack of correlation, but largely due to the directional spread of the waves (see Isobe *et al.* (1984) who provide analytical results for a theoretical spectrum).

A better approach to the coherency is to consider the whole data set together, i.e., considering the coherence between every pair of instruments to generate a more meaningful result. The analysis then becomes multivariate analysis, of which spectral analysis is a special case, as it assumes a known deterministic (sinusoidal) relationship between the data. In general, this relationship is to be found from the data.

Several multivariate analyses were used in this study, specifically, principal component analysis (or EOF), multiple and partial coherence, and canonical coherence. All of these techniques were developed to be used in the real domain; however, we used them in the complex (frequency) domain. The resulting equations are identical to their real counterparts, except that the variables are complex. The significance test is slightly different; its derivation is based on the results of Goodman (1964), Khatri (1965), Giri (1965) and Krishnaiah (1976). Only a brief sketch of the various methods will be provided here; further details are found in books on multivariate analysis (e.g., Anderson, 1958, and Press, 1982).

10B.4.1 Empirical Orthogonal Eigenfunction Analysis

It is generally desirable to extract one or more dominant components from a set of data. These components can be interpreted with physical models or used as a base, with greatly reduced number of variables, for further analysis. The technique is based on two requirements: to choose the dominant components of the data and to determine how many of the components are statistically meaningful, such that the rest can be considered noise. The technique is generally known as principal component analysis, or EOF. [Wallace and Dickenson (1972) in the analysis of meteorological data (complex domain); Winant *et al.* (1975) analysis of seasonal beach profile changes (real domain); Katoh (1981), Holman and Bowen (1984) analysis of wave data to detect edge waves (real domain)].

The criterion for picking the dominant component is to extract from the data the component associated with the maximum variance. After the first component is extracted, the same principle is used to extract the next dominant component. This procedure is continued until the data can be represented by n components, corresponding to n instruments. Analytically, the process is equivalent to finding the eigenvectors and eigenvalues of the cross-spectral matrix.

The EOF is based on a discrete representation of the data (only n orthogonal eigenfunctions are found). For data consisting of a continuous spectrum, the results may not be applicable, but the encouraging results of Katoh (1981) and Winant *et al.* (1975) seem to indicate that the method is useful. Some insightful discussion of the resolving power of the EOF technique is given by Wallace and Dickenson (1972).

Each EOF component represents a possible type of wave structure (plus noise). Often the structure can be interpreted physically. For standing waves, the amplitude of the component would vary periodically along the longshore instrument array and the phases would have a 180 degree jump at the nodal points. A progressive wave form would have constant amplitude in space and the phases would decrease linearly in the propagation direction with a 360 degree phase change after one wave length. More complicated wave forms may not be resolvable by the technique.

A statistical test of significance was developed based on the results of Krishnaiah (1976) and Anderson (1958), which involves a χ-square distribution.

10B.4.2 Multiple and Partial Coherency Analysis

Multiple and partial coherency analyses study the relationship between two sets of data, each containing a different number of instruments. The details of the techniques can be found in Jenkins and Watts (1968) and Priestley (1981). Examining the coherence between two data sets indicates the amount that the output of one instrument in a data set is explained by the other data set. Partial coherence is used to determine the relationship between any pair of instruments in one data set after statistically deleting the portion of the data which is correlated to the other data set.

If the first set of data is forced by the second data set (including noise), the multiple coherencies will be high, while the partial coherencies in the first set, given the second set, will be low. If some free wave components exist in the first set, in addition to forced components, the partial coherencies will be high as well. The statistical significance test (Jenkins and Watts, 1968) is used to determine the dependencies between two sets of data.

10B.4.3 Canonical Coherence

Canonical coherency analysis treats two data sets together to determine the maximum possible coherence between them. The procedure involves finding pairs of components in each data set which have the maximum possible coherence between them. The maximum number of pairs is determined by the smallest number of instruments in the data sets.

The canonical coherency analysis is used for two reasons. The first is to examine the components of the two data sets: the offshore wave groups and the nearshore currents, and the second is to test the dependency of the two data sets. If dependency is found, then the pair with the maximum coherency can be subtracted and the dependency can be examined again. In this way, the dependency of the two data sets on numerous forcing functions can be examined. A criterion developed by Tatsuoka (1971) was used to test independence.

10B.5 Results and Discussion

Before a detailed analysis of the data using statistical techniques is discussed, it is necessary to show that the low frequency energy in the rip current band is meaningful, instead of a resolution problem or the result of tidal changes over the measurement period. The wave number resolution of the field data is not high, due to the short array length; however, the frequency resolution is adequate. The spectral resolution is 0.00049 Hz, but there are only 10 degrees of freedom assuming longshore homogeneity. One hundred degrees of freedom were obtained by averaging the frequency spectrum of the ten sensors over the longshore array. After examining the 99 percent confidence limits, there is no doubt that the lowest frequency bands of the longshore flows are the most energetic.

The effect of the tidal change during the duration of the data gathering can be assessed by noting that, at a fixed point offshore, for a stationary wave field, the signals collected at that point will be nonstationary, due to the continuous change in water depth. To recover the stationary spectrum desired, it is necessary to shift the instrument horizontally, keeping the water depth constant. For all practical purposes, we shall confine our attention to fluid motion with length and times scales much smaller than the corresponding tidal scales. We can transform the coordinate system by a simple translation, so that the origin of the coordinate system is placed on the mean shoreline with the equations of motion governing the fluid motion remaining virtually the same since the shoreline accelerates much slower than the fluid motion. Now the stationary spectrum obtained at the fixed position of the new coordinate is desired, while the instrument, moving with the speed of the shoreline relative to the new coordinates, is where the data are collected. We can heuristically examine the tidal effects on two types of motion. For a standing wave, such as surf beat, the motion can be described by $F(x) \cos(\sigma t)$, where σ is the angular frequency of the infragravity motion. A Taylor series expansion allows us to approximate the signal at $x + \partial(t)$ as

$$\left[F(x) + F'(x)\, \partial(t) \right] \cos(\sigma t)$$

for small shoreline excursions. As $\partial(t)$ is periodic with the tidal frequency, side bands at $\sigma \pm$ (tidal frequency) are produced but no low frequency signals are produced. The frequency of a progressive incoming wave train will be doppler-shifted by an amount proportional to kc, where k is the wave number of the wave train and c is the speed of the moving observor, who is moving with the speed of the shoreline tidal migration speed. With a change in tide of 20 cm over the data gathering period on November 20, the effects are very small.

The possiblity that the low frequency peak is a random error can be excluded by noting that a low frequency peak followed by a valley in the frequency spectrum is often found among other published data: Holman and Bowen (1984, peak at 0.002 Hz, for a beach slope of 0.04), Huntley (1976, <0.005 Hz, slope of 0.022), and Holman (1979, 0.005 Hz and slopes of 0.03 to 0.04). These peaks are more pronounced for the longshore velocity and run-up spectra. While the second and higher peaks and valleys in their spectra are often explained by edge waves, the first spectral peak and valley were never examined, partially due to the long time and space length scales involved.

For edge waves of such long period, say 1000 sec, the length scale will be about 30 km based on the dispersion relationship and a beach slope of 0.02. The longshore velocities should be smaller than or equal to the cross-shore velocities, which is not the case for the November 20 data. The long length scale would also make such waves detectable far offshore, but they are not detectable at the offshore pressure sensors. Furthermore, the amplitude of such an edge wave, corresponding to the measured velocities at this frequency would be much larger (16 cm) than the measured amplitudes in the surf zone. Therefore, the conclusion is that the major portion of the low frequency motion is locally induced, decaying rapdily offshore, such as is characteristic of nearshore circulation systems.

10B.5.1 MLM Spectral Analysis Results

Transformation of the cross-spectral matrices of the longshore arrays using the MLM results in the wave number-frequency spectrum shown in Figure 10B-4a, b, c. The contour levels are the enhancement factor, which is defined as the ratio of the calculated energy to the reference energy, obtained by assuming the observed energy in the range of 0.00049 to 0.04 Hz is uniformly distributed over the frequency range and the wave number space between $\pm$ 0.025 m^{-1} for longshore and cross-shore currents and $\pm$ 0.0085 m^{-1} for the offshore wave groups (corresponding roughly to the Nyquist wave number for the respective array configurations). The edge wave dispersion curves for a plane beach of slope 0.02 are also shown.

The highest peak in the longshore current spectrum is located at 0.00049 Hz and at a longshore wave number of $-$ 0.001 m^{-1} which represents a northward propagating wave. The second highest peak is located at the same frequency with the opposite wave number as before. There are several minor low frequency peaks (at less than 0.005 Hz spread out at larger wave numbers up to the Nyquist wave number (not shown on these plots). These peaks may be due to the higher harmonics of a nonlinear spatial wave structure rather than caused by aliasing because the peaks are less than one tenth the amplitude of the major peak. This is revealed by plotting the same spectrum with a different reference scale as shown in Figure 10B-5, where the contour levels are the band-by-band enhancement factors with the calculated energy normalized by the energy which would occur if the observed energy at the frequency band were evenly spread over the wave number axis. The plot shows that aliasing is not a serious problem; in addition, the local peaks at each frequency are seen to cluster around the edge wave dispersion relationship as first shown by Huntley *et al.* (1981) for frequencies between 0.005 and 0.025 Hz. The lower frequency peak contains much more energy than the edge wave peaks (factor of 15).

It was initially expected that the wave number-frequency spectra of the offshore wave groups (Figure 10B-4b) would show the same structure as observed in the longshore currents (Figure 10B-4a). There is no obvious correspondence. This may be due to two reasons: the responses of the surf zone velocities to offshore wave groups are very complicated functions of wave number and frequency (Gallagher, 1971) and the mismatching or inadequate resolving power of the MLM. In addition the statistical instability associated with the low number of degrees of freedom (10) may change the appearance of the spectra.

The cross-shore current wave number-frequency spectrum (Figure 10B-4c) is drastically different from the longshore results. Besides the peak at (0.00049 Hz, 0.00175 m^{-1} there are five peaks between 0.005 and 0.015 Hz.

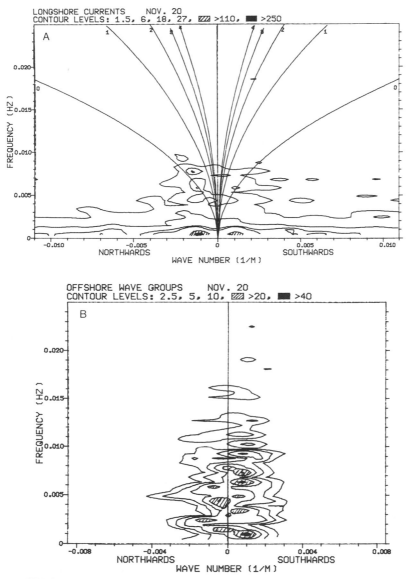

Figure 10B-4. W.-F. spectra measured by the sensors at approximately the same offshore position. The contour levels are the enhancement factors defined as the ratio of calculated energy to the reference energy which would occur if the observed energy between 0.00049 Hz to 0.04 Hz were uniformly distributed over the wave number space ranging from ±0.025 (1/m). The curves are edge wave dispersion curves; mode numbers are indicated under the curves. a) Longshore currents, on the main longshore line. b) Offshore wave groups, on the deepwater longshore line.

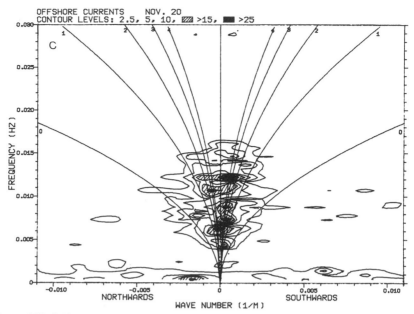

Figure 10B-4. W.-F. spectra measured by the sensors at approximately the same offshore position. The contour levels are the enhancement factors defined as the ratio of calculated energy to the reference energy which would occur if the observed energy between 0.00049 Hz to 0.04 Hz were uniformly distributed over the wave number space ranging from ±0.025 (1/m). The curves are edge wave dispersion curves; mode numbers are indicated under the curves. c) On-offshore currents, on the main longshore line.

10B.5.2 Empirical Orthogonal Eigenfunction Analysis

For the EOF analysis, 50 degrees of freedom were used by combining the data within five frequency bands, with the final resolution of 0.0024 Hz.

In most of the cases, the results of the EOF compare well with the MLM spectra, as they should. The first orthogonal functions of the lowest frequency band (0.00146 Hz with a bandwidth of 0.00244 Hz) for the offshore wave groups and longshore currents are shown in Figure 10B-6a. For longshore currents, the linear change in phase and the small variation of the amplitudes along the array indicates a progressive profile moving to the north; the wave length is estimated to be about 750 m (based on the 180 degree change of phase over about 375 m), compared to the peak of the MLM at 1000 m in Figure 10B-4a. The first component of the offshore wave group is interpreted to be an almost normally incident wave group, which is not in disagreement with the corresponding peak in the wave number frequency diagram (Figrue 10B-4b), when effects of the band averaging are considered. The percentages given in the figure are the amount of the variance explained by the eigenfunction. The

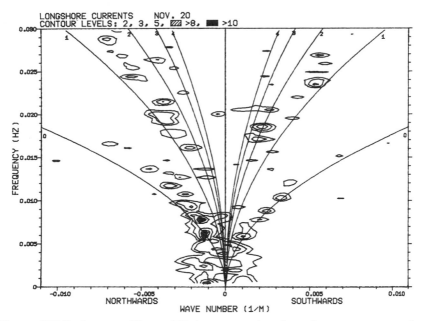

Figure 10B-5. Same as Figure 10B-4a, except that the reference energy used to determine the enhancement factor is different at each frequency line, which is defined as the energy would be observed if the energy at a given frequency band were smeared in the wave number space between ±0.025 (1/m).

eigenfunctions with lower percentages of the total variance are not very meaningful and are not shown. Another example is shown in Figure 10B-6b for 0.0039 Hz, which is not an energetic frequency band. The figure shows another nearly normally incident wave group and a far different response in the longshore current. The longshore variation in the longshore current and the constant phase indicates a standing wave motion at this frequency. The longshore EOF structure suggest only one-half a wave length over the array, or a wave length of 0 (1000 m).

A typical EOF analysis is shown in Figure 10B-6c ($f = 0.01367$ Hz), suggesting cross-shore currents may have a standing wave structure and that the performance of the MLM W-F spectral analysis may be less satisfactory.

The EOF technique was used to test the significance of the data. The cumulative percentage of the energy associated with the significant components at the 0.05 significance level was calculated for each frequency band and is summarized in Tables 10B-1 and 10B-2 for the longshore and cross-shore velocities. The test shows that the lowest frequency band for both components of velocity are significant (and not noise); the 1000 m length scale shown in Figure 10B-6a further exemplifies the fact that this motion is not tidal.

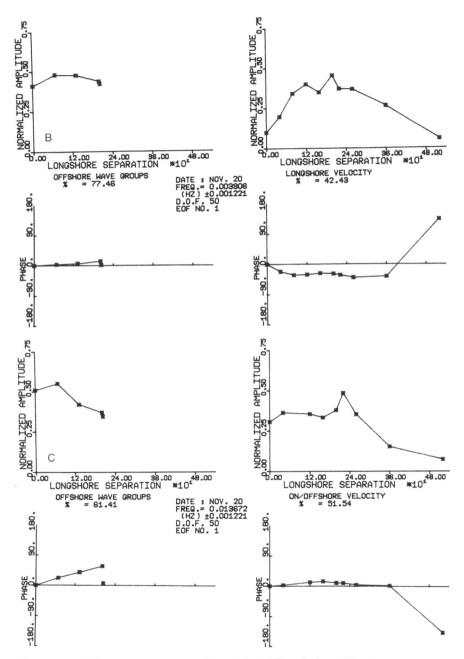

Figure 10B-6. The wave structure profiles of the EOF analysis. a) The first components for v and groups, f = 0.00147 Hz b) The first components for v and groups, f = 0.00391 Hz c) The first components for u and groups, f = 0.00391 Hz.

Table 10B-1. Summary of Multivariate Analysis
for Longshore Currents

Frequency[1]	Energy Level $10^3(cm^2/s)/Hz$	%Energy Significant[2] (E.O.F.)	%Energy Explained by Offshore Groups[3]		Average Squared Multiple Coherence
0.00147	88.96	98.8	44.5		0.445
0.00391	12.38	92.0	38.9		0.472
0.00635	9.27	93.4	37.3		0.398
0.00879	4.99	94.9	19.4		0.236
0.01123 *	1.93	54.8	0.0	(I)	0.193
0.01367 *	1.452	63.7	25.1		0.250
0.01611	1.453	59.7	26.5	(I)	0.246
0.01856	1.47	71.1	5.89	(I)	0.207
0.02100	1.17	69.1	9.9	(I)	0.219
0.02343	1.14	61.6	20.6		0.287
0.02588	0.93	62.2	9.3		0.274
0.02832	0.8	65.6	18.0	(I)	0.214
0.03076	0.7	45.7 (I)	10.4	(I)	0.195
0.03256	0.578	44.5 (I)	8.2	(I)	0.222
0.03565	0.618	41.7 (I)	9.5	(I)	0.172
0.03809	0.59	81.3	10.1	(I)	0.230

(1)   The symbols * denote edge wave zero crossing.

(2)   At 0.05 significant level; (I) denotes independence at 0.01 significant level.

(3)   The results of canonical coherency analysis at 0.1 significant level; (I) denotes independence on groups at 0.05 significant level.

The coherencies between the first EOF components of the velocities and the offshore wave groups is small in general, meaning that the first component of each group are not directly related. Multivariate analysis was then used to examine the relationships in more detail.

10B.5.3 Multivariate Analysis

Offshore wave groups were calculated at five sensors (P1, P2, P3, P4, and P7), which are treated as a set, set G. The longshore and cross-shore currents were treated as two different sets (set u and set v), with 9 sensors (excluding C26) for u and 10 sensors for v. Fifty degrees of freedom were used in this analysis as well.

For each of the low frequency bands canonical coherency analysis was used for the sets (v, G) and (u, G). Five pairs of components, arranged in descending order of canonical coherence, were obtained. In Figure 10B-7a, b, the first and the fourth pair in the longshore current-wave groups analysis are shown for 0.001447 Hz, the lowest frequency band. The results show a northerly propagating wave motion in the longshore current (as was found by the EOF,

Table 10B-2. Summary of Multivariate Analysis
for Cross-Shore Currents

Frequency[1]	Energy Level $10^3(cm^2/s)/Hz$	%Energy Significant[2] (E.O.F.)	%Energy Explained by Offshore Groups[3]		Average Squared Multiple Coherence
0.00147	28.3	96.6	9.4	(I)	0.2
0.00391	12.7	66.4	27.9		0.3
0.00635	22.2	94.3	39.4		0.45
0.00879	22.19	94.2	11.0	(I)	0.23
0.01123	27.2	98.2	16.13	(I)	0.2
0.01367	18.22	95.9	21.89	(I)	0.28
0.01611	8.753	94.2	9.63		0.36
0.01856 *	3.2	70.5	26.6		0.29
0.02100 *	1.83	64.9	11.1	(I)	0.23
0.02343	2.46	94.5	19.5	(I)	0.25
0.02588	4.49	92.7	31.4		0.31
0.02832	4.6	98.0	8.7	(I)	0.18
0.03076	3.43	84.5	16.34		0.29
0.03256	2.728	98.9	18.03	(I)	0.26
0.03565 *	2.05	33.3	44.0		0.31
0.03809 *	1.95	99.1	32.6		0.36

(1) The symbols * denote edge wave zero crossing.

(2) At 0.05 significant level.

(3) The results of canonical coherency analysis at 0.1 significant levels; (I) denotes independence on groups at 0.05 significant level.

Figure 10B-6a), with a high coherence to a highly spatially dependent structure in the offshore wave groups. This structure, with a strong longshore variation, may be a northerly propagating wave group or a standing wave group pattern (more pressure sensors would have been needed to resolve this); however, both would be able to force the migration of the nearshore circulation system. It should be noticed that the component of the wave group picked out by the canonical coherency analysis does not correspond to the normally incident wave train, picked first by the EOF technique (which explained more of the variance in the wave group signal). The forcing of the longshore currents appears to be a result of a less energetic wave group structure, which however has a longshore dependency. The first EOF wave group structure corresponds to the fourth pair of the canonical coherency analysis (Figure 10B-7b). This normally incident wave structure provides almost no forcing of the longshore current (7.4 percent of the variance).

The above results demonstrate that the canonical coherency analysis can be interpreted reasonably well physically. Furthermore, the analysis regroups the

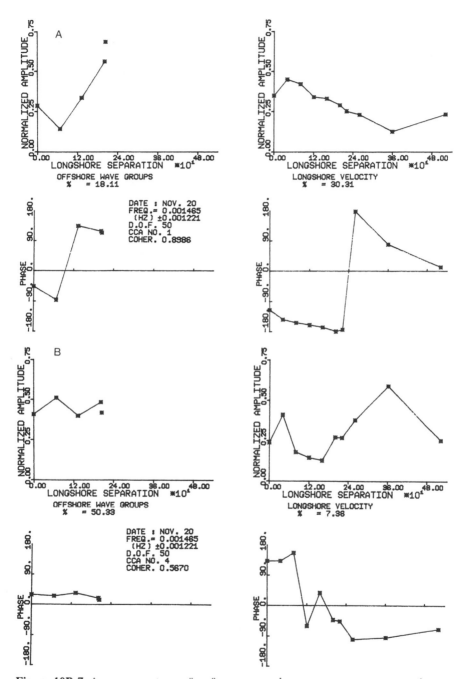

Figure 10B-7. Average spectrum of surf zone cross-shore current groups, averaged over the nine sensors on the main longshore line.

EOF components so the coherences between sets is higher -- but the energy dominance of the first canonical coherence component is usually lower than the first component resulting from the EOF.

The null hypothesis that the currents and groups are independent was tested. The result at 0.05 significance level for each frequency band is listed as the fourth column in Tables 10B-1 and 10B-2 for v and u, respectively. Listed in the same column are the cumulative percentages of energies of significant canonical components at the 0.1 level. The results of the significance test of the EOF analysis (percentages of significant energies, Column 3), canonical coherency analysis and the multiple coherency analysis (the average of the squared multiple coherence at each current meter given the offshore wave group, G, Column 5) are listed in the tables.

At the lowest frequency band, 0.00147 Hz, which contains the migrating rip current, the longshore velocities are highly correlated to the offshore wave groups (at the 0.001 significance level). Figure 10B-7a demonstrated that the forced longshore currents have a progressive wave structure moving to the north, but it is not immediately clear why there is no cross-shore flow component associated with this wave group (only 9 percent of the energy in the cross-shore currents is explained by the groups). It is likely that the nearshore motion at this frequency is not composed of free waves or resonantly forced edge waves, as these types of waves would have strong cross-shore velocities, which are not evident in the data. It is suggested that the longshore currents are therefore forced motion, due to the wave groups.

At the frequencies 0.00391 and 0.00635 Hz, both u and v are correlated with the offshore wave groups. It is probable that the forced motion is of non-resonant type, because there is a spectral valley at 0.00391 Hz for u (existing theories for edge wave forcing (Holman, 1979) do not predict such a spectral valley for zero mode edge waves).

Huntley et $al.$ (1981) identified the frequency band ranging from 0.00879 to 0.01367 Hz to be a zero or first mode edge wave band using the same data set. There is a spectral peak for u at 0.01125 Hz and a valley for v at the same frequency, presumably due to the zero crossing of the first mode edge waves. The complete lack of correlation at the frequency band between v and G indicates that there is only resonantly forced motion besides first of higher mode edge waves and the low coherency between u and G suggests that the edge waves may be free. This is supported by the high partial coherencies (usually >0.46, the significant partial coherence at 0.01 significance level) of cross-shore currents, given the offshore wave groups.

At slightly higher frequencies (0.0161 to 0.021 Hz), u and v again show different behavior. It is possible that both free motion and forced motion (of resonant and/or non-resonant type) coexist at these bands. The situation is further clouded by a possible mode 1 or higher edge wave u component zero crossing at frequencies between 0.019 and 0.21 Hz.

For frequencies 0.02343 to 0.038 Hz, the upper limit of the frequency bands we examined, the coherencies for the longshore current is low, and the energy levels are low with respect to the cross-shore flow, indicating that the motion at this frequency is likely to be one-dimensional surf beat, which has no longshore variability or the combination of several wave structures. The higher coherencies between u and G at some frequencies may indicate preferential frequencies for the surf beat (Wright et al., 1979).

10B.6 Conclusions

The water motions in the surf zone are comprised of numerous motions: surf beat, infragravity waves, forced long waves and nearshore circulation, consisting of rip currents, longshore currents and feeder currents. These motions can coexist in the surf zone, either at the same frequencies or different frequencies.

This study has shown that the nearshore circulation at Torrey Pines appears in part to be driven by the offshore wave groups, based on a canonical coherency analysis. Wave groups, with strong longshore variation, force a response in the surf zone. The frequency of this motion is very low, 0.00147 Hz and less (longer than 680 sec period). Field observations of the rip currents on November 20 support the wave length resulting from the statistical methods.

The number of current meters was not sufficient to resolve a true rip current using the statistical methods; however, their utility in showing a correlation of the nearshore motion to the offshore wave groups has been demonstrated.

Further analyses are necessary to generalize our results. Other days of the Torrey Pines data set need to be analyzed as well as other data sets, for different beach slopes.

10B.7 References

Anderson, T. W., 1958, An Introduction to Multivariate Analysis, John Wiley, New York.

Arthur, R. S., 1962, A note on the dynamics of rip currents, Journal of Geophysical Research, 67: 2777-2779.

Bowen, A. J., 1969, Rip currents 1: theoretical investigations, Journal of Geophysical Research, 74: 5479-5490.

Bowen, A. J. and R. T. Guza, 1978, Edge waves and surf beats, Journal of Geophysical Research, 83: 1913-1920.

Bowen, A. J. and D. L. Inman, 1969, Rip currents 2: laboratory and field observations, Journal of Geophysical Research, 74: 5479-5490.

Capon, J., 1969, High resolution frequency-wave number spectrum analysis, Proceedings, IEEE, 57: 1405-1418.

Capon, J., R. J. Greenfield and R. J. Kolker, 1967, Multidimensional maximum-likelihood processing of a large aperture seismic array, Proceedings, IEEE, 55: 192-211.

Cox, H., 1973, Resolving power and sensitivity to mismatch of optimum array processors, Journal of Acoustic Society of America, 54: 771-785.

Dalrymple, R. A., 1975, A mechanism for rip current generation on an open coast, *Journal of Geophysical Research*, 80: 3485-3487.

_____. 1978, Rip currents and their causes, *Proceedings*, Sixteenth Coastal Engineering Conference, August 27-September 3, 1978, Hamburg, Germany, American Society of Civil Engineers, New York, 2: 1414-1427.

Dalrymple, R. A. and G. A. Lanan, 1976, Beach cusps formed by intersecting waves, *Geological Society of America Bulletin*, 87: 57-60.

Dalrymple, R. A. and C. J. Lozano, 1978, Wave-current interaction models for rip currents, *Journal of Geophysical Research*, 83: 6063-6071.

Davis, R. E. and L. A. Regier, 1977, Methods for estimating directional wave spectra from multi-element arrays, *Journal of Marine Research*, 35: 453-477.

Ebersole, B. A. and R. A. Dalrymple, 1979, A numerical model for nearshore circulation including convective accelerations and lateral mixing, *Ocean Engineering Report*, 21, Department of Engineering, University of Delaware, Newark, Delaware.

_____. 1980, Numerical modelling of nearshore circulation, *Proceedings*, Seventeenth Coastal Engineering Conference, March 23-28, 1980, Sydney, Australia, American Society of Civil Engineers, New York, 3: 2710-2725.

Gallagher, B., 1971, Generation of surf beat by non-linear wave interactions, *Journal of Fluid Mechanics*, 49: 1-20.

Giri, N., 1965, On the complex analogues of T and R tests, *Annals Mathematical Statistics*, 36: 664-670.

Goodman, N. R., 1964, Analysis based on a certain multivariate complex Gaussian distribution, an introduction, *Annual Mathematical Statistics*, 34: 152-177.

Guza, R. T. and D. C. Chapman, 1979, Experimental study of the instabilities of waves obliquely incident on a beach, *Journal of Fluid Mechanics*, 95: 199-208.

Guza, R. T. and D. L. Inman, 1975, Edge waves and beach cusps, *Journal of Geophysical Research*, 80: 2997-3012.

Hansen, J. B. and I. A. Svendsen, 1984, A theoretical and experimental study of undertow, *Proceedings*, Nineteenth Coastal Engineering Conference, September 3-7, 1984, Houston, Texas, American Society of Civil Engineers, New York, 3: 2246-2262.

Hino, M., 1974, Theory on formation of rip current and cuspidal coast, *Proceedings*, Fourteenth Coastal Engineering Conference, June 24-28, 1974, Copenhagen, Denmark, American Society of Civil Engineers, New York, 2: 901-919.

Holman, R. A., 1979, Infragravity waves on beaches, Ph.D. thesis, Dalhousie University, Halifax, N.S., Canada.

Holman, R. A. and A. J. Bowen, 1984, Longshore structure of infragravity wave motions, *Journal of Geophysical Research*, 89 (C4): 6446-6452.

Huntley, D. A., 1976, Long period waves on a natural beach, *Journal of Geophysical Research*, 81: 6441-6449.

Huntley, D. A. and A. J. Bowen, 1973, Field observations of edge waves, *Nature*, 243: 160-162.

Huntley, D. A., R. T. Guza and E. B. Thornton, 1981, Field observation of surf beats; 1, progressive edge waves, *Journal of Geophysical Research*, 86: 6451-6460.

Isobe, M., K. Kondo and K. Horikawa, 1984, Extension of MLM for estimating directional wave spectrum, *Proceedings*, Symposium on Description and Modeling of Directional Seas, Technical University of Denmark, Chapter A6: 1-15.

Iwata, N., 1978, Rip current spacing as an eigenvalue, *Proceedings*, Sixteenth Coastal Engineering Conference, August 27-September 3, 1978, Hamburg, Germany: 828-843.

Jenkins, G. M. and D. G. Watts, 1968, *Spectral Analysis and Its Applications*, Holden-Day Inc.

Katoh, K., 1981, Analysis of edge waves by means of empirical eigenfunctions, Report of Port and Harbor Institute, 20: 2-51.

Khatri, C. G., 1965, Classical statistical analysis based on a central multivariate complex Guassian distribution, *Annals Mathematical Statistics*, 36: 98-114.

Krishnaiah, P. R., 1976, Some recent developments on complex multivariate distribution, *Journal of Multivariate Analysis*, 6: 1-30.

LeBlond, P. H. and C. L. Tang, 1974, On the energy coupling between waves and rip currents, *Journal of Geophysical Research*, 79: 811-816.

Liu, P. L.-F. and C. C. Mei, 1976, Water motion on a beach in the presence of a breakwater, 2, mean currents, *Journal of Geophysical Research*, 81: 3085-3094.

Longuet-Higgins, M. S. and R. W. Stewart, 1962, Radiation stress and mass transport in gravity waves, with application to "surf-beats," *Journal of Fluid Mechanics*, 13: 81-504.

_____. 1964, Radiation stress in water waves, a physical discussion with applications, *Deep-Sea Research*, 11: 529-563.

McDonough, R. N., 1982, Application of the maximum-likelihood method and the maximum-entropy method to array processing, *Nonlinear Methods of Spectral Analysis*, Haykin, S., ed.

Mei, C. C. and D. Angelides, 1977, Longshore circulation around a conical island, *Coastal Engineering*, 1: 31-42.

Melville, W. K., 1983, Wave modulation and breakdown, *Journal of Fluid Mechanics*, 128: 489-506.

Miller, C. and A. Barcilon, 1978, Hydrodynamic instability in the surf zone as a mechanism for the formation of horizontal gyres, *Journal of Geophysical Research*, 83: 4107-4116.

Munk, W., 1949, Surf beats, EOS Transactions, American Geophysical Union, 30: 849-854.

Munk, W., F. Snodgrass and F. Gilbert, 1964, Long waves on the continental shelf: an experiment to separate trapped and leaky modes, *Journal of Fluid Mechanics*, 20: 529-554.

Noda, E. K., 1974, Wave-induced nearshore circulations, *Journal of Geophysical Research*, 79: 4097-4109.

Press, S. J., 1982, *Applied Multivariate Analysis: Using Bayesian and Frequentist Methods of Inference*, Robert E. Krieger Publishing Co.

Priestley, M. B., 1981, *Spectral Analysis and Time Series, V.2 Multivariate Series, Prediction and Control*, Academic Press.

Rice, S. O., 1945, Mathematical analysis of random noise, Bell System Technology Journal, 46-156; reprinted in *Selected Papers on Noise and Stochastic Process*, edited by N. Wax, 133-294, Dover Publishing Inc.

Sasaki, T., 1975, Simulation on shoreline and nearshore current, *Proceedings*, Speciality Conference on Civil Engineering in the Ocean III, University of Delaware, Newark, Delaware, June 9-12, American Society of Civil Engineers, 1: 176-196.

Shepard, F. P. and D. L. Inman, 1950, Nearshore circulation, *Proceedings*, First International Conference on Coastal Engineering, Council on Wave Research, October, 1950, Berkeley, California, 1: 50-59.

Sonu, C. J., 1972, Field observation of nearshore circulation and meandering currents, *Journal of Geophysical Research*, 77: 3232-3247.

Symonds, G. D., D. A. Huntley and A. J. Bowen, 1982, Two dimensional surf beat: long wave generated by a time-varying breakpoint, *Journal of Geophysical Research*, 87: 492-498.

Tatsuoka, M. M., 1971, *Multivariate Analysis: Techniques for Educational and Psychological Research*, John Wiley.

Tucker, M. J., 1950, Surf beats: sea waves of 1 to 5 min period, *Proceedings, Royal Society*, Series A, 202: 565-573.

Ursell, F., 1952, Edge waves on a sloping beach, *Proceedings, Royal Society*, Series A, 214: 79-98.

Wallace, J. M. and R. E. Dickenson, 1972, Empirical orthogonal representation of time series in the frequency domain, part I: theoretical considerations, *Journal of Applied Meteorology*, 11: 887-892.

Winant, C. D., D. L. Inman and C. E. Nordstrom, 1975, Description of seasonal beach changes using empirical eigenfunctions, *Journal of Geophysical Research*, 80: 1979-1986.

Wright, L. D., J. Chappell, B. G. Thom, M. P. Bradshaw and P. Cowell, 1979, Morphodynamics of reflective and dissipative beach and inshore system, *Southeastern Australia, Marine Geology*, 32: 105-140.

Chapter 11

SUSPENDED SEDIMENT MEASUREMENTS

A. Continuous Measurements of Suspended Sediment

Richard W. Sternberg
Contribution No. 1702 of the School of Oceanography
University of Washington

N. C. Shi
Northern Technical Services
Redmond, Washington

John P. Downing
Batelle Memorial Institute
Washington, D.C.

11A.1 Introduction

Field measurements of suspended sediment in the nearshore zone have been carried out with a variety of instruments, including pumps operated from ocean piers (Watts, 1953; Fairchild, 1972, 1977) and from sleds pulled through the surf zone (Coakley, 1980); self-siphoning samplers (Wright *et al.*, 1982a); light-scattering devices (Thornton and Morris, 1978; Downing, 1983); light transmission devices (Brenninkmeyer, 1974, 1976a, b); and diver-operated samplers [Kana, 1976, 1978, 1979; Kana and Ward, 1980 (modified for use from a pier); Inman *et al.*, 1980]. While each of the above studies have limitations, the combined information presented gives a general view of suspended sediment movement in the nearshore zone. Those studies that provide background data directly applicable to the present investigation are summarized briefly.

Brenninkmeyer (1974, 1976a, b), who used a light transmission device (almometer) at three locations across the surf zone, concluded that in the outer surf zone suspension of sand to elevations 15 cm above the seabed was almost non-existent. Suspended sediment transport increased in magnitude in the inner surf zone and was the predominant transport mode in the transition region (region of interaction of backwash and incoming bores) with concentrations as high as 380 g/l occurring in association with episodic suspension events. The lower limit of detection by the almometer was approximately 10 gm/l which limited the detection of suspension events at higher elevations (>15 cm) in the central and outer surf zone.

As summarized by Kana (1978) for data collected by diver-operated samplers, suspended sediment concentration in the surf zone primarily depends on elevation above the bed, breaker type, wave height, and distance from the breakpoint. Concentration decreases exponentially above the bottom to approximately 60 cm elevation as a function of the intermittent suspension of sediment from the bed. In spilling waves, concentration rapidly increases shoreward of the breakpoint, then remains relatively constant under the bore as it propagates toward the beach. In plunging waves, concentration peaks within a few meters of the breakpoint, then decreases gradually toward shore.

Komar (1978) in an extensive analysis of existing field and laboratory data evaluated the relative contribution of suspended load transport to the total transport. He concluded that suspended load transport could account for up to 25% of the total littoral transport rate.

Kana and Ward (1980) made suspended sediment and current measurements from the CERC field research facility at Duck, North Carolina, and compared the measured rates of longshore transport of suspended sediment with the total transport rate determined from the CERC equation (CERC, 1973). The suspended sediment samplers collected samples from 5 cm off the seabed to the surface and this comparison suggested that during storm conditions (H_s = 1.99 m) the magnitude of the longshore transport rate of suspended sediment was equal to the total transport predicted by the CERC equation. During low wave conditions (H_s = 0.88 m) the suspended load transport rate was 30% of the rate predicted by the CERC equation. Inman et al. (1980) reported that the immersed weight suspended sediment rate at Torrey Pines Beach was about 15-20% of the total transport. The lowest suspended sediment sample was collected 10 cm off the seabed and bedload was defined as sediment transport occurring within 10 cm of the seabed.

Using a vertical array of optical backscatter sensors (lowest sensor at z = 3.5 cm) at Twin Harbors Beach, Washington, Downing (1983) found that the mean suspended load was high in the mid-surf zone position (X/X_b = 0.4-0.5) and decreased toward the breaker region. Total suspended load was also very high in the transition zone. His measured longshore transport rate of suspended sediment was 47% of the total transport rate predicted by the method outlined in Komar and Inman (1970) and Komar (1983, equation 22).

The measurements described herein were also directed towards the characterization of the suspended sediment field in the surf zone at Leadbetter Beach in Santa Barbara. Miniaturized optical backscatter (OBS) sensors which have fast response (10 Hz) and can be operated continuously within the surf zone were packaged in two types of vertical arrays. The OBS array has five sensors mounted in a 51 cm tube with a logarithmic spacing between sensors of

3, 6, 12, and 30 cm and the MOBS, a miniature array, has five sensors mounted within 16 cm with a spacing of 3, 3, 5, 5 cm between sensors. (See also Downing *et al.* (1981) and Chapter 5C for a more detailed description.)

Continuous measurements of suspended sediment on time and space scales used in this study, in conjunction with flow and water level measurements, have not been made before, so these data provide a unique view of the role of suspended sediment in the surf zone.

11A.2 Methods

11A.2.1 Data Collection

The basic strategy of the sampling program was to maintain a cross-shore transect of OBS and MOBS arrays over as wide a range of wave conditions as possible. Up to four sensor arrays were deployed along the main transect; however, due to various problems with electronics, instrument damage under storm conditions, severe beach erosion, etc., data recovery varied from day to day. (See Table 5C-3 for a summary of sensor array locations, operational times, and dates.) Figure 11A-1 gives a graphic representation of the data series collected by individual sensor arrays relative to significant wave height (H_s), which varied from about 0.3 to 1.8 m over the duration of the experiment. The OBS arrays (AO, BO, and CO) were deployed in the central and outer surf zone and the MOBS arrays (AMO, BMO) in the inner surf zone. The arrays were mounted vertically with the lowest sensor adjusted to 3.5 cm off the seabed and the upper sensor located at 54.5 cm and 19.5 cm above the seabed for OBS and MOBS arrays, respectively. The elevation of each array was monitored before and after each data series.

The deployment of each MOBS and OBS array is discussed in detail in Chapters 4 and 5. Generally, the OBS arrays were located 6 m west of the main rangeline and on the same isobath as an EM current meter ($z \approx 45$ cm) and a pressure sensor ($z \approx 20$ cm). MOBS arrays were deployed from a shallow water tripod and were located within 50 cm of an EM current meter positioned at $z = 4$ cm.

11A.2.2 Data Analysis

Three categories, high frequency data analysis, averaged data analysis, and sediment transport calculations and predictions, were employed. The techniques involved are outlined in Table 11A-1.

Initially, the raw data from all sensor outputs were block averaged to a 2 Hz sampling rate. For the high frequency data analysis the outputs from optical backscatter sensors, offshore slope array pressure sensors (west array), depth, and u and v velocity components were converted to scientific units. (See Chapter 5C for the calibration information.) Then the total immersed weight suspended sediment (G_s), immersed weight suspended sediment longshore

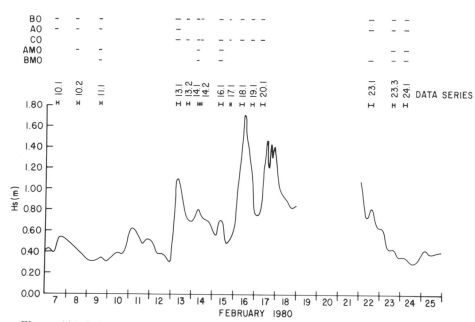

Figure 11A-1. Sensor status during the field experiment at Leadbetter Beach. Individual sensor arrays are designated AO, BO, CO (OBS arrays) and AMO, BMO (MOBS arrays). The length of each sensor status line shows the approximate sampling time within each day. Data series designations and significant wave heights are also shown.

transport per unit width of seabed (i_{ls}), and depth of erosion (b) necessary to supply the measured suspended load were computed every 1/2 sec. An example of a 9-minute data series analyzed according to the above procedures is illustrated in Figure 11A-2.

The second category of analysis involved the computations of mean values of various sediment parameters. The averaging period for these computations was taken as 2048 sec or approximately 34 minutes. This period is thought to be sufficiently long to give stable averages of most parameters and yet short enough to minimize the influence of tidal variations during the averaging period. The mean suspended sediment concentration at each level above the bed ($\overline{C}$), the mean immersed weight suspended sediment load ($\overline{G}_s$), and the mean erosion depths ($\overline{b}$, $\hat{b}$) were computed by equations shown in Table 11A-1B. The average depth of erosion required to supply the suspended load was estimated by (i) time averaging values of b (designated $\overline{b}$), which represent the hypothetical thickness of a sand layer equivalent to $\overline{G}_s$, and by (ii) computing the mean value of the maximum erosion depth associated with each suspension event in a 34-minute period. For example, if the individual peaks in b shown in

Table 11A-1. Summary -- Data Analysis

Procedure	Sensors Analyzed	Equation
A. High Frequency Data Analysis ($2Hz$ or $\Delta t = 0.5\ sec$)		
1. Convert sensor outputs to scientific units (C, $kg\ m^{-3}$)	Each optical sensor, depth, u, v and offshore slope array	Use various calibration equations
2. Compute immersed weight suspended sediment load (G_s, $nt\ m^{-2}$) **where upper 2 sensor outputs are extrapolated to the surface. If C(h)>0, then C(h) set=0, and output of lower 2 sensors extrapolated linearly to the seabed.**	Each OBS and MOBS array	$G_s = \dfrac{\rho_s^- \rho}{\rho_s}\, g\displaystyle\int_0^h C\ dz$ (1)
3. Compute immersed weight longshore transport of suspended sediment per unit width of seabed (i_{ls}, $nt\ m^{-1}sec^{-1}$)	Each OBS and MOBS array	$i_{ls} = G_s v$ (2)
4. Compute depth of erosion required to support measured suspended load (b, cm)	Each OBS and MOBS array	(3) $b = \dfrac{\displaystyle\int_0^h C\ dz}{\rho_s\, C_b}$ where C_b = **volume concentration at the bed** (= 0.6)
B. Averaged Data Analysis (2048 sec = 4096 data points)	Each optical sensor	(4)
1. Compute mean immersed weight suspended sediment concentration ($\bar{C}$, $kg\ m^{-3}$)		
2. Mean immersed weight suspended sediment load ($\bar{G}_s$, $nt\ m^{-2}$)	Each OBS and MOBS array	$\bar{G}_s = \dfrac{1}{4096}\displaystyle\sum_{n=1}^{4096} G_s$ (5)
3. Mean erosion depth required to support suspended load	Selected OBS and MOBS array in 24.1 series	$\bar{b} = \dfrac{1}{4096}\displaystyle\sum_{n=1}^{4096} b$ (6)
a) nominal thickness of sand sheet in continuous motion ($\bar{b}$, cm) **b) mean depth of maximum erosion during individual suspension events** ($\hat{b}$, cm)	Each OBS and MOBS array	$\hat{b} = \dfrac{1}{n}\displaystyle\sum_{i=0}^{n} b\max$ (7) where b max = **maximum depth of erosion during each suspension event (see Fig. 2)**
C. Longshore Transport Calculations and Predictions		
1. Compute local immersed weight longshore transport rate of suspended sediment (per unit width of seabed) ($\bar{i}_{ls}$, $\bar{i}_{ls}'$, $nt\ s^{-1}\ m^{-1}$)	Each OBS and MOBS array	a) $\bar{i}_{ls} = \bar{G}_s v$ (8) b) $\bar{i}_{ls}' = \bar{G}_s \bar{v}$ (9)
2. Compute total immersed weight longshore transport rate of suspended sediment (I_{ls}, $nt\ s^{-1}$)	$\bar{i}_{ls}$ summed over each data series	$I_{ls} = \sum \bar{i}_{ls}\,\Delta x$ (10)
3. Predict total immersed weight longshore transport rate [using equations of Komar and Inman (1970) and Komar (1983) (I_{ls}, $nt\ s^{-1}$)]	Over width of surf zone	c) $I_l = KP_l$ where $K = 0.77$, and $P_l = (ECn)_b \sin\alpha_b \cos\alpha_b$ (11) d) $I_l = K'(ECn)_b \dfrac{\bar{v}_l}{u_m}$ where $K' = 0.28$ (12)

Figure 11A-2I are designated as b_{max} then $\hat{b}$ is the average of the b_{max} events (equation 7, Table 11A-1B). The number of individual suspension events roughly equals the number of waves suspending sediment over a 2048-second period.

The third category of analysis included computations of immersed weight suspended sediment transport rate from the field measurements and the prediction of the total longshore transport rate using the equations of Komar and Inman (1970) and Komar (1983). The local immersed weight longshore transport rate of suspended sediment $(\overline{i}_{ls}, \overline{i}_{ls}')$ occurring at each sensor array was computed by two methods (equations 8 and 9, Table 11A-1C). The total longshore transport rate of suspended sediment (I_{ls}) was determined by the summation of the individual measurements of $\overline{i}_{ls}$ made across the surf zone (equation 10, Table 11A-1C). Because of the small number of sensor arrays deployed, measurements of $\overline{i}_{ls}$ made at different stages of the tide (and hence different values of X/X_b) during a data series were combined to increase data coverage across the surf zone. This assumes stationary surf zone conditions over the duration of a data series. The constancy in significant wave height and wave period discussed below suggests that this is a reasonable assumption. For computations of I_{ls}, the total width of the surf zone (X_b) had to be determined (see Appendix). Equations 11 and 12 (Table 11A-1C) were used to predict the total immersed weight longshore transport rate I_l for each data series. The wave information required for these computations was collected by other NSTS investigators. (See the Appendix for the treatment of wave data.)

11A.3 Results

During the experiment 15 suspended sediment data series were collected (Figure 11A-1). Each data series consisted of continuous sampling of all operational sensor arrays over a 2- to 4-hour period. Concurrent measurements of velocity, water level, and beach profiles were made by other NSTS investigators.

For a variety of reasons, it was not possible to analyze all of the data collected. In some cases sensor problems limited the value of the data either due to electronic noise in the sensors (e.g., Series 19.1) or damage to electromagnetic current meters (16.1). Several data series were collected during periods of very low tides where a significant number of sensors, including current meters, were not submerged (e.g., Series 11.1). The latter data will be valuable for investigating sediment movement in the swash-backwash zone; however, special techniques will be required for analyses. Data collected during storms also cannot be used (e.g., Series 18.1). The quantity of very fine sediment suspended as wash load in the nearshore zone due to storm drain flooding in Santa Barbara increased the background signal levels so that individual suspension events of nearshore sand were frequently obscured.

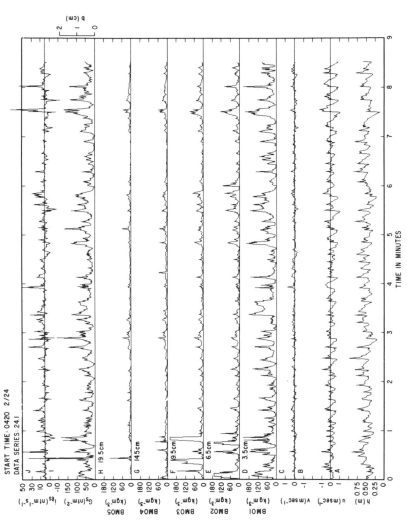

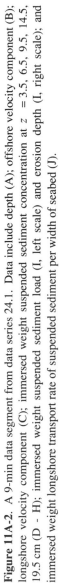

Figure 11A-2. A 9-min data segment from data series 24.1. Data include depth (A); offshore velocity component (B); longshore velocity component (C); immersed weight suspended sediment concentration at z = 3.5, 6.5, 9.5, 14.5, 19.5 cm (D - H); immersed weight suspended sediment load (I, left scale) and erosion depth (I, right scale); and immersed weight longshore transport rate of suspended sediment per width of seabed (J).

After inspecting all of the data records a data quality index was used to choose the data series that were analyzed. The primary considerations were that the data series spanned a range of wave conditions, had low background signals from wash load, provided reasonable coverage across the surf zone, and required a minimum amount of editing. Data runs that did not meet the above conditions but still contain interesting or valuable information (e.g., from the swash-backwash zone) will be analyzed subsequently.

Six data series (13.1, 13.2, 14.1, 14.2, 23.3, 24.1) were chosen for analysis. Within these six data series, thirty segments representing different tide levels were chosen arbitrarily for averaged data analyses (34-minute averages). These thirty segments represent the data set.

11A.3.1 Waves and Currents

Wave information needed for this study includes wave height, maximum oscillatory velocity, depth of breaking, and sea level variations and flow conditions at the various sensor locations. Depth of breaking, when combined with the beach profiles and tide level, can be used to estimate the surf zone width (X_b).

Wave and flow parameters have been computed from outputs of current meters and pressure sensors outputs located adjacent to each OBS and MOBS array. These data (Table 11A-2) consist of the values of depth (h), significant wave height (H_s), significant oscillatory velocity (U_s), mean shore-normal velocity ($\bar{u}$), mean shore-parallel velocity ($\bar{v}$), mean immersed weight suspended sediment load ($\bar{G}_s$), mean local immersed weight longshore transport of suspended sediment ($\bar{i}_{ls}$), and relative position of the instruments in the surf zone (X/X_b). The sediment parameters $\bar{G}_s$ and $\bar{i}_{ls}$ are discussed in the next section. Values of u are positive in an offshore direction and those of v are positive in the easterly direction, which is toward Santa Barbara Harbor from the field site. Missing data points in Table 11A-2 indicate that instruments were inoperative or not deployed.

Various characteristics of the waves at breaking are required to calculate the littoral transport rate (equations 11-11, 11-12, Table 11A-1C). These include breaker height, breaker depth, breaking angle, and bottom oscillatory velocity. As all of these variables were not measured throughout the experiment (e.g., angle of breaking), predictions of breaker characteristics for each data series were made using offshore pressure data. Waves measured at the offshore wave array (west) were subjected to computer analysis that included considerations of refraction and breaking. Where possible the predicted breaking characteristics were compared to the observed breakers to test the methods; however, the predicted breaker characteristics were used for all calculations of littoral transport. The detailed procedure for this analysis is given in the Appendix. A summary of wave period (T), predicted significant breaker height (H_{sb}), direction at breaking (α_b), breaking depth (h_b) is given in Table 11A-3. Also

Table 11A-2. Longshore Suspended Load Transport Flux Calculations

TIME/ DATA SERIES	SENSOR ID	h (m)	H_s (m)	U_s (m/s)	$\bar{u}$ (m/s)	$\bar{v}$ (m/s)	$\bar{G}$ $(\frac{nt}{m^2})$	$\bar{i}_{ls}$ $(\frac{nt}{s/m})$	$\frac{X}{X_b}$
0756/13.1	AO	1.13	0.39	1.50	0.17	−0.95	26.89	−26.44	0.79
0756/13.1	BO	1.44	0.57	1.50	0.21	−0.71	30.48	−22.22	0.97
1915/13.2	CO	0.32	----	1.59	0.15	−0.74	8.27	−6.13	0.16
1957/13.2	CO	0.38	----	1.50	0.16	−0.67	7.16	−4.42	0.20
1915/13.2	BO	0.96	0.58	1.75	0.26	−0.73	15.10	−11.28	0.45
1957/13.2	BO	1.02	0.59	1.70	0.26	−0.63	26.10	−17.14	0.53
0805/14.1	BO	1.59	0.77	1.67	0.11	−0.11	11.05	−0.73	0.74
0605/14.2	CO	1.20	----	1.94	0.14	−0.13	30.43	−3.56	0.45
0705/14.1	CO	1.31	----	1.79	0.08	−0.10	27.1	−2.88	0.46
0805/14.1	CO	1.25	----	1.98	0.15	−0.18	25.50	−3.60	0.46
0905/14.2	BO	1.31	0.68	1.89	0.14	−0.22	47.18	−10.03	0.62
1005/14.2	BO	0.88	0.61	1.90	0.26	−0.08	41.06	−3.16	0.41
1051/14.2	BO	0.50	0.49	1.99	0.18	−0.12	32.16	−4.78	0.25
0904/14.2	CO	0.84	----	1.92	0.19	−0.30	13.24	−3.48	0.36
1010/14.2	CO	0.41	----	1.84	0.20	−0.12	18.06	−2.01	0.20
1528/23.3	AO	1.06	----	1.12	0.10	0.16	7.13	1.23	0.74
1628/23.3	AO	1.17	----	1.04	0.06	0.13	10.13	1.12	0.96
1728/23.3	AO	1.19	----	1.03	0.07	0.12	11.33	0.99	0.88
1428/23.3	AMO	0.43	----	----	----	----	11.16	----	0.28
1528/23.3	AMO	0.60	----	1.16	0.12	0.15	8.22	1.48	0.35
1628/23.3	AMO	0.71	----	1.06	0.10	0.23	3.66	0.91	0.50
1728/23.3	AMO	0.73	----	1.01	0.08	0.22	9.64	2.51	0.46
0720/24.1	AO	1.22	----	1.01	0.04	0.03	2.61	0.18	1.40
0300/24.1	AMO	1.13	0.36	1.04	0.04	0.06	7.24	0.36	0.97
0525/24.1	AMO	1.16	0.37	1.01	0.07	0.09	3.92	0.30	0.96
0625/24.1	AMO	0.98	0.33	0.97	0.09	0.05	6.48	0.41	0.84
0720/24.1	AMO	0.74	0.20	0.90	0.10	0.15	2.05	0.33	0.65
0300/24.1	BMO	0.76	0.29	1.74	0.12	0.07	37.01	2.92	0.47
0425/24.1	BMO	0.89	0.32	1.35	0.11	0.04	45.98	2.22	0.60
0525/24.1	BMO	0.79	0.30	1.41	0.09	0.08	31.37	2.73	0.49

listed are the measured values of mean longshore velocity $(\bar{v}_l)$, P_l and $(ECn)_b \bar{v}_l / u_m$ (equations 11-11 and 11-12, Table 11A-1C), and relative state of the tide for each averaging period. The measured longshore velocity was estimated by plotting values of $\bar{v}$ from all of the current meters placed across the surf zone and interpolating between points to $X/X_b = 0.5$. The values of P_l, and $(ECn)_b \bar{v}_l / u_m$ were computed using rms rather than significant wave conditions to be consistent with other studies.

The beach profiles for each of the data series are shown in Figure 11A-3. Also included are the positions of each sensor array, tide range during the sampling time, and the estimated breaker position.

11A.3.2 Sediment Movement

Analysis of suspended sediment includes mean concentration at each level above the bed $(\bar{C})$, mean suspended sediment load $(\bar{G}_s)$, local immersed weight longshore transport rate of suspended sediment per unit width of the surf zone computed by two methods $(\bar{i}_{ls}, \bar{i}_{ls}')$, and the total immersed weight longshore

Table 11A-3. Breaker Predictions

TIME DATA SERIES	T (sec)	$*H_{sb}$ (m)	$*\alpha_b$ (°)	h_b (m)	$\bar{v}_\ell$ (m/sec)	P_ℓ (nt/s)	$(ECn)_b$ (nt/s)	$\dfrac{\bar{v}_\ell}{u_m}$	** TIDE
0756/13.1	4.8	0.62	-15.1	1.48	-0.90	-203	-1470		H
1915/13.2	8.7	0.90	-11.8	2.16	-0.80	-445	-2760		LH
1957/13.2	8.0	0.82	-10.5	1.95	-0.70	-304	-1970		LH
0605/14.1	8.2	0.80	-6.8	2.07	-0.12	-200	-353		H
0705/14.1	8.0	0.84	-6.7	2.17	-0.10	-219	-322		H
0805/14.1	8.0	0.90	-6.8	2.34	-0.19	-270	-711		H
0905/14.2	8.4	0.90	-6.5	2.15	-0.22	-248	-758		EBB
1005/14.2	8.8	1.01	-6.9	2.41	-0.14	-346	-601		EBB
1051/14.2	9.2	0.88	-6.6	2.11	-0.15	-242	-496		EBB
1428/23.3	12.5	0.54	1.8	1.28	0.14	19.4	170		FL
1528/23.3	12.5	0.56	1.8	1.33	0.18	21.4	236		FL
1628/23.3	11.8	0.50	1.7	1.20	0.22	15.6	233		H
1728/23.3	10.9	0.53	1.8	1.28	0.21	19.0	253		H
0300/24.1	12.5	0.47	2.4	1.13	0.07	19.5	66.1		H
0425/24.1	13.1	0.50	2.7	1.20	0.04	25.2	42.8		H
0525/24.1	13.0	0.50	2.1	1.20	0.08	20.1	85.6		EBB
0625/24.1	13.5	0.45	1.6	1.06	0.10	11.0	83.9		EBB
0720/24.1	14.9	0.44	1.2	1.04	0.12	7.79	96.5		EBB

*predicted values, H = High, LH = Lower High, EBB = Ebb, FL = Flood

transport of suspended sediment ($I_{\ell s}$). Tables summarizing the values of $\bar{C}$ and $\bar{i}_{\ell s}'$ are not presented in this chapter because of length limitations. The values of G_s and $\bar{i}_{\ell s}$ are listed in Table 11A-2.

11A.4 Discussion

11A.4.1 Suspended Sediment

A relatively minor effort was devoted to the characteristics of suspended sediment profiles (Figure 11A-2). However, some observations of the time series (e.g., 24.1, Figure 11A-2) are summarized below:

(1) The suspended sediment concentration near the seabed ($z = 3.5$ cm) varied over a wide range and reached 180 kgm^{-3} during individual suspension events (Figure 11A-2D).
(2) The concentration decreased systematically away from the seabed (Figure 11A-2D-H).
(3) Individual suspension events were in phase with bores propagating across the surf zone. Where the offshore directed flow prior to an incoming bore attained significant magnitude (i.e., the threshold of grain motion is exceeded), a sediment suspension event was initiated and then reinforced by the passage of a bore propagating shoreward.
(4) The frequency and duration of sediment suspension events were strongly correlated with incident wave conditions. Low frequency oscillations of

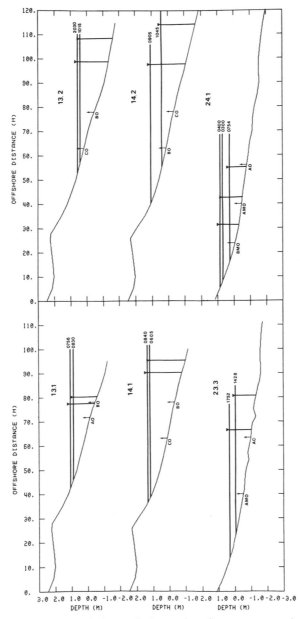

Figure 11A-3. Beach profiles during each data series. Sensor arrays and maximum and minimum tide level at beginning and end of the data series is also shown. Breaker position for each tide level is designated (∇).

cross-shore velocity and water level were observed in many data records. Large temporal variations in the magnitude of suspension events also occurred; they appear to be of greatest magnitude in relation to major offshore flow; of water corresponding to troughs in the low frequency signals.

Thirty-four minute averages of suspended sediment concentration at each OBS sensor have been computed. Four representative suspended sediment profiles from data series 24.1 are shown in Figure 11A-4. Two are from the mid-surf zone (X/X_b = 0.49, 0.60) and two are from the breaker region (X/X_b = 0.84, 0.96). The position of the lowest sensor in these profiles is 3.5 cm above the seabed which is over 150 grain diameters above the boundary (Md = 0.23 mm); thus any particles measured by the sensors are considered as suspended load (Smith and Hopkins, 1972; Francis, 1973). The dominance of suspended load transport at z = 3.5 cm is also suggested by the fact that the maximum measured volume concentration of solids was 0.07 which according to Bagnold (1963), Inman (1963), and Bailard and Inman (1979) is below the minimum concentration (0.08 or 0.09) considered necessary to produce a bedload transport layer supported by intergranular collisions. Also included in Figure 11A-4 for comparison is Kana's (1978, 1979) profile of suspended sediment, collected shoreward of the breaking point, which he considered to be representative of suspended sediment profiles associated with plunging breakers.

The profiles plotted from data series 24.1 generally are similar to the results summarized by Kana (1978, 1979). The slopes of the concentration profiles in the region of z = 10 − 20 cm for data series 24.1 are similar to the slope of the Kana data. Also the mean concentration measured at z = 10 cm for data series 24.1 ranges from approximately 2 − 12 $\mathrm{kgm^{-3}}$ while the concentration in the Kana profile is 3 $\mathrm{kgm^{-3}}$. Thus at z = 10 cm the mean concentrations observed in this study are of the same order as those of Kana's study. The explanation of the higher concentration measured in this study are twofold: instantaneous suspension events occurring higher in the water column that were not sampled consistently by other instruments and high concentrations that frequently occur close to the seabed that were not sampled by other investigators.

11A.4.2 Sediment Transport

11A.4.2.1 Local Transport Rate

Two methods (equations 11-8 and 11-9, Table 11A-1C) were used to compute the mean longshore transport of suspended sediment per unit width of seabed. The first was to compute the time average of the instantaneous suspended sediment transport rate ($\overline{i_{ls}}$) for each 1/2 sec over a 2048-sec period. The second method was to compute the product of the mean total suspended load and the mean longshore velocity ($\overline{i_{ls}}'$). A comparison of $\overline{i_{ls}}$ and $\overline{i_{ls}}'$ (Figure

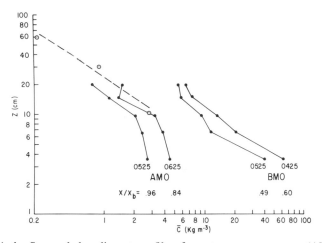

Figure 11A-4. Suspended sediment profiles from two sensor arrays (AMO, BMO) during data series 24.1. $\overline{C}$ is mean concentration (34 min), z = level above the seabed, times indicate beginning of averaging period, X/X_b shows relative position within the surf zone where X_b = breaking point. The open circles are data points of Kana (1978) representative of plunging breaker conditions. The dashed line shows the trend of Kana's data.

11A-5) shows that they have equal values for all data series regardless of wave conditions or position in the surf zone. This indicates that a systematic correlation does not exist between the fluctuating components of longshore currents and suspended sediment and that it is only necessary to measure mean suspended load ($\overline{G}_s$) and mean longshore current ($\overline{v}$) to estimate the local immersed weight longshore transport rate of suspended sediment per unit width of seabed.

11A.4.2.2 Erosion Depth

Calculations of the mean depth of erosion ($\overline{b}$) necessary to supply the measured values of G_s and the mean of the maximum erosion depths associated with each suspension event ($\hat{b}$) were made from the high frequency data (Tables 11A-1A and 11A-1B). This determination assumes that the measured suspended loads are caused by local erosion and that fluctuations in erosion depth are accompanied by instantaneous changes in suspended load. Both $\overline{b}$ and $\hat{b}$ are artifices of an analytical procedure and the assumptions made; however, they may provide insights into the processes of resuspension and mixing of sediment in the surf zone.

Figure 11A-6A shows number-frequency plots of b_{max} values observed for three segments (inner, central, and outer surf zone) of data series 24.1 over 34-

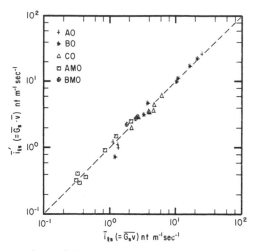

Figure 11A-5. Comparison of the local immersed weight longshore transport rate of suspended sediment computed as the mean of the instantaneous transport ($\bar{i}_{ls}$) and as the product of the mean total suspended load and the mean longshore velocity ($\bar{i}_{ls}{}'$).

min periods. The values of $\bar{b}$ and $\hat{b}$ are also shown except for the inner surf zone where $\bar{b}$ could not be determined because the OBS sensors were out of the water much of the time. The suspension peaks were measured, however, because they are associated with bore passage or times of deeper water and thus $\hat{b}$ could be estimated.

The results of this analysis show that the mean suspended sediment load in the central and outer surf zone during periods of low wave conditions ($H_s = 0.3$ m) is equivalent to a layer of sand ($\bar{b}$) of 0.48 and 0.11 cm thickness, respectively. The depths of erosion causing the suspended sediment peaks are considerably greater. Near the breaker region ($X/X_b = 0.96$), most values of b_{max} were relatively small although one erosion event reached a depth of 4.4 cm. The average value of $\hat{b}$ was 0.4 cm. In the mid-surf zone ($X/X_b = 0.60$) the modal value of b_{max} is 0.7 cm, $\hat{b}$ equals 1 cm, and the maximum erosion depth is 3.6 cm. In the inner surf zone ($X/X_b = 0.20$) the b_{max} mode is at 1.5 cm, $\hat{b}$ equals 1.8 cm, and the maximum erosion depth is 4.0 cm.

These estimates of $\hat{b}$ are similar to estimates of mixing depth of dyed sand carried out in various tracer studies such as those of Inman and Flick (1980) as shown in Figure 11A-6B. This suggests that further analysis of suspension events of sediment may provide an independent means of interpreting mixing depths in the nearshore zone.

An estimate of the net advective velocities experienced by the suspended sediment load has also been made, with the assumptions that sediment grains move with the fluid and that the maximum erosion depth (b_{max}) is achieved quickly. The data series shown in Figure 11A-6A ($X/X_b = 0.60$) was chosen for

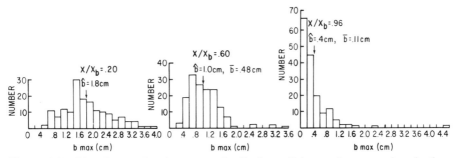

Figure 11A-6A. The number-frequency distributions of the maximum erosion depths (b_{max}) causing the peaks in suspension events at inner ($X/X_b = 0.20$), middle ($X/X_b = 0.60$), and outer ($X/X_b = 0.96$) surf zone positions during data series 24.1, $\hat{b}$ equals the mean of the b_{max} events, $\bar{b}$ equals the mean erosion depth.

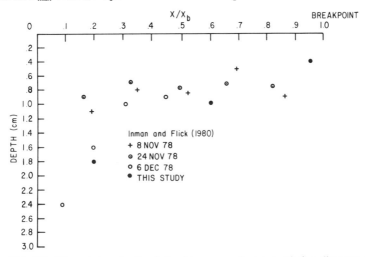

Figure 11A-6B. The mixing depth of dyed tracer sand versus relative distance offshore (after Inman *et al.*, 1980).

this analysis because its surf zone position approximates the regions where sand advection velocities have been measured (e.g., Komar and Inman, 1970; Inman *et al.*, 1980; Kraus *et al.*, 1981), thus making a gross comparison possible. The seabed was sliced into 1 mm levels and the number of data points (out of 4096) in which the erosion depth exceeded each level was determined. The ratio of NPTS/4096 × 100 represents the percent of time that grains at any particular depth were travelling with the suspended load. The product of the fraction of time suspended and the mean longshore velocity ($\bar{v}_\ell$) is an estimate of the average advective velocity of the suspended load (V_a)

$$V_a = \frac{\%}{100} \, \bar{v}_\ell \tag{11A-13}$$

Table 11A-4. Estimates of Erosion Depth

The percent of time that particles from a particular depth are suspended, and the advection velocity $(\overline{V}_a)$ for a mid-surf position during Data Series 24.1. The mean longshore current $(\overline{v}_l)$ is 10 cm/sec.

EROSION DEPTH (cm)	TIME SUSPENDED (%)	$\overline{V}_a$ (cm/sec)
0.2	85.5	8.6
0.5	18.1	1.8
1.0	5.4	.5
1.5	1.0	.1
2.0	0.2	.02
2.5	0.05	.005

The result of this analysis suggests that particles lying near the surface are frequently suspended by waves and are advected downcoast with nearly the rate of the longshore velocity (Table 11A-4). Particles lying deeper within the seabed are eroded less frequently. Thus their average advection rate is proportionally reduced. For example, a particle lying at a depth of 1.5 cm (the depth at which a sharp decrease is indicated in Figure 11A-6A, $(X/X_b = 0.60)$ is in suspension about 1% of the time and would have a mean advection rate of about 0.1 cm/s for the conditions observed during data series 24.1 $(\overline{v}_l = 10$ cm/s). These advection rates are of the same order as estimates from tracer studies of Komar and Inman (1970), Komar (1978), and Kraus *et al.* (1981), which show advection velocities associated with the movement of the center of gravity of a dyed sand mass equal to $0.5 - 1.25\%$ of the mean longshore velocity.

11A.4.2.3 Total Longshore Transport of Suspended Sediment

The longshore transport of suspended sediment (I_{ls}) was determined for each of the six data series. The values of $\overline{i}_{ls}$ from each data series were plotted according to their relative position within the surf zone (X/X_b) and the area encompassed by the data points is proportional to I_{ls} (Figure 11A-7). The curves have been extended to zero at the positions of $X/X_b = 0$ and 1.0, which neglects any swash zone transport or transport beyond the breaker point. The shapes of the curves in Figure 11A-7, although irregular, do tend to follow a general pattern, with maximum longshore transport in the mid-surf position. The data points tend to be grouped on either the shoreward or the seaward side of $X/X_b = 0.5$ and no single data series spanned the width of the surf zone. Nevertheless, visual inspection of the six data series appears to justify the way in which the curves are extrapolated to zero. This general distribution of $\overline{i}_{ls}$ is suggested by the field measurements of Kana (1979) and Downing (1983), the laboratory measurements of Sawaragi and Deguchi (1978), and the theoretical

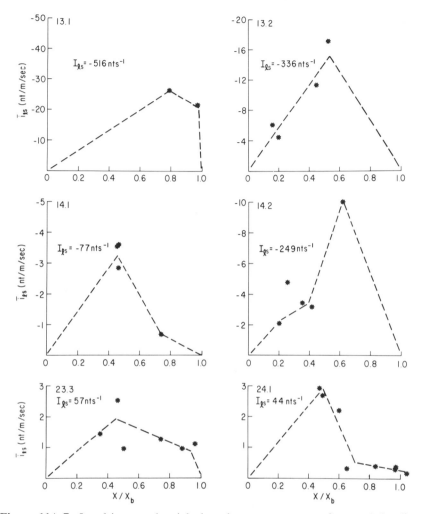

Figure 11A-7. Local immersed weight longshore transport rate of suspended sediment per unit width of beach ($\bar{i}_{ls}$) versus relative position in the surf zone (X/X_b) for each data series. The total immersed weight longshore transport rate of suspended sediment (I_{ls}) is noted for each curve.

predictions of Komar (1977). The area under each curve in Figure 11A-7 multiplied by the width of the surf zone (Table 11A-5) represents the measured value of I_{ls}. The values of I_{ls} are noted for each data series in Figure 11A-7 and are summarized in Table 11A-5.

Table 11A-5. Longshore Transport Calculations

DATA SERIES	AVERAGE P_l $(\frac{nt}{s})$	AVERAGE $(ECn)_b$ $\frac{\bar{v}}{u_m}$ $(\frac{nt}{s})$	WIDTH $\times$ (m)	BEACH SLOPE	MEASURED I_{ls} $(\frac{nt}{s})$	K	K'
13.1	-203	-1470	34	0.041	-516	2.54	0.35
13.2	-374	-2360	48	0.041	-336	0.90	0.14
14.1	-229	-462	55	0.037	-77	0.34	0.17
14.2	-278	-618	60	0.037	-249	0.89	0.40
23.3	19	207	55	0.022	57	3.00	0.28
24.1	17	75	35	0.021	44	2.59	0.59
A----				0.033		1.71	0.32

It is also of interest to know how the longshore transport rate of suspended sediment compares with the estimated total littoral transport rate. Assuming that (11A.11) and (11A.12) of Table 11A-1C predict the total longshore transport rate (I_l) as discussed by Komar (1983), the ratio of I_{ls} to I_l represents the contribution of suspended sediment to the total. In Figures 11A-8A and 11A-8B the values of I_{ls} from this study (solid circles) have been superimposed on a plot drawn by Komar (1983) which compares the equations of Komar and Inman (1970) with values of I_l determined by previous field studies.

In Figure 11A-8A the data from this study tend to fall above the predictive line while the agreement in Figure 11A-8B is excellent. The magnitude of K and K' measured for each data series (Table 11A-5) shows that the average value of K is 1.71. This is approximately twice the 0.77 value determined from the previous studies reported by Komar (1983). The average value of K' is 0.32 which is within 14% of the value determined from previous studies (0.28).

According to Komar (1983), equation 12 (Figure 11A-8B) is the more fundamental of the two sand transport relationships and should be applicable regardless of the origin of the longshore current $(\bar{v}_l)$, e.g., tide- or wind-generated, currents of the cell circulation, and from oblique wave approach. Further, Kraus *et al.* (1981) demonstrated the applicability of equation 12 under a condition where the longshore current was the net result of two opposing currents. Komar also pointed out that it is usually easier to measure the longshore current than the wave breaker angle needed in the evaluation of P_l. The close agreement between the results of this study and previous studies (Figure 11A-8B) suggests that all of the longshore transport at Leadbetter Beach can be accounted for by suspended sediment transport and that equation 12 adequately predicts the total immersed weight longshore transport rate of suspended sediment in the surf zone.

The ratio of I_{ls} to I_l is significantly higher in this study than has been generally measured or estimated by other investigators. A brief summary is listed below for comparison. Although it would be expected that physical processes in the surf zone would be a major influence on sediment suspensions

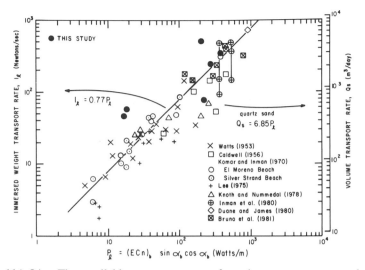

Figure 11A-8A. The available measurements of sand transport rates on beaches compared with the wave conditions expressed as P_l of (11A-11). The sand transport rate is expressed as either the immersed-weight transport rate, I_l, or as the volume transport rate, Q_s, with the two empirical relationships shown (after Komar, 1983). Data from this study are superimposed. (I_{ls} from suspended sediment and longshore velocity measurements, P_l from rms wave calculations).

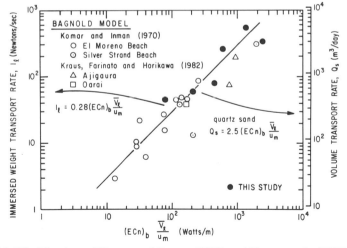

Figure 11A-8B. The data of Komar and Inman (1970) and Kraus *et al.* (1982) testing the Bagnold model of (11A-12) with $K' = 0.28$ for this relationship which yields the immersed-weight sand transport rate. Alternatively, the relationship can be expressed as the volume transport rate, Q_s (after Komar, 1983). Data from this study are superimposed. (I_{ls} from suspended sediment and longshore velocity measurements, $(EC_n)_b \, \bar{V}_l/u_m$ from rms wave calculations).

and hence might account for the differences observed above, there are also major differences in sampling techniques that play an important role.

Komar (1978), who analyzed the relative quantities of suspension transport versus bedload transport, developed an equation (equation 4) to estimate the ratio I_{susp}/I_{total} $(=I_{ls}/I_l)$. The important parameters in this equation are the mean volume concentration $(\overline{c})$, beach slope (tan$b$), and ratio of breaker height to breaker depth (γ). Komar used the pump data collected by Watts (1953) and Fairchild (1972) and the diver collected data of Kana (1976) to evaluate his equation. Sediment samples were collected no closer than 7.6 cm from the seabed for the studies of Watts and Fairchild and 10 cm from the seabed for the Kana study. None of these investigations sampled the higher concentrations of suspended sediment near the bed (e.g., see Figure 11A-4). Kana, who collected instantaneous profiles sporadically over time using diver-operated samplers, would tend to undersample many of the individual suspension events extending to higher elevations (lasting only several seconds). The analysis of Komar suggests that the value I_{ls}/I_l is ≤ 0.25.

$$\frac{I_{ls}}{I_l} \text{ COMPARISON}$$

STUDY	LOWEST MEASUREMENT (cm)		$\dfrac{I_{ls}}{I_l}$
Komar (1978)	7.6 - 10		.07 - .26
Inman et al. (1980)	10		.15 - .20
Kana and Ward (1980)	5	storm	1.0
		post-storm	.30
Downing (1983)	3.5		.47
This Study	3.5		1.0

In the study by Inman et al. (1980) also using diver-operated samplers, the suspended load would be undersampled for the same reasons discussed above for Kana's study. The data of Kana and Ward (1980) were collected with bulk water samplers that were mounted from a pier and positioned down to 5 cm above the seabed. During storm conditions when one might expect sediment suspension events to occur for longer periods of time and to extend to greater distances from the seabed, their results show a high ratio of I_{ls} to I_l, similar to the results of this study. During post-storm conditions, however, there would be a tendency to undersample the suspended load for the same reasons as discussed above.

Downing (1983) used sensors and analytical methods similar to those of the present study. His lowest sensor was deployed at $z = 3.5$ cm and the concentration gradient was extrapolated linearly to the seabed. Downing's measurements were carried out on a wide dissipative beach rather than on a narrow, reflective beach. Since the sampling and analytical methods of the two studies were similar, the difference in I_{ls}/I_l is quite possibly related to dynamic processes.

For the present study the mean values of $\overline{c}$ for all data (average of $\overline{G}_s$ from Table 11A-2 converted to volume concentration) is 1.46×10^{-3}, $\tan b = 0.033$, and $\gamma = 0.53$. Using these values in Komar's (1978) equation 4 gives an estimate of $I_{ls}/I_l = 0.92$ which is in agreement with the results shown in Figure 11A-8B. This further suggests that the observations from this study are consistent with Komar's analysis and that I_{ls}/I_l is predictable on the basis of mean volume concentration, beach slope, and the breaker height to breaker depth ratio.

11A.5 Conclusions

Using a miniature optical backscatter probe designed for this study, continuous measurements of suspended sediment concentrations at five levels between either 3.5 and 54.5 cm or 3.5 and 19.5 cm of the seabed and at different positions across the surf zone have been used to describe the characteristics of the suspended sediment concentration field. Suspended sediment concentration data have also been combined with concurrent flow and water level measurements to determine both local and total immersed weight longshore transport rates. Interpretation of these results in light of past studies of suspended sediment in the nearshore zone leads to the following conclusions:

(1) Sediment transport in the nearshore zone occurs as individual suspension events associated with bores propagating landward. Frequently, when the offshore-directed backwash flow exceeds the threshold condition for grain motion a resuspension event is initiated that is quickly reinforced by the next incoming bore. Sediment concentrations as high as 180 kgm^{-3} are measured at 3.5 cm off the seabed under passing bores and may be greater than 50 kgm^{-3} at 55 cm elevation. Individual suspension events occur abruptly and may last only seconds. Temporal variations of the magnitude of suspension events are observed in most data records. These variations in magnitude appear to be related to the bore amplitude and low frequency oscillations of the velocity component. The largest suspension events tend to be associated with strong low frequency offshore currents.

(2) Mean concentration profiles show an approximately logarithmic decrease away from the bed.

(3) Maximum values of the mean suspended load occur within the mid-surf region and decrease toward both the breaking point and the shore.

Suspended load is high in the swash-backwash region of the shore; however, detailed analysis has not been completed.

(4) During a period of low waves ($h_{sb} = 0.5$ m) the mean depth of erosion required to supply the measured suspended load is on the order of 0.1 to 0.5 cm. The mean of the maximum erosion depths during each suspension event is approximately 1 cm and is comparable to the mixing depth evaluated in various sand tracer studies. Assuming that the sediment moves with longshore velocity when suspended, the advection velocity is approximately 1% of the longshore velocity. The results suggest that further analysis of suspension events of sediment may provide an independent means of interpreting mixing depth used in tracer studies.

(5) A comparison of the longshore transport of suspended sediment using computational methods employing both mean and instantaneous analytical techniques shows that the mean of the instantaneous values of longshore transport equals the product of the mean values ($\overline{G_s v} = \overline{G_s}\,\overline{v}$). This implies that the fluctuating components of the suspended load and the longshore current are not correlated. Thus it is only necessary to measure the mean suspended load and the mean longshore current to estimate the immersed weight longshore transport rate of suspended sediment per unit width of seabed.

(6) The total immersed weight longshore transport rate of suspended sediment accounts for all of the total longshore transport rate as predicted by the equations of Komar and Inman (1970). The suspended sediment transport rate as measured during the experiments at Leadbetter Beach can be predicted by the equation

$$I_{ls} = 0.28 \, (Ecn)_b \, \frac{\overline{v}_l}{u_m} \qquad\qquad (11A\text{-}14)$$

11A.6 Acknowledgement

We would like to thank Stan Woods and Rex Johnson for their expert advice and fine craftsmanship in fabricating the instrumentation used for this study. Mr. H. Smith provided energetic assistance in gathering the field data and Mr. B. Jaffe assisted in the early stage of data analysis.

11A.7 Appendix

11A.7.1 Breaker Calculations

Estimation of the longshore sediment transport requires a knowledge of breaker height (H_b), angle (α_b), and depth (h_b), and width of the breaker zone (X_b). Since no direct measurements of these variables were performed during the experiment, wave data collected by the offshore slope array were used to hindcast the breakers.

Table 11A-A1. Conditions Governing
Wave Propagation Seaward of the Surf Zone

Conditions	Equations
Dispersion Relation	$gk \tanh(kh) = \text{constant}$
Refraction	$\dfrac{s}{\cos}\alpha = \text{constant}$ (Snell's law)
Energy Conservation	$sECn = s'E'cn'$ $+ \Delta E \, \Delta r$
Breaking Condition	$\dfrac{H_b}{h_b} = \gamma_b$

s	=	distance between wave rays
α	=	angle between wave ray and contour normal
E	=	wave energy density
ΔE	=	energy dissipation rate per bed area
Δr	=	distance along ray path
H_b	=	breaker height
h_b^*	=	depth under wave trough $(= h_b - \dfrac{H_b}{2})$
h_b	=	mean water depth
γ_b	=	breaking wave height to depth ratio

The prime " ′ " denotes value at the next Δr distance.

Seaward of the breaker zone, wave propagation is controlled by the dispersion relation, refraction, energy conservation, and the constraint on wave breaking. These conditions are tabulated in Table 11A-A1 together with the corresponding equations. Dissipation of wave energy before breaking is mainly produced by bottom friction. The rate of dissipation (ΔE) per unit area of the seabed can be expressed as $1/2\tau_m u_m \cos\phi'$ (Kajiura, 1968), where τ_m is the maximum wave boundary shear stress, u_m is the maximum bottom orbital velocity, and ϕ' is the phase lead of the oscillatory boundary shear stress relative to the orbital velocity. Assuming small phase lead and $\tau_m = \rho/2 \, f_w \, u_m^2$ with f_w representing the wave friction factor of Jonsson (1966), $\Delta E = \rho/4 \, f_w \, u_m^3$.

Theoretically predicted values of the wave height to depth ratio (γ_b) for solitary waves vary between 0.73 and 1.03 (Galvin, 1972). The most frequently used value is 0.78 (McCowan, 1894). Field tests of the magnitude of γ_b are lacking; thus, instead of applying a theoretically derived value, γ_b was determined from the wave data measured by a series of pressure sensors. The greatest number of bottom pressure sensors were deployed across and beyond the breaker zone on the 14th and 16th of February and thus provide a data base for estimating the breaker position and the breaker height. By converting the measured pressure into surface wave height through linear theory and using the Fourier transform method to obtain the significant wave height (with wave period, $4 < T < 30$ sec), the variation of the significant wave height with distance offshore was plotted (Figure 11A-A1). The breaker position is chosen at the point where the significant wave height starts to decrease (points 1, 2, 3, in Figure 11A-A1). The wave height to depth ratio (γ_b) obtained from these three curves are 0.48, 0.53 and 0.59, respectively. Based on these values, the average of 0.53 is chosen as the general breaking criterion for the data series analyzed except for series 14.1 and 14.2 for which their specific values were determined (see Figure 11A-A1). Although this method of determining the breaker position involves some uncertainty because of the spacings between pressure sensors, it is statistically stable, gives the breaker position to within ±5 m, depending on the distance between sensors, and appears to be more reliable than visual determination. Pressure data from the offshore slope array were converted into wave heights that were then Fourier transformed to obtain significant wave height and the wave period of the spectral peak.

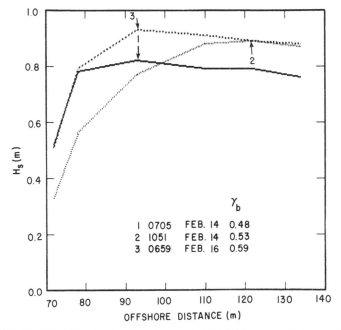

Figure 11A-A1. Variation in significant wave height (H_s) versus distance offshore for three data series (14.1, 14.2, 16.1). The breaking point is indicated by the arrows and the corresponding ratio of breaker height to depth (γ_b) is listed.

Table 11A-A2. Predicted and Observed
Breaker Height and Breaker Depth

TIME/DAY/NO.	PREDICTED		'OBSERVED' from Figure 11A-A1		
	H_{sb} (m)	h_b (m)	H_{sb} (m)	h_b (m)	γ_b
0705/14.1	0.84	2.17	0.82	2.10	0.48
1051/14.2	0.88	2.11	0.89	2.12	0.53

Directions for significant waves from the NSTS Leadbetter Beach experiment report (Gable, 1981) were used as the incoming wave direction at the slope array. Using the four conditions listed in Table 11A-A1, an iterative scheme was developed to find the breaker height, breaker angle, and breaker depth. Various values of the wave friction factor were tested in the calculation. A value of 0.15 was chosen based on a comparison between the calculated breaker height and breaker depth the measured values shown in Figure 11A-A1. Table 11A-A2 shows the comparison between the calculated and observed breaker heights and breaker depths which are within 2%. Using the above methods, breaker predictions were made for each data series (see Table 11A-3).

It should be mentioned that the wave friction factor of 0.15 as estimated represents a dissipation coefficient that may have included energy dissipation in addition to bottom friction. Although the value appears to be high, it is of the same order of magnitude as those observed by Wright *et al.* (1982b) in the nearshore zone and by Carstens *et al.* (1969) in laboratory studies.

11A.8 References

Bagnold, R. A., 1963, Mechanics of marine sedimentation, In: Hill, M. N. (ed.), *The Sea*, 3: 507-528, Interscience Publishers, a division of John Wiley & Sons, New York, London.

Bailard, J. A. and D. L. Inman, 1979, A reexamination of Bagnold's granular-fluid model and bedload transport equation, *Journal of Geophysical Research*, 84: 7827-7833.

Brenninkmeyer, B. M., 1974, Mode and period of sand transport in the surf zone, *Proceedings*, Fourteenth Coastal Engineering Conference, June 24-28, 1974, Copenhagen, Denmark, American Society of Civil Engineers, New York: 812-827.

_____. 1976a, In situ measurements of rapidly fluctuating, high sediment concentrations, *Marine Geology*, 20: 117-128.

_____. 1976b, Sand fountains in the surf zone, In: Davis, R. A. and R. L. Ethington (eds.), *Beach and Nearshore Sedimentation*, Society of Economic Palaeontologists and Mineralogists Special Publication No. 24: 69-91.

Carstens, M. R., R. M. Neilson and H. D. Altinbilek, 1969, Bed forms generated in the laboratory under oscillatory flow: analytical and experimental study, U. S. Army Coastal Engineering Center Technical Memo 28, 78 pp.

Coastal Engineering Research Center, 1973, *Shore Protection Manual*, U. S. Army Coastal Engineering Research Center, U. S. Government Printing Office, Washington, D.C. 20402 (3-part set).

Coakley, J. P., 1980, Field measurements of suspended sediment transport in the littoral zone of Lake Ontario, Canadian Coastal Conference 1980 Proceedings/Comptes-Rendus Conference Canadienne sur le Littoral de 1980, April 22-24, 1980, Burlington, Ontario: 31-46.

Downing, J. P., Jr., 1983, Field studies of suspended sand transport, Twin Harbors Beach, Washington, Ph.D. dissertation, University of Washington, Seattle, Washington.

Downing, J. P., R. W. Sternberg and C.R.B. Lister, 1981, New instrumentation for the investigation of sediment suspension processes in the shallow marine environment, In: Nittrouer, C. A. (ed.), Sedimentary Dynamics of Continental Shelves: 19-34, *Marine Geology*, 42 (Special Issue).

Fairchild, J. C., 1972, Longshore transport of suspended sediment, *Proceedings*, Thirteenth Coastal Engineering Conference, July 10-14, 1972, Vancouver, B.C., Canada, American Society of Civil Engineers, New York, 2: 1069-1088.

_____. 1977, Suspended sediment in the littoral zone at Vetnor, New Jersey, and Nags Head, North Carolina, U. S. Army Coastal Engineering Research Center, Technical Paper No. 77-5: 97 pp.

Francis, J. R. D., 1973, Experiments on the motion of solitary grains along the bed of a water-stream. *Proceedings*, Royal Society of London, Series A, 332: 443-471.

Gable, C. G. (ed.), 1981, Report on data from the Nearshore Sediment Transport Study experiment at Leadbetter Beach, Santa Barbara, California, January-February, 1980, 314 pp.

Galvin, C. J., Jr. 1972, A gross longshore transport rate formula, *Proceedings*, Thirteenth Coastal Engineering Conference, July 10-14, 1972, Vancouver, B.C., Canada, American Society of Civil Engineers, New York, 3: 953-970.

Inman, D. L., 1963, Sediments: physical properties and mechanics of sedimentation, In: Shepard, F. P., *Submarine Geology*: 101-151, Second Edition.

Inman, D. L. and R. E. Flick, 1980, Sediment transport studies in the nearshore environment - Task 4B, In: The NSTS Program for 1981 and 1982: 72-93, A Sea Grant National Program, Scripps Institution of Oceanography, La Jolla, California.

Inman, D. L., J. A. Zampol, T. E. White, D. M. Hanes, B. W. Waldorf and K. A. Kastens, 1980, Field measurements of sand motion in the surf zone, *Proceedings*, Seventeenth Coastal Engineering Conference, March 23-28, 1980, Sydney, Australia, American Society of Civil Engineers, New York: 1215-1234.

Jonsson, I. G., 1966, Wave boundary layers and friction factors, *Proceedings*, Tenth Conference on Coastal Engineering, September, 1966, Tokyo, Japan, American Society of Civil Engineers, New York: 127-148.

Kajiura, K., 1968, A model of the bottom boundary layer in water waves, Bulletin of the Earthquake Research Institute 46: 75-123.

Kana, T. W., 1976, A new apparatus for collecting simultaneous water samples in the surf zone, *Journal of Sedimentary Petrology*, 46: 1031-1034.

_____. 1978, Surf zone measurements of suspended sediment, *Proceedings*, Sixteenth Coastal Engineering Conference, August 27-September 3, 1978, Hamburg, Germany, American Society of Civil Engineers, New York: 1725-1743.

_____. 1979, Suspended sediment in breaking waves, Technical Report No. 18-CRD, Coastal Research Division, Department of Geology, University of South Carolina, Columbia, South Carolina: 153 pp.

Kana, T. W. and L. G. Ward, 1980, Nearshore suspended sediment load during storm and post-storm conditions, *Proceedings*, Seventeenth Coastal Engineering Conference, March 23-28, 1980, Sydney, Australia, American Society of Civil Engineers, New York, 2: 1158-1174.

Komar, P. D., 1977, Beach sand transport: distribution and total drift, *Journal of the Waterway, Port, Coastal and Ocean Division*, American Society of Civil Engineers, New York, 103(WW2): 225-239.

_____. 1978, Relative quantities of suspension versus bed-load transport on beaches, *Journal of Sedimentary Petrology*, 48(3): 921-932.

_____. 1983, Nearshore currents and sand transport on beaches, In: Johns, B. (ed.), *Physical Oceanography of Coastal and Shelf Seas*: 67-109, Elsevier Science Publishers B. V., Amsterdam, Netherlands.

Komar, P. D. and D. L. Inman, 1970, Longshore sand transport on beaches, *Journal of Geophysical Research*, 75: 5914-5927.

Kraus, N. C., R. S. Farinato and K. Horikawa, 1981, Field exeriments on longshore sand transport in the surf zone, *Coastal Engineering in Japan*, 24: 171-194.

McGowan, J., 1894, On the highest wave of permanent type, *Philosophical Magazine*, 5(38): 351-358.

Sawaragi, T. and I. Deguchi, 1978, Distribution of sand transport rate across a surf zone, *Proceedings*, Sixteenth Coastal Engineering Conference, August 27-Setpebmer 3, 1978, Hamburg, Germany, American Society of Civil Engineers, New York: 1596-1613.

Smith, J. D. and T. S. Hopkins, 1972, Sediment transport on the continental shelf off of Washington and Oregon in light of recent current measurements, In: Swift, D. J. P., D. B. Duane and O. H. Pilkey (eds.), *Shelf Sediment Transport: Process and Patterns*: 143-180, Dowden, Hutchinson and Ross, Inc., Stroudsburg, Pennsylvania.

Thornton, E. B. and W. D. Morris, 1978, Suspended sediments measured within the surf zone, *Coastal Sediments '77*, November 2-4, 1977, Charleston, South Carolina, American Society of Civil Engineers, New York: 655-668.

Watts, G. M., 1953, Field investigation of suspended sediment in the surf zone, *Proceedings*, Fourth Conference on Coastal Engineering, October, 1953, Chicago, Illinois, Council on Wave Research, J. W. Johnson, ed., Berkeley, California: 181-199.

Wright, L. D., P. Nielsen, A. D. Short, F. C. Coffey and M. O. Green, 1982a, Nearshore and surfzone morphodynamics of a storm wave environment: eastern Bass Strait, Australia, Coastal Studies Unit Technical Report No. 82/3, June 1982, University of Sydney, Coastal Studies Unit, Sydney, N.S.W. 2006, 154 pp.

Wright, L. D., P. Nielsen, A. D. Short and M. O. Green, 1982b, Morphodynamics of a macrotidal beach, *Marine Geology*, 50: 97-128.

Chapter 11

SUSPENDED SEDIMENT MEASUREMENTS

B. Discrete Measurements of Suspended Sediment

James A. Zampol and Douglas L. Inman

Center for Coastal Studies

Scripps Institution of Oceanography

La Jolla, California

11B.1 Introduction

Theories and empirical studies of longshore sediment transport rates within the surf zone have generally assumed or have estimated in some *ad hoc* manner, the contribution of suspended-load sediment to total-load transport. However, there has been broad disagreement upon the importance of suspended-load transport; ranging from dominant (Watts, 1953, Fairchild, 1972; Dean, 1973; Galvin, 1972; Brenninkmeyer, 1976; and Kana, 1977) to less than one-half of the bedload transport (Komar and Inman, 1970; Komar, 1976, 1978; and Inman *et al.*, 1980). Fairchild (1977), Kana (1977, 1978, 1979), Kana and Ward (1980) and Inman *et al.* (1980) suggest that the ratio of suspended load to total load is not constant, but that the importance of suspended load varies for different surf zone conditions (e.g., breaker type) and wave intensities.

Suspended sediment concentrations have been measured in a number of field investigations using a variety of sampling methods (Chapter 5A). However, few field measurements of longshore suspended-load transport have been made in conjunction with measurements of other important parameters such as waves, currents, and bedload transport rate. The intent of this study was to determine the vertical and cross-surf distribution of suspended sediment, relate suspended-load transport to wave and current parameters, and compare suspended-load transport rates with the concurrently measured rates of bedload transport (Chapter 13).

Classic treatments of the subject define suspended load as that portion of the total sediment load that is supported by fluid turbulence (O'Brien, 1933; Inman, 1949), while bedload is material that is placed in motion by the tangential shear stress of the fluid over the bottom and has a vertical dispersion maintained as a result of grain-to-grain contact and by the lift forces on the bed material (Bagnold, 1954, 1966; Inman, 1963). For sufficiently intense flows a region of high sediment concentration exists well above the at-rest surface of the bed (Einstein and Chien, 1955; Inman, 1963; Coleman, 1969). The motion of grains in this near-bed layer appears to involve elements of both grain dispersive pressure and fluid turbulence.

This leads to some confusion in estimates of suspended-load transport since some workers have assumed that all sediment above the level of the at-rest bed is suspended load (Watts, 1953; Thornton, 1972; Kana, 1977, 1978; Fairchild, 1972, 1977) while others have included the near-bed high sediment concentration region with the bedload (Komar, 1976, 1978) or specified a level above which sediment is considered to be suspended (Brenninkmeyer, 1974; Inman *et al.*, 1980). For practical purposes, suspended load in the present study has been arbitrarily chosen as that portion of the load that occurs 10 cm or more above the at-rest sand bed.

11B.2 Study Sites

The data for this study were collected at Torrey Pines Beach during the March 1977 and November-December 1978 experiments and at Leadbetter Beach during the February 1980 experiments, usually as a concurrent part of the sand tracer experiments. The median diameter of the beach sand was about 200 μ in 1977 and 150 μ in 1978; Leadbetter Beach contained sand of about 240 μ. During the experiments, Torrey Pines Beach had a low tide terrace slope, tanβ, of about 0.01 to 0.02 and a beach face slope of about 0.04. Beach slopes at Leadbetter Beach during the 5 to 9 February 1980 experiments were about 0.03 for the low tide terrace and 0.07 for the beach face. Following the storms of mid-February the surf zone widened and the beach slope decreased. For the 21 to 23 Feburary 1980 experiments the low tide terrace had a slope of 0.02 and had increased in width while the beach face nearly disappeared.

Wave conditions ranged from low energy to moderately high energy during storms (refer to Table 13-1). The dominant breakers during sampling on most days were visually classified as weakly to strongly plunging except on 22 February 1980 when they were visually classified as spilling.

11B.3 Results

During the NSTS experiments, 465 suspended sediment concentration determinations were made using the discrete water-core type samplers (Chapter 5A), providing 144 vertical concentration profiles. In addition, 19 plastic bag samples (Chapter 5A) were collected in the swash. These vertical suspended sediment concentration profiles were collected at stations on shore-normal ranges and usually traversed about three quarters of the surf zone. In the surf zone, suspended sediment concentrations ranged from less than 0.005 gm/liter to a high of 11 gm/liter; a sample of 110 gm/liter was collected in the swash.

The number of data points in each vertical profile of suspended sediment concentration is not always adequate for a thorough evaluation of the variation of concentration with height above the bottom (Figure 11B-1). However, concentration profiles of more than two points have been categorized according to the shape of the best fit curve to the data. The dominant category (47%) shows a depth-concentration curve that increases at a growing exponential rate (i.e., the curve is concave upward on a plot of sample elevation versus the logarithm of sediment concentration). Of the remainder, 16% are best approximated as a constant exponent, 11% are linear, and 26% are either irregular or inverted. The concentration data at each suspended sediment sample station have been interpolated and extrapolated, using the best fit to the data, from the arbitrarily chosen lower boundary of 10 cm above the bed to the water surface. The water surface was measured at the instant of sampling with a hand-held wave staff.

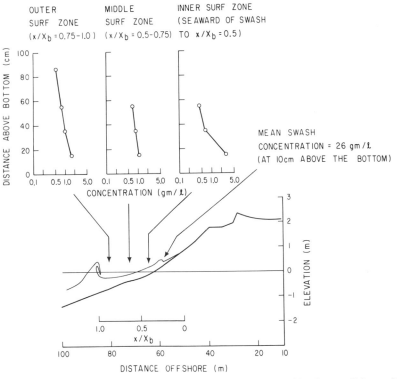

Figure 11B-1. Averaged suspended-sediment concentration profiles for conditions of low to moderate wave energy and steep beach (5-9 February 1980). Concentration data from different model samplers have been interpolated or extrapolated to common elevations of 15, 35, 55 and 85 cm above the bottom. Beach profile and surf zone location are for sampling period on 7 February 1980 and is representative of the conditions during 5-9 February 1980.

The suspended-sediment concentration profiles have been summed vertically to obtain local suspended load in units of mass per unit area of bed, $m_s (x, t/T)$. Suspended load also varies with time relative to the passage of a bore. Most of the data presented here were taken under the crest of the advancing bore which was found to provide the most consistent results.

The cross-shore distributions of locally averaged suspended-sediment load for data collected under the bore crest for each day are shown in Figure 11B-2a, b. There was, in general, a bimodal distribution of suspended load across the surf zone. The lowest suspended loads occurred in the middle of the surf zone and higher values were measured near the breakpoint and in the inner surf zone and swash. The largest and most variable suspended load was observed in the swash zone and in the region of interaction between the bore and the backwash. Similar suspended sediment distributions have been reported by Aibulatov (1958), Aibulatov et al. (1960) on a barred coast, Brenninkmeyer (1974, 1976) for sediment plumes of high concentration, and Inman et al. (1980) for a portion of the data presented here. Kana (1978, 1979) reports a peak in suspended-sediment concentration close to the breakpoint for plunging breakers, for samples collected approximately two seconds after passage of the bore crest.

All of the data shown in Figure 11B-2a, b are for days with primarily plunging breakers except that for 22 February 1980 when spilling breakers dominated. The measured distribution of suspended load is consistent with the fluid motions that have been observed for plunging breakers. Fairchild (1977), Kana (1977, 1978), and Inman et al. (1980) report that plunging breakers suspend significantly more sediment than do spilling breakers. This is intuitively expected from the different character of the bottom stresses associated with the two break types (Sawaragi and Iwata, 1974; Miller, 1976). However, the measured suspended load for the seaward-most sample on 22 February 1980, the only day of dominantly spilling breakers, is very high. This was a fairly high energy, post-storm day, and only one sample profile is available in the region of the breakpoint. Therefore, it is difficult to assess the degree to which this profile represents typical conditions. It is also worth noting that, using the break type criteria, c_{rb} (refer to discussion Chapter 13), the conditions on 22 February 1980 are in the same range as some other days described as weakly plunging and are within the plunging breaker range (Galvin, 1972; Inman and Guza, 1976).

11B.4 Suspended-Load Transport

Assuming that the suspended sediment travels along the shore at the same velocity as the longshore current, $v (x)$, and that $v (x)$ is approximately constant above 10 cm from the bed, then the local immersed weight longshore transport rate of suspended sediment is

$$i_s(x) = g \frac{(\rho_s - \rho)}{\rho_s} \overline{m}_s (x) \cdot v (x) \qquad (11B\text{-}1)$$

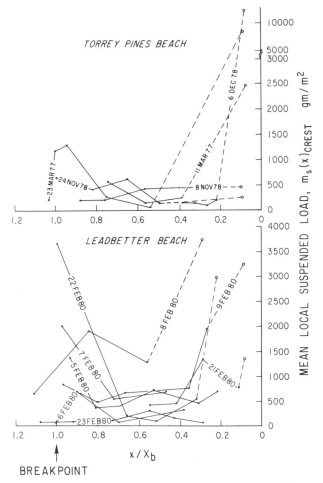

Figure 11B-2. Distribution of averaged suspended sediment load ($m_s(x)_{crest}$) relative to position in the surf zone (closed circles) and swash zone (open circles): (a) wave (bore) crest samples at Torrey Pines State Beach; (b) wave (bore) crest samples at Leadbetter Beach, Santa Barbara.

where ρ_s and ρ are the densities of the sediment and fluid respectively and g is the acceleration of gravity. Ideally, the average suspended load over a wave period would be used in calculating $i_s(x)$. Unfortunately, there is insufficient data available from which to estimate an average suspended load for most of the days analyzed here. Therefore, in order to be consistent, only data from the crest of the bore are used to calculate $i_s(x)$ even though it will, in general, cause $i_s(x)$ to be overestimated. In part, this is because the suspended load under the crest of the bore is summed to the maximum water surface elevation of the bore which is given as the depth of water in the trough (h) plus the height of the bore above the trough (H). The mean water surface elevation is, in general, closer to the elevation of the water surface in the trough of the wave (h). Therefore, suspended load and suspended-load transport have also been calculated for suspended-sediment concentration under the crest of the bore using h as an upper summation boundary. However, since this does not include the variation of suspended-sediment concentration during passage of the bore, even this lower value may be an overestimate of the average suspended-load transport.

The longshore velocity data are from nearby electromagnetic current meters (Chapter 10A), which provide a cross-surf distribution of $v(x)$. However, on 11 March 1977, 6 December 1978, 21 and 22 February 1980 the current meter coverage was sparse. On these days the cross-surf distribution of $v(x)$ was also measured with drift bottles. These data have been adjusted to fit the available current meter data. The local immersed weight suspended transport rate was calculated using local longshore current velocities averaged over time periods from tracer injection to either the end of temporal tracer sampling or to the mid-time of each spatial tracer sampling run. These time intervals were used so that the suspended-load transport rates pertain to the same time intervals as the bedload tracer experiments reported in Chapter 13.

For all six experimental days the strongest longshore currents occurred in the central portion of the surf zone (Inman *et al.*, 1980, Figures 4-3, 5, 6). Thus, the average bi-modal cross-surf pattern of suspended load distribution (highest near the breakpoint and in the inner surf zone and swash zone) is to varying degrees counter-balanced by the cross-surf pattern of longshore velocity distribution (highest in mid-surf zone). There does not appear to be a consistent position in the surf zone where suspended-load transport is a maximum for the six days presented here (Figure 11B-3a, b, c).

The total immersed-weight longshore transport rate of suspended sediment in the surf zone is given as

$$I_s = \sum_{i=1}^{n} i_{s_i}(x)\Delta x(x) \tag{11B-2}$$

where $\Delta x(x)$ is the interval about each determination of $i_s(x)$ and over which $i_s(x)$ is assumed to apply. I_s has been calculated both with the upper vertical summation boundary at the crest $(H + h)$ and at the trough (h). This was done for the temporal sampling time interval and for the two spatial sampling time intervals associated with the fluorescent tracer data (Table 11B-1). For all calculations of I_s, the breakpoint has been used as the seaward summation boundary. However, the landward summation boundary differed, being the seaward limit of the swash for the trough (h) summation level, and the landward limit of the uprush $(x/X_b = 0)$ for the crest $(H + h)$.

High energy surf conditions on 6 December 1978 and on 21 February 1980 prevented acquisition of adequate data from the outer surf zone. Since the

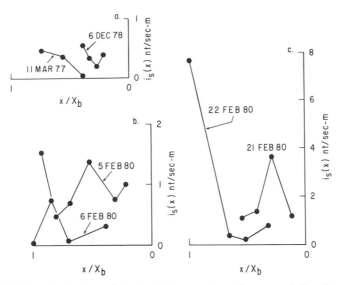

Figure 11B-3. The distribution of the local immersed-weight suspended-sediment transport rate ($i_s(x)$) relative to position in the surf zone. The values of $i_s(x)$ are calculated for averaged suspended load ($m_s(x)_{crest}$) summed to $H + h$ with current meter and drift bottle longshore current data averaged over the temporal sampling time interval. Swash zone values of suspended load have been excluded in calculating $i_s(x)$.

distribution of suspended load across the surf zone is commonly bimodal for plunging breakers (as shown in Figure 11B-2a, b), the lack of data in the outer surf zone could lead to underestimating the suspended load and the suspended-load transport. As previously mentioned, the most seaward suspended load value on 22 February 1980 is larger than might be expected for the prevailing surf conditions, but there is no *a priori* reason to discount it.

11B.5 Discussion

The ratio of suspended-load transport I_s to total longshore transport I_{ℓ} ($I_{\ell} = I_b + I_s$) has been calculated (Table 11B-2) for the six days discussed here and in Chapter 13. As discussed in Chapter 13, the bedload transport rate, I_b, used in the calculation was the rate obtained when using the *preferred estimator* of the thickness of the bedload layer, Z_o (Tables 13-3a, b, c thickness estimator 1 and temporal sampling using x-variation in sand velocity). The suspended-load transport data used for this comparison were calculated with current meter and drift bottle data averaged over the temporal time interval. The 11 March 1977 ratio is an exception in that only data for the spatial tracer method are available. Estimates of bedload transport rates determined by the spatial tracer method tend to be lower than temporally determined transport rate

Table 11B-1. Immersed-Weight Longshore
Suspended-Sediment Transport Rates

Date	Upper Summation Limit	Sample Time (PST)	I_s (nt/sec)		
			Temporal	Spatial Run #1	Spatial Run #2
11 March 1977	crest (H+h)*	1308	-	34	-
		1414	-	144	-
		average	-	89	-
	crest (H+h)**	1308	-	17	-
		1414	-	55	-
		average	-	36	-
	trough (h)*	1308	-	29	-
		1414	-	87	-
		average	-	58	-
	trough (h)**	1308	-	15	-
		1414	-	36	-
		average	-	26	-
6 December 1978	crest (H+h)*	1310	82	88	82
	crest (H+h)**	1310	86	94	87
	trough (h)*	1310	57	61	57
	trough (h)**	1310	60	65	60
5 February 1980	crest (H+h)*	0945	56	50	55
		1051	19	17	18
		1104	50	45	48
		average	42	37	40
	trough (h)*	0945	35	29	35
		1051	9.6	8.7	9.4
		1104	31	27	36
		average	25	22	27
6 February 1980	crest (H+h)*	1022	8.4	10	8.4
		1104	1.5	1.9	1.5
		1112	1.7	2.0	1.7
		average	3.9	4.6	3.9
21 February 1980	crest (H+h)*	1226	-	182	-
		1530	-	274	-
		average	-	228	-
	crest (H+h)*	1226	160	161	170
		1530	206	200	209
		average	183	181	190
	trough (h)*	1226	-	113	-
		1530	-	123	-
		average	114	112	117
	trough (h)**	1226	110	110	117
		1530	118	113	117
		average	114	112	117

* Velocity data from current meters
** Velocity data from current meters and drift bottles

Table 11B-1 (continued). Immersed-Weight Longshore
Suspended-Sediment Transport Rates

Date	Upper Summation Limit	Sample Time (PST)	I_s (nt/sec) Temporal	Spatial Run #1	Spatial Run #2
22 February 1980	crest (H+h)*	1259	287	252	261
		1319	29	22	24
		average	158	139	144
	crest (H+h)**	1259	255	227	240
		1319	26	22	24
		average	141	125	132
	trough (h)*	1259	156	137	142
		1319	22	19	20
		average	89	78	81
	trough (H+h)**	1259	134	118	125
		1319	19	16	18
		average	77	67	72

* Velocity data from current meters
** Velocity data from current meters and drift bottles

Table 11B-2. Ratios of Suspended-load to
Total-load Longshore Transport Rates

I_s/I_l, where $I_l = I_b + I_s$

	11 Mar 1977*	6 Dec 1978	5 Feb 1980	6 Feb 1980	21 Feb 1980	22 Feb 1980	Average Ratio: all days
I_b (nt/sec)	72	296	405	119	1090	273	
I_s(H+h) (nt/sec)	36	86	42	8.9	183	144	
I_s(h) (nt/sec)	26	60	25	3.9	114	77	
I_s(H+h)/(I_b + I_s(H+h))	0.33	0.23	0.094	0.070	0.14	0.33	0.20
I_s(h)/(I_b + I_s(h))	0.27	0.17	0.058	0.032	0.095	0.21	0.14

*Uses spatial estimates

estimates (Inman *et al.*, 1980; Kraus *et al.*, 1982; Chapter 13, Figure 13-4, 6), therefore the ratio of suspended-load to total-load transport may be overestimated when the total load is based on spatial sampling. The average ratios of suspended-load to total-load longshore transport for all six days were 0.20 and 0.14 using vertical summation to $H + h$ and to h, respectively.

These ratios have been calculated with the assumption that mostly bedload transport is measured in the bed tracer experiments (refer to Chapter 13). If, as others have assumed (Komar, 1976, 1978; Kraus *et al.*, 1982), the tracer experiments measure something close to the total-load transport rate, then the above ratios could be as large as 0.27 and 0.19 using vertical summation to $H + h$ and h, respectively.

Inspection of the estimates of suspended-load transport rate I_s in Table 11B-1 and the longshore sand transport potential $[CS_{xy}]_b$ values in Table 13-1 suggest a relationship of the form

$$I_s = K_s \, [CS_{xy}]_b \tag{11B-3}$$

where K_s is an empirically determined coefficient. A least squares best fit to these data, forced to pass through the origin, give $K_s = 0.14$ and 0.08 for the estimates of I_s which were calculated with vertical summation to $H + h$ and to h, respectively (Figure 11B-4). However, the range of data presented here is

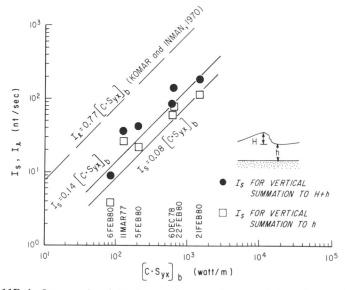

Figure 11B-4. Immersed-weight transport rate of suspended sediment I_s versus longshore sand transport potential $[CS_{xy}]_b$. The dashed line is from Komar and Inman (1970) with coefficient 0.77.

insufficient to preclude some other relationship between I_s and $[CS_{xy}]_b$. The coefficient, K_s, is expected to depend in some fashion on breaker type, beach slope, and grain size. However, any dependence on these parameters cannot be resolved with the present data.

In contradiction to our findings here and in Chapter 13, Sternberg *et al.* in Chapter 11A conclude that suspended-load transport can account for the total longshore transport of sediment. In part these contradictions may result from different sampling devices and the nature of their deployment, and from the different concepts of bedload and suspended load which lead to different methods of separating the two transport modes. Ten cm is used here as a practical lower limit of suspended load, while Sternberg *et al.* extrapolate to the bed surface.

Our sensors were designed to avoid artificial suspension by *coring* the water column from above (see Chapter 5A). They were deployed far from the turbulence caused by other instruments and were held up-current from the operator. The optical sensors of Sternberg *et al.* were attached to bottom-mounted platforms and at Santa Barbara were placed near NSTS *0* range which contained a high concentration of bottom-mounted instruments. Visual observations showed a higher concentration of suspended sediment in the waters near this range, presumably associated with the instrument mountings and with the frequent occurrence of rip currents which increase suspension. The occurrence of rip currents near this range was shown by the pattern of dye dispersion (e.g., see Figure 13-1), motivating the placement of the temporal sampling grid up-current from *0* range. In some cases, this decreased the recovery of tracer-sand from the down-current portion of the spatial sampling grids (Table 13-2b).

The data from our water-core samplers compare favorably with data collected in other studies that have employed both continuous (Watts, 1953; Fairchild, 1977; Coakley *et al.*, 1978; Nielsen *et al.*, 1982) and discrete (Kana, 1979) sampling methods. The water-core sampler data are in the same range as these other data for similar elevations. In contrast to these data, some of the suspended-sediment concentrations presented by Sternberg *et al.* are high, especially for the wave and surf zone conditions which existed during sampling (refer to instrument BMO in Chapter 11A, Figure 4).

11B.6 Conclusion

The vertical distribution of suspended-sediment concentration was measured at a series of stations across the surf zone. These were summed vertically to obtain local suspended load and in turn combined with local longshore current measurements to obtain longshore suspended-load transport. The data were collected in low energy to moderately high energy wave conditions with dominantly plunging breakers. Simultaneous measurements were made of the wave parameters and of bedload transport.

The results of this study indicate that:

(1) Suspended-sediment load shows a bimodal distribution across the surf zone with maxima near the breakpoint and in the swash zone (Figure 11B-2a, b).

(2) Within the constraints imposed by the experimental conditions, the suspended-load longshore transport rate is roughly 10 to 30% of the total longshore transport rate of sand.

(3) The longshore suspended-load transport rate data can be fit to an equation of the form

$$I_s = K_s [CS_{xy}]_b$$

The empirically determined values of K_s are 0.08 and 0.14 for the two methods used to estimate I_s.

11B.7 References

Aibulatov, N. A., 1958, New research on the longshore displacement of sand sediment in the sea, *Byull. Okeanogr. Kom. Akad. Nauk SSSR*, 1: 72-80 (in Russian).

Aibulatov, N. A., V. L. Boldyrev and V. P. Zenkovitch, 1960, Some new data on alongshore sediment streams, In: *21st International Geological Congress, Reports of Soviet Geologists, Problem 10*, (Akad. Nauk. SSSR) (in Russian with English summary).

Bagnold, R. A., 1954, Experiments on a gravity-free dispersion of large solid spheres in a Newtonian fluid under shear, *Proceedings*, Royal Society London, Series A, 225: 49-63.

_____. 1966, An approach to the sediment transport problem from general physics, U. S. Geological Survey Professional Paper 422-I, 37 pp.

Brenninkmeyer, B. M., 1974, Mode and period of sand transport in the surf zone, *Proceedings*, Fourteenth Coastal Engineering Conference, June 24-28, 1974, Copenhagen, Denmark, American Society of Civil Engineers, New York, 2: 812-827.

_____. 1976, Sand fountains in the surf zone, In: *Beach and Nearshore Sedimentation*, R. A. Davis and R. L. Ethington (eds.): 69-91, (Soc. Econ. Paleont. Mineral. Spec. Pub. n. 24), 187 pp.

Coakley, J. P., H. A. Savile, M. Pedrosa and M. Laroque, 1978, Sled system for profiling suspended littoral drift, *Proceedings*, Sixteenth Coastal Engineering Conference, August 27-September 3, 1978, Hamburg, Germany, American Society of Civil Engineers, New York: 1764-1775.

Coleman, N. L., 1969, A new examination of sediment suspension in open channels, *Journal of Hydraulic Research*, 7: 67-82.

Dean, R. G., 1973, Heuristic models of sand transport in the surf zone, *Proceedings*, Conference on Engineering Dynamics in the Surf Zone, Sydney, Australia: 208-214.

Einstein, H. A. and N. Chien, 1955, Effects of heavy sediment concentration near the bed on the velocity and sediment distribution, University of California Institute of Engineering Research, Berkeley, Series 33, Issue 2 (Water Resources Center Archives), 78 pp.

Fairchild, J. C., 1972, Longshore transport of suspended sediment, *Proceedings*, Thirteenth Coastal Engineering Conference, July 10-14, 1972, Vancouver, B.C., American Society of Civil Engineers, New York, 2: 1069-1088.

_____. 1977, Suspended sediment in the littoral zone at Ventnor, New Jersey and Nags Head, North Carolina, U. S. Army Corps of Engineers, Coastal Engineering Research Center, Technical Paper No. 77-5.

Galvin, C. J., 1972, A gross longshore transport rate formula, *Proceedings*, Thirteenth Coastal Engineering Conference, July 10-14, 1972, Vancouver, B.C., American Society of Civil Engineers, New York: 953-970.

Inman, D. L., 1949, Sorting of sediments in the light of fluid mechanics, *Journal of Sedimentary Petrology*, 19(2): 51-70.

_____. 1963, Sediments: physical properties and mechanics of sedimentation, Chapter 5, In: F. P. Shepard, *Submarine Geology* (second edition): 101-147, Harper and Row Publishers, New York, 557 pp.

_____. 1978, Status of surf zone sediment transport relations, *Proceedings*, Workshop on Coastal Sediment Transport, with Emphasis on the National Sediment Transport Study, University of Delaware, 2-3 December, 1976, University of Delaware, Sea Grant Report DEL-SG-15-78: 9-20.

Inman, D. L. and R. T. Guza, 1976, Application of nearshore processes to the design of beaches, *Abstracts*, Fifteenth International Conference on Coastal Engineering, American Society of Civil Engineers, New York, 2: 1215-1234.

Inman, D. L., J. A. Zampol, T. E. White, D. M. Hanes, B. W. Waldorf and K. A. Kastens, 1980, Field measurements of sand motion in the surf zone, *Proceedings*, Seventeenth Coastal Engineering Conference, American Society of Civil Engineers, New York, 2: 1215-1234.

Kana, T. W., 1977, Suspended sediment transport at Price Inlet, S. C., *Coastal Sediments '77*, American Society of Civil Engineers, New York: 366-382.

_____. 1978, Surf zone measurements of suspended sediment, *Proceedings*, Sixteenth Coastal Engineering Conference, August 27-September 3, 1978, Hamburg, Germany, American Society of Civil Engineers, New York, 2: 1725-1743.

_____. 1979, Suspended sediment in breaking waves, *Technical Report No. 8-CRD*, Coastal Research Division, Department of Geology, University of South Carolina, Columbia, South Carolina, 153 pp.

Kana, T. W. and L. G. Ward, 1980, Nearshore suspended sediment load during storm and post-storm conditions, *Proceedings*, Seventeenth Coastal Engineering Conference, March 23-28, 1980, Sydney, Australia, American Society of Civil Engineers, New York, 2: 1158-1174.

Komar, P. D., 1976, *Beach Processes and Sedimentation*, Prentice-Hall, Inc., Englewood Cliffs, NJ, 429 pp.

_____. 1978, Relative quantities of suspension versus bed-load transport on beaches, *Journal of Sedimentary Petrology*, 48(3): 921-932.

Komar, P. D. and D. L. Inman, 1970, Longshore sand transport on beaches, *Journal of Geophysical Research*, 75(30): 5914-5927.

Kraus, N. C., M. Isobe, H. Igarashi, T. O. Sasaki and K. Horikawa, 1982, Field experiments on longshore sand transport in the surf zone, *Proceedings*, Eighteenth Coastal Engineering Conference, November 14-19, 1982, Cape Town, Republic of South Africa, American Society of Civil Engineers, New York, 2: 969-988.

Miller, R. L., 1976, Role of vortices in surf zone prediction: sedimentation and wave forces, In: R. A. Davis and R. L. Ethington (eds.), *Beach and Nearshore Sedimentation*: 92-114, (Soc. Econ. Paleont. Mineral. Spec. Pub. n. 24), 187 pp.

Nielsen, P., M. O. Green and F. C. Coffey, 1982, Suspended sediment under waves, *Coastal Studies Unit Technical Report No. 82/6*, Department of Geography, The University of Sydney, Sydney, N.S.W. 2006, 157 pp.

O'Brien, M. P., 1933, Review of the theory of turbulent flow and its relation to sediment transportation, *American Geophysical Union, Transcript*, 14th Annual Meeting: 487-491.

Sawaragi, T. and K. Iwata, 1974, Turbulence effect on wave deformation after breaking, *Coastal Engineering in Japan*, 17: 39-49.

Thornton, E. B., 1972, Distribution of sediment transport across the surf zone, *Proceedings*, Thirteenth Coastal Engineering Conference, July 10-14, 1972, Vancouver, B. C., American Society of Civil Engineers, New York: 1049-1068.

Watts, G. M., 1953, Field investigation of suspended sediment in the surf zone, *Proceedings*, Fourth Conference on Coastal Engineering, October, 1953, Chicago, Illinois, American Society of Civil Engineers, New York: 181-199.

Chapter 12

CROSS-SHORE TRANSPORT

Richard J. Seymour
Scripps Institution of Oceanography

12.1 Measurement Problems

Chapters 13 and 14 describe two relatively straightforward techniques for measuring longshore transport, one involving short-term tracer experiments and the second longer-term trap experiments. Unfortunately, cross-shore transport is not readily amenable to either of these techniques. Cross-shore transport is driven by oscillatory velocities, near the bottom, which are nearly perpendicular to the shoreline. Net cross-shore transport results from a local or general assymmetry in these oscillations. In many cases, this bias is very small compared to the magnitude of the oscillatory component.

Attempting to measure the net transport caused by this small difference results in significant complications with the use of tracers. The diffusion of tracer is very rapid [Ingle (1966), Price (1968)] compared with the rate of cross-shore progression of the tracer centroid. Also, most models for cross-shore transport predict significant gradients in transport rate in the shore normal direction during intervals of great change to the profile. Both these factors tend to smear the tracer distribution, making it much more difficult to interpret, compared to a well-executed experiment on longshore transport.

However, if changes in the volume of sand on an area of straight, uninterrupted beach contained between two shore-normal lines extending into deep water can be measured, and the change is satisfactorily close to zero for some time interval, it can be concluded that there is no gradient in the longshore transport. Therefore, any observed rearrangement of the sand volume must be caused by cross-shore transport. It is relatively easy to insure that all of the net sand volume change has been accounted for in the shallow water portion of the experiment area. In deep water the elevation changes are smaller and require greater measurement accuracy. Because of the characteristically shallower slopes in deep water, even small changes in elevation can result in significant sediment volume changes over a given depth interval. The benchmarks for establishing the vertical references are always landward of the beach area to be measured. This results in the requirement that measurement accuracy increase with increasing distance from the benchmark, the reverse of the way in which most practical measurement systems work. The problem is further compounded by the inability, in the NSTS program, to make measurements in the deepwater portion of the site during storm wave conditions when offshore transport is likely to be greatest.

In an attempt to alleviate some of the measurement problems, investigators have followed one of two courses. The first is directed towards improving the accuracy of the deepwater measurements. These efforts under NSTS are detailed in Chapter 3. A number of other approaches using measuring systems that rest on the bottom are described in Seymour and Bothman (1984). The second method eliminates deep water measurements, or reduces the need for extreme accuracy in them, by using circumstantial evidence that no significant longshore gradient exists in the longshore transport. This can be accomplished from an analysis of the offshore and nearshore bathymetry, for example. If the contours are sufficiently straight and parallel, it can be assumed that the longshore forcing function is uniform. It is also possible for non-parallel contours to provide uniform longshore transport. It can be argued that quasi-stable shorelines must arrange themselves so that this is true. Therefore, measurements of wave directional properties at different longshore locations could be expected to test this premise. Also, if the change in the foreshore is documented over a significant longshore distance, synchrony in volume changes over the entire reach could be accepted as evidence that longshore transport gradients were not important.

Regardless of the rationale for accepting that the transport is not aliased by longshore gradients, under these conditions it is possible to study the effects of cross-shore transport on the observable portion of the beach. Net gain or loss, or a redistribution of sediment volume on any portion of the nearshore profile can be interpreted as cross-shore transport.

12.2 Cross-shore Transport at Torrey Pines

The offshore bathymetry at the Torrey Pines experiment site is remarkably regular (Figure 1A-1). Analyses of the linear wave array (see Chapter 2A), require the assumption of a homogeneous wave field over its 363 m length. The beach length used to measure cross-shore transport response is of the same order (500 m). The assumption of longshore homogeneity in the incident wave field was exhaustively verified in Pawka (1982). Therefore, it appears reasonable to assume that the forcing function for longshore transport will not vary significantly along the study area in a systematic fashion. Observations of beach face volume changes at Torrey Pines have been assumed to be caused solely by cross-shore transport.

A total of ten range lines were used to measure cross-shore transport at Torrey Pines, although all ten were not measured each day. Because the longshore gradient in the wave field was assumed small, it was concluded that

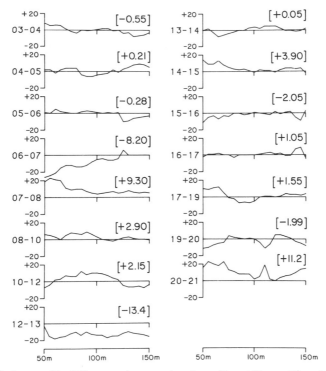

Figure 12-1. Interprofile differences in mean beach profiles at Torrey Pines Beach, CA. Dates are shown in pairs at left of each plot. Mean sea level is approximately at 90 m from origin. Elevation changes are in cm. Quantities in square brackets are volume changes in m^3 /m.

variation between survey lines was noise superimposed on the basic response of the beach to the waves. The sources of this noise might include cusp formation or decay, rip currents, local differences in the sediment characteristics or measurement errors. Seymour and King (1982) describe a scheme for averaging all of the available profiles for a single day to yield a mean beach profile. The intent was to cancel the noise as much as possible and reveal the underlying beach morphology. It should be noted that rip currents are a strong candidate for involvement in the cross-shore transport mechanism. However, since they have been observed to migrate longshore at Torrey Pines, their effects on the beach geometry at their instantaneous positions can be considered noise, even if their integrated effect is significant in driving the mean beach response.

Changes between successive mean beach profiles represented the response of the beach to cross-shore transport forcing during the inter-survey interval. Fifteen of these difference determinations were made at Torrey Pines and the volumetric changes calculated (Figure 12-1). Volume changes varied from -13.4 to $+11.2$ m^3/m during the inter-survey intervals (usually one day). The maximum erosionary event (12-13 November) was marked by a rather uniform 10 to 16 cm loss of material over the whole profile. The 6-7 November change involved a greater than 20 cm reduction in elevation of the beach near the berm. The next day, almost all of this material was restored. None of the other inter-survey differences have any clearly defined spatial patterns. As described above, all of the observed differences were interpreted as net cross-shore transport.

An objective analysis technique for beach profile variation has been demonstrated using empirical orthogonal functions [for example, Aubrey *et al.* (1980)]. Simply stated, this approach is analogous to Fourier decomposition, except that the components (called functions) are allowed to assume any shape rather than being constrained to be sinusoids. The functions are selected so that the minimum number are required to reconstruct the original data set in space and time. In this analysis, spatial functions are found. The analysis typically produces one or two functions, which when scaled and added to the mean profile for the whole time interval, will adequately predict the beach profile on any day. A scaling factor for each day is determined for each function and each value in the composite function curve is multiplied by the appropriate factor. When this method was applied to the Torrey Pines Beach data two significant functions were found. The first function [Figure 12-2(a)] shows that 62% of the variance from the mean in the horizontal beach position was involved with changing the concavity of the beach face with the mean sea level position remaining unchanged. The second function [Figure 12-2(b)], which accounts for almost 25% of this variance, erodes or accretes the beach above a null point at a depth of about 70 cm relative to MSL. The erosion or accretion is slightly more pronounced at MSL. In the figures, the mean of all the beach profiles is shown as a solid line flanked by the envelope (dotted) of the maximum and minimum measured positions. There is a time series that accompanies each function which gives the appropriate scaling factor for each function for each day in the record. Each function was multiplied by the maximum and minimum scale factors and these were added to the mean to produce the points that are plotted as crosses and stars for each elevation. In the case of Torrey Pines, adding the first two functions together produces an estimate of the beach profile which explains 97.1% of the variance.

Seymour and King (1982) attempted to predict beach face erosion and accretion at Torrey Pines using the models then available in the literature. None of the models showed any significant skill in prediction. When compared with the predictability of the other three data sets, as shown in Chapter 17, the Torrey Pines data set appears noisy, lacking in significant erosionary events, and unlikely to provide general insights into cross-shore transport mechanisms. Some of the noise may be attributed to the variable number of profiles involved

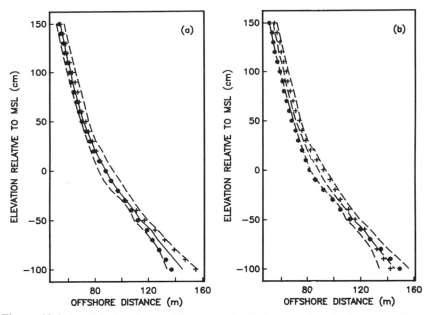

Figure 12-2. Empirical orthogonal functions (solid lines) for beach changes at Torrey Pines. The first function (a), explains 62.2% of the variance. The second function (b), explains 24.9%. The solid line is the mean position of the beach. The crosses (+) show the maximum position predicted by the function and the stars (*) the minimum. The envelope of maximum and minimum positions are shown by the dotted lines.

in determining the mean beach profile for a given day and to several data gaps when no profiles were obtained.

12.3 Cross-shore Transport at Santa Barbara

Based upon the experience with analyzing the Torrey Pines profile data, a different approach was adopted for the experiment at Santa Barbara. Only five range lines were employed, but these were all measured at least daily, eliminating the problems of variable data density. The measurement scheme, including a series of longshore profiles to define longshore variability, has been described in Chapter 3 and elsewhere [Gable (1981), Seymour and Aubrey (1985)].

The offshore bathymetry at Santa Barbara does not exhibit the straight and parallel contours seen at Torrey Pines. Therefore, an assumption of no longshore transport contribution to changes in profiles is not warranted without further information. Chapter 2B describes the two arrays that were installed to measure wave direction and intensity. Their longshore separation was approximately 200 m. The signals from these arrays were recorded four times per day for a period of several years. The data from the first two weeks of the

experiment were converted to directional spectra following the method of
Longuet-Higgins *et al.* (1963). These spectra, using central angles for each
frequency interval, were refracted by linear theory into a depth of about 3 m. At
this depth, the contours were straight and parallel. Therefore, converting the
local directional spectra to a spectrum of S_{xy}, allowed the estimation of
longshore transport as described in Seymour and Higgins (1978). The transports
predicted by each of the two arrays were summed algebraically for a two week
period, yielding a net transport value. The variation in the predicted net
transport between the locations of the two arrays was less than 5%. The total
transport difference, if spread evenly over the width of the surf zone, would
have resulted in an elevation difference of about 1 cm over two weeks between
the two ends of the beach cell considered. Thus, it can be assumed that, under
normal conditions, there is negligible gradient in longshore transport at Santa
Barbara.

The cross-shore transport during the experiment can be divided conveniently
into two time regimes. From 28 January to 15 February, 1980 was a period of
modest wave activity with small changes to the profiles on the order of those
observed at Torrey Pines. During this interval, during a time of low wave
steepness, a cusp formation and destruction event was recorded. As described in
Seymour and Aubrey (1985), this provided one of the few complete data sets
against which models for cusp formation mechanisms could be tested.

From 16 to 22 February was a period of intense storm waves and very
significant and rapid changes to the profiles. An erosion event occurred during
an episode of intense longshore currents and strong longshore transport resulted.
Evidence for significant longshore transport under the breakers is contained in
the displacement of the bar formed offshore by this storm. Figure 12-3 shows
beach contours before and after the storm and also contours of the differences.
The bar can be seen to have shifted to the east of the beach with its greatest
extent opposite the breakwater. The discontinuity in longshore transport along
the bar may have been caused by the headland to the west. The figure also
shows that, in addition to the erosion of the beach, there was a realignment in
the form of a clockwise rotation of the mean sea level (MSL) line and the
contours just seaward of it. This major, three-dimensional adjustment of the
whole nearshore zone to the very strong eastward thrust of the storm, rules out
any simple assumptions about longshore transport gradients during the second
time period. However, the beach profiles were remarkably uniform throughout
its length after the storm. Allowing for a rotation of the reference baseline to
correspond to the observed beach axis shift, there was far less profile-to-profile
variation than during the prestorm period. The existence of the eastward-shifted
bar, shown in Figure 12-3(c), suggests a longshore gradient in wave intensity
that might well have produced the beach rotation entirely by variation in the
intensity of cross-shore transport. It is not necessary to invoke longshore
transport gradients to explain the observed conditions. The data from Virginia

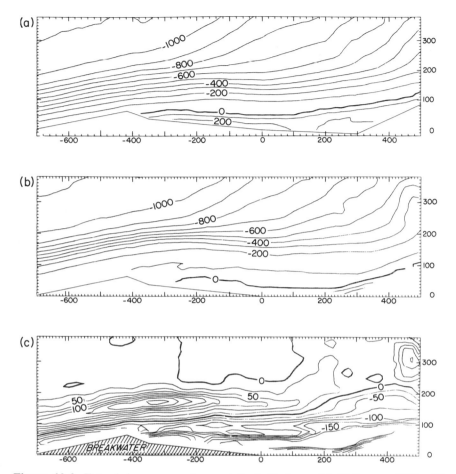

Figure 12-3. Extended survey contours at Santa Barbara. (a) 22 January 1980. (b) 25 Feburary 1980. Distances are in meters. Contours are in cm relative to MSL with intervals of 100 cm. (c) Difference between (a) and (b). Contours are in cm with intervals of 25 cm.

Beach, discussed later in this chapter, provide circumstantial evidence that the existence of this type of offshore bar may have a very significant influence on the rate of cross-shore transport. The conclusion is that the condition was indeterminate. The presence of the large offshore bar clearly indicates that there was massive offshore movement. Its displacement, because of the unusually strong longshore currents, prevents the traditional method of accounting for all of the sediment with the study area boundaries. Because the effects of beach rotation are averaged out in the mean beach scheme, treating the profile change as all cross-shore transport produced seems the best choice to salvage this rare data set of severe erosion. The contours as functions of time are shown in Figure 12-4.

The complete Santa Barbara record was examined using the objective function technique described above. Figure 12-5 shows the result using the same conventions described for Figure 12-2. The horizontal excursion of nearly 60 m that occurred as a result of the storm wave regime is shown by the dotted envelopes of maximum and minimum positions. The first function describes 97% of the variation for this data set. The stars and crosses show clearly that the principal element in the erosion process at Santa Barbara was a shoreward shift of the profile, without substantial change in shape or slope.

12.4 Other Cross-shore Transport Experiments

As described in Seymour (1986), two other experiments were conducted to study cross-shore transport. One was on Scripps Beach in La Jolla, California (8 October to 5 November 1980.) The second was at Virginia Beach, Virginia (29 September to 20 October 1981.) In both cases a single survey line was profiled each day and waves were measured several times per day. At Virginia Beach, a prototype hydrostatic profiler [see Seymour and Bothman (1984)] was employed that allowed long, accurate daily profiles to a depth of at least 2 m.

Scripps Beach is protected on the south by Point La Jolla and the La Jolla submarine canyon and on the north by the Scripps submarine canyon so that it functions as a pocket beach. Because of refraction away from the canyons,

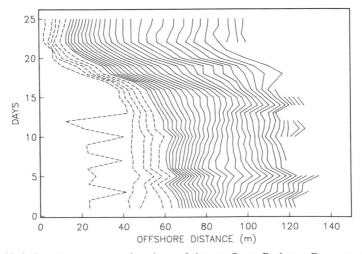

Figure 12-4. Beach contours as functions of time at Santa Barbara. Days are dates in February 1980. Dashed lines are contours from +2.0 m to +1.2 m in 0.1 m intervals. Solid lines are contours from +1.1 m to −2.0 m in 0.1 m intervals. All elevations are relative to MSL.

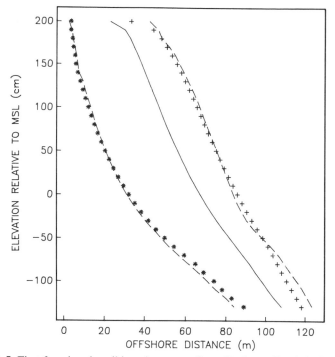

Figure 12-5. First function describing changes at Santa Barbara. Symbols follow Figure 12-2. This function explains 97% of the variation.

waves approach the beach at low angles. Longshore transport is slight at any time and approaches a zero net over long periods. The nearshore contours are quite regular in the vicinity of the survey line and gradients in longshore transport can be reasonably assumed to be small.

Virginia Beach has very smooth contours across the shelf, comparable to those at Torrey Pines. Therefore, it was assumed that all observed variation was caused by cross-shore transport. The beach contours as functions of time are shown in Figure 12-6 for Scripps Beach and Figure 12-7 for Virginia Beach.

As before, the predictive functions were found for each of these data sets that explained the maximum amount of the variance of horizontal position. These are shown, with the mean and the extreme positions of the beach, in Figures 12-8 and 12-9. In the Scripps Beach experiment (Figure 12-8) 85% of the variance is explained by the function illustrated. It can be seen to alter the beach slope or concavity about a null point at an elevation of about 150 cm above MSL. The Virginia Beach experiment (Figure 12-9) produced almost purely horizontal translation of the beach face. The first function explains 92%

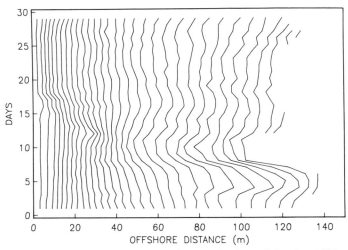

Figure 12-6. Beach contours at Scripps Beach. Days begin on 8 October 1980. Contours are from +1.9 m to −2.0 m in 0.1 m intervals, relative to MSL.

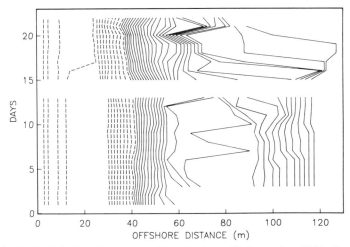

Figure 12-7. Virginia Beach contours. Days begin on 29 September 1981. Dashed lines are elevation contours from +2.8 m to +0.6 m in 0.2 m intervals. Solid lines are contours from +0.5 m to −2.0 m in 0.1 m intervals. All elevations are relative to MSL.

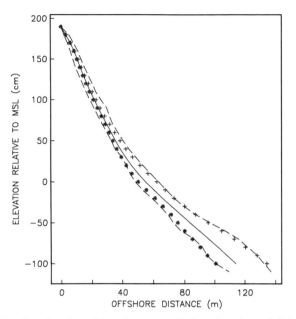

Figure 12-8. First function describing changes at Scripps Beach, explaining 85% of the variation. Symbols follow Figure 12-2.

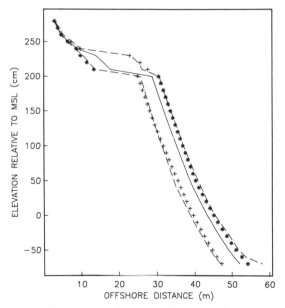

Figure 12-9. First function at Virginia Beach for beach face change only (bar not included). This explains 92% of the variance. Symbols follow Figure 12-2.

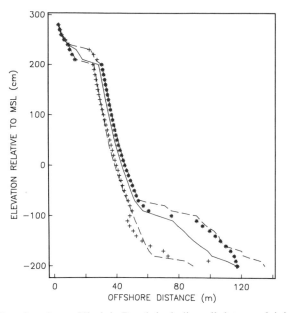

Figure 12-10. First function at Virginia Beach including all data, explaining 80% of the variance. Symbols follow Figure 12-2.

of the variance in this regime. When the complete profiles are considered (Figure 12-10), the major variation at Virginia Beach can be seen to be a substantial shift in the position of the offshore bar (on the order of 40 m). For the complete profile, the first function explains 80% of the variance.

Mason *et al.* (1984) describe a similar occurrence at Duck, North Carolina, approximately 60 km south of Virginia Beach. During an episode of large waves, there was little response of the beach face but rapid movement of the bar, on the order of 1 m/hr. The sediment, shelf slope and wave exposure are similar at these two sites.

It is interesting to compare the variations at the four sites. Torrey Pines and Scripps are low slope beaches with small breaker angles. Their principal mode of variation, as expressed by the dominant function, is a change in concavity. The beach face becomes steeper and less concave as material is eroded from it. Santa Barbara and Virginia Beach have steeper slopes and larger breaker angles. In these two, the variation is principally a horizontal retreat of the beach with little change in slope or concavity as material is eroded. Virginia Beach also has a bar at a depth of about 1 m, and Santa Barbara formed a bar during the major erosion event, while the other two beaches remained planar.

The Scripps Beach, Santa Barbara and Virginia Beach data sets, with the corresponding times series of waves and tides, are contained in Seymour (1986).

12.5 References

Aubrey, D. G., D. L. Inman and C. D. Winant, 1980, The statistical prediction of beach changes in Southern California, *Journal of Geophysical Research*, 85(C6): 3264-76.

Gable, C. G., Ed., 1981, Report on data from the Nearshore Sediment Transport Study Experiment at Leadbetter Beach, Santa Barbara, California, January-February, 1980, University of California, San Diego, Institute of Marine Resources, IMR Reference Number 80-5, 314 pp.

Ingle, J. C., 1966, *The Movement of Beach Sand*, Elsevier, Amsterdam, 221 pp.

Longuet-Higgins, M. S., D. E. Cartwright and N. D. Smith, 1963, Observations of the directional spectrum of sea waves using the motions of a floating buoy, *Ocean Waves Spectra, Proceedings of a Conference*, Prentice-Hall, Inc., Englewood Cliffs, New Jersey: 111-136.

Mason, C., A. H. Sallenger, R. A. Holman and W. A. Birkemeier, 1984, DUCK82--a coastal storm processes experiment, *Proceedings*, Nineteenth Coastal Engineering Conference, September 3-7, 1984, Houston, Texas, American Society of Civil Engineers, New York, 2: 1913-1928.

Pawka, S. S., 1982, Wave directional characteristic on a partially sheltered coast, Ph.D. dissertation, Scripps Institution of Oceanography, University of California, San Diego, 246 pp.

Price, W. A., 1968, Variable dispersion and its effects on the movements of tracers on beaches, *Proceedings*, Eleventh Coastal Engineering Conference, London, American Society of Civil Engineers, New York: 329-34.

Seymour, R. J., 1986, Results of cross-shore transport experiments, *Journal of the Waterway, Port, Coastal and Ocean Engineering*, American Society of Civil Engineers, 112(1): 168-173.

Seymour, R. J. and D. G. Aubrey, 1985, Rhythmic beach cusp formation: a conceptual synthesis, *Marine Geology*, 65: 289-304.

Seymour, R. J. and D. P. Bothman, 1984, A hydrostatic profiler for nearshore surveying, *Coastal Engineering*, 8: 1-14.

Seymour, R. J. and A. L. Higgins, 1978, Deepwater wave direction from an intensity array, *Proceedings*, Sixteenth Coastal Engineering Conference, August 27-September 3, 1978, Hamburg, Germany, American Society of Civil Engineers, New York, 1: 305-311.

Seymour, R. J. and D. B. King Jr., 1982, Field comparisons of cross-shore transport models, *Journal of the Waterway, Port, Coastal and Ocean Division*, Proceedings of the American Society of Civil Engineers, 108(WW2): 163-179.

Chapter 13

MEASURING LONGSHORE TRANSPORT WITH TRACERS

Thomas E. White and Douglas L. Inman

Center for Coastal Studies
Scripps Institution of Oceanography
La Jolla, California

13.1 Introduction

This chapter describes the results of experiments using sand tracers to measure the bedload response to forcing by waves and currents. Previous field studies of the longshore transport of sand have used wave arrays to measure the wave field and fluorescent sand tracer to measure the longshore transport of sand (e.g., Inman, Komar and Bowen, 1968; Komar and Inman, 1970). The former field procedures have been modified and improved to include arrays of pressure sensors and electromagnetic current meters and injections of dye in the water to measure the wave and current field. Fluorescently dyed sand is employed as a tracer to measure the motion of the sediment. The methods and assumptions involved in our use of sand tracer are detailed in Chapter 6B of this volume.

13.2 Driving Forces

The deepwater incident wave field was measured at Torrey Pines with a linear longshore array of pressure sensors and current meters approximately 400 m in length. The Santa Barbara wave field was measured by two six-meter square slope arrays of four pressure sensors each. The compactness of the Santa Barbara arrays resulted in significantly less directional accuracy of the estimated wave spectra. The directional wave spectra at both sites were estimated using maximum-likelihood techniques detailed by Pawka, Inman and Guza (1983) and then refracted band-by-band to approximate breaking depth. Wave and current data are listed in Table 13-1 for the six sand-tracer experiments described in this chapter (11 March 1977 and 6 December 1978 at Torrey Pines; 5, 6, 21, 22 February 1980 at Santa Barbara). The wave energy, E, was summed over the spectra at breaking depth. Various methods of determining the limiting wave frequencies for the spectral summation were used, resulting in the range of

Table 13-1. Wave and Current Data

	11 March 1977	6 Dec 1978	5 Feb 1980	6 Feb 1980	21 Feb 1980	22 Feb 1980
Surf Zone Location (m)						
Number of Estimates			6	5	9	11
Swash Location (NSTS coordinate)						
Mean of Observations	60	55	45	50	5	20
Standard Deviation			1.8	1.1	12.5	13.0
Breaker Location (NSTS coordinate)						
Mean of Observations	200	235	95	85	115	115
Standard Deviation			2.8	4.7	4.2	3.4
Significant Surf Zone Width						
Mean of Observations	140	180	50	35	110	95
Standard Deviation			4.9	4.6	14.4	6.0
Depth at Breaker Location, h_b (m)	1.43	1.65	1.40*	0.91*	2.56*	2.22*
Wave Angle of Approach of Spectral Peak, $\alpha_b(°)$	0.8-3	3-4	7-8	8-10	7-11	9-14
Spectral Wave Energy, E_b $\left[\frac{nt}{m} \text{ or } \frac{J}{m^2}\right]$	1506	1863-1910	270-293	107-118	1496-1881	682-787
Radiation Stress, S_{sy_b} $\left[\frac{nt}{m} \text{ or } \frac{J}{m^2}\right]$	28.0	118-125	38.3-40.9	15.6-17.6	256-322	118-131
Transport Potential $(C \cdot S_{xy})_b$ $\left[\frac{nt}{s} \text{ or } \frac{W}{m}\right]$	128.2	609-646	204-218	82.1-91.8	1365-1711	626-692
Average Mid-surf Longshore Current, $v\left[\frac{cm}{s}\right]$	45	65	33	14	46	32
Beach Reflection Coefficient for Peak Spectral Frequency c_{rb} (Eq. 13-4)	.024	.031	.418	.141	.044	.052

* At Santa Barbara wave data were shoaled to 3m depth. Radiation stress was assumed to have undergone no significant change between 3m and breaking depth (i.e., plane parallel beach contours in that depth range).

energy estimates in Table 13-1. The methods of selecting these limits, including the shallow water criterion ($L_\infty > 20\ h_b$ where L_∞ and h_b are the deepwater wavelength and depth at breaking point), are described as follow: one technique was to sum the radiation stress over all measured frequencies. A second method deleted all bands from the summation which failed to satisfy the aforementioned shallow water criterion. A third method deleted each band containing less than one percent of the radiation stress contained in the peak band. These methods resulted in the range of values in Table 13-1.

Both the frequency and directional spectra were refracted to a depth of 3 m. The 3-meter depth and the actual breaking depth were assumed to be sufficiently close that S_{xy} would not vary between them. The onshore flux of longshore directed momentum, S_{xy}, as defined by the radiation stress tensor S_{ij}, was computed as

$$S_{xy} = [En \cos \alpha \sin \alpha]_b \tag{13-1}$$

α is the angle of wave approach, n the ratio of group to phase velocities, and the subscript b denotes evaluation at the breakpoint of the waves. The longshore transport relations of a number of workers (e.g., Eaton, 1951; Inman and Bagnold, 1963; Komar and Inman, 1970) assume that a certain portion of the flux of longshore radiation stress drives the longshore transport of sediment

$$I = K[C \cdot S_{xy}]_b \tag{13-2}$$

where K is a dimensionless coefficient, and C is the wave phase velocity. $C_b \cdot S_{xy_b}$ with units of power per unit length of beach, is a measure of the potential for the longshore transport of sand that has a common occurrence in longshore transport relations (Inman, 1978). The phase velocity was computed from $C = \sigma/k$, where σ is the wave radian frequency and k the wave number computed at the breakpoint. The wave number was in turn determined from the linear-theory dispersion relation

$$\sigma^2 = gk \tanh kh \tag{13-3}$$

As expected, the angle of approach for the peak frequency at breaking (Table 13-1), was much higher at Santa Barbara than at Torrey Pines. The Santa Barbara experiments had high breaker angles of 7-14 degrees. During two days of high energy and low breaker angles, the very long sensor array at Torrey Pines gave excellent directional resolution, even for the low angles of approach. The compact Santa Barbara arrays provide lower directional resolution.

The reduced directional resolution of the compact arrays used at Santa Barbara led us to investigate the accuracy of the measured angles. Peak frequencies on several days were refracted outward using Snell's law. Errors at the west array were less than two degrees in the measured angle. The waves of peak energy on the four sand-tracer study days all approached from the west, and thus passed through that array on their approach to the section of the beach used for the sand-tracer study. Therefore for this paper we used the wave data as measured at the west array as input for the shoreward refraction analysis.

In addition to Eulerian currents obtained by the arrays of electronic sensors, dye was released into the water during the tracer experiments to obtain a Lagrangian picture of the current system. The dispersion of the dye (Figure 13-1) during the Santa Barbara experiments showed the presence of rip currents, which had considerable effect on at least three of the sand-tracer sampling grids

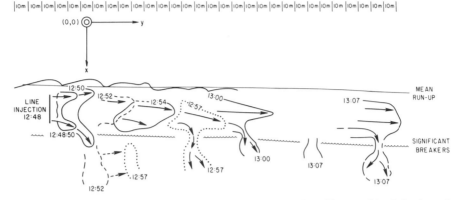

Figure 13-1. Water circulation patterns on 6 February 1980 as illustrated by injection of dye at 12:48:50 PST.

(spatial grids 1 and 2 on 21 Feburary 1980 and the second spatial grid on 22 February, 1980). In each of the four Santa Barbara experiments a rip current was present on the cross-shore line of sensors and apparently carried some sand tracer offshore, affecting the spatial sampling grids, but not the temporal grid, which was purposely placed up-current of the line of instruments. The 6 December 1978 Torrey Pines experiment also took place in the presence of rip currents. The rip currents at the 11 March 1977 experiment were all observed to be outside the sampling grids. (The dye dispersion for the Torrey Pines site is described in Inman *et al.*, 1980).

Rip currents and water circulation in beach cusps produced a significant effect on tracer distribution and recovery. The concentration of instruments in the relatively narrow surf zone at Santa Barbara formed a partial cross-shore barrier that favored the occurrence of rip currents along zero range. Rip currents were clearly observed by the seaward flow of dyed water injected to measure longshore currents (Figure 13-1). Since the temporal sampling station was updrift of the rip current, it invariably showed a higher recovery of tracer than the spatial sampling that extended down current of the rips. Swash cusps caused a pronounced variation in the pattern of spatial tracer sand on 5 February 1980, with concentrations related to integrals of one-half cusp length. However, the spatial grid with longshore dimensions of 135 m was large compared with the cusp wavelength, so that the smaller scale variations tended to be averaged out of the spatial transport calculations. Even so, the lower transport calculations from the spatial grid probably reflect loss of tracer in the cusp circulation.

13.3 Bedload Velocity

The longshore velocity of sand in the bedload was determined by monitoring the motion of the sand-tracer centroid with the spatial and temporal sampling grids described in Chapter 6B. Each day's tracer experiment utilized two spatial sampling grids and one temporal grid, except on 11 March 1977 when just one spatial grid was sampled. Maps of spatial distribution of sand tracer on 11 March 1977 and temporal distribution on 22 February 1980 are presented in Figures 13-2 and 13-3. Neither of these grids were significantly affected by rip currents, as the grids were between positions of semi-stationary rip currents. However, many of the spatial distributions at Santa Barbara were observed to have abrupt decreases in tracer concentration at the cross-shore instrument line where a semi-stationary rip current was observed to occur (Figure 13-1). Also, some of the spatial concentration distributions at Santa Barbara and in the December experiment at Torrey Pines were skewed in the offshore direction by rip currents.

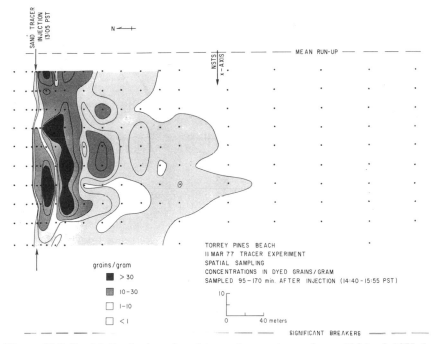

Figure 13-2. Spatial distribution of sand-tracer from grab samples on 11 March 1977. Injection of tracer was performed at 10 locations on the indicated cross-shore line.

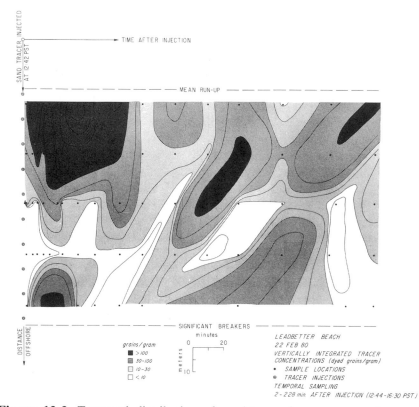

Figure 13-3. Temporal distribution of sand-tracer from core samples on 21 February 1980. The faster velocities in the swash can be observed from the cross-shore slope of lines of high concentration.

The cross-shore variation of the measured longshore bed velocity is illustrated in Figure 13-4a for the 6 February 1980 experiment. The results from the temporal grid are the more reliable as rip currents had little effect, and tracer recovery rates were high as described below under transport rates. A bimodal variation of velocity that is large in the swash and near the breakers, is observed in Figure 13-4a and confirmed by the other Santa Barbara experiments. The drop in velocity magnitude at the outer sampling station on the temporal grid is due to failure to obtain the earliest bed samples at that station. As seen in (6B-1), the principal contribution in the summation that determines bed velocity is from samples at times shortly after tracer injection. Thus the most seaward point in Figure 13-4a cannot be considered reliable. Accordingly we have extended the data from the next shoreward data location ($x/X_b = 0.86$) out to the breakers, as indicated by the dotted line in Figure 13-4a.

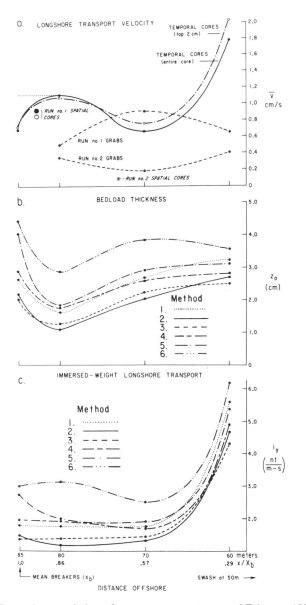

Figure 13-4. Cross-shore variation of transport parameters on 6 February 1980. (a) Sand transport velocity from (6B-1). (b) Bedload thickness in the temporal grid computed by the methods listed in Table 13-3b. (c) Longshore sand transport from the temporal grid in (a) and bedload thickness from (b).

13.4 Bedload Thickness

To compute bedload transport rates, both the velocity and the thickness of the moving bedload layer, Z_o, are needed, as indicated in (6B-3). The eight estimators described in Chapter 6B were used to estimate this thickness at each sampling station in the cross-shore direction. At each sampling station in the temporal grid several core samples were used to yield an average transport thickness for each estimator at each station. The cross-shore variations of these thicknesses are shown in Figure 13-4b for 6 February 1980. A generally bimodal distribution is observed. Noting the bimodality in the other NSTS experiments, we suggest that this variation in thickness across the surf zone is related to the characteristic motions of the fluid under plunging breakers and overturning bores in the swash. It is readily observed that these two types of fluid motion have great influence on the sand bed, while mid-surf motions do not generally involve fluid surface masses that plunge directly into the bed. The expected greater vertical velocities and stresses in plunging breakers and overturning bores should lead one to expect a thicker moving bed layer in those regions.

The spatial grids were sampled primarily by grab samplers which do not yield information on vertical variations in tracer concentration. However, in addition to the grab samples, one longshore line of core samples was obtained to provide a longshore distribution of bedload thickness. These values were then averaged, which resulted in one average bedload thickness (for each estimator) at one cross-shore location in the spatial grid. The variation of thicknesses from the temporal sampling were then used to provide cross-shore definition to this average thickness from the spatial grid. However, this procedure was not followed for the 11 March 1977 experiment, since there was no temporal sampling grid on that day. The longshore average of thicknesses at the cross-shore location was used as the estimate of Z_o on that day.

One estimator of Z_o which was used, the maximum-penetration estimator, does not appear in Fiugre 13-4b, as the values obtained from this estimator are so large that they dwarf the others (Tables 13-3 through 13-5). The 0.5 grain/gram penetration estimates usually gave the next largest layer thicknesses, while the remaining estimators generally provided quite comparable results.

The maximum-penetration estimator is one which has often been used in the past (King, 1951; Komar and Inman, 1970). It is based on the maximum penetration of significant amounts of tracer, so that the resulting estimates of tracer layer thickness are subjective, depending on the estimate of significant amounts. Furthermore, what is desired in tracer experiments is some temporal mean of the thickness of the moving bed. Use of an estimator which relies on extreme events is unlikely to be the most meaningful measure when a time-averaged process is involved.

13.5 Tracer Recovery and Transport Rate

The traditional method of estimating the quality of the transport rates obtained from a sampling grid is to estimate the amount of the injected tracer that has been recovered in the samples, assuming the concentration in each sample to be representative of the surrounding area of the bed. This method of estimating the budget of tracer and using it as a measure of quality control was introduced by Inman and Chamberlain (1959), and followed by Komar and Inman (1970) and Inman *et al.* (1980). The methods of computing transport rates and tracer recovery rates are detailed in Chapter 6B for each of the two types of grids.

The recovery estimates for our experiments are listed in Table 13-2 in units of percent of the injected tracer mass. The entire volume of sand in transport as indicated by tracer is considered in estimating the tracer grain recovery rate for the spatial grid [see (6B-5)]. Thus Z_o is critical to this estimate, and there are eight different estimates of recovery rate listed in Table 13-2 for the spatial grid sampling. On the other hand, estimates of tracer recovery rates for temporal grid sampling are based on the flux of tracer across a vertical plane in the beach containing all of the tracer grains over all depths. The cores in temporal sampling all extended beyond the maximum thickness of tracer penetration. Thus Z_o does not enter the calculation for tracer recovery for temporal grid

Table 13-2a. Tracer Recovery 1978

	Run #1	Run #2
Spatial Grid Recovery (%) Thickness estimator used for Z_o:		
Crickmore profile with		
*1) Abrupt decay (Eq 6B-7)	20.8	63.8
2) Crickmore estimator (Eq 6B-6)	21.9	59.1
Original profile with		
3) 80% cutoff	18.9	54.3
4) 90% cutoff	20.4	61.8
5) 1 grain/gm cutoff	5.3	39.5
6) ½ grain/gm cutoff	19.1	60.4
7) Abrupt decay (Eq 6B-7)	30.6	76.4
8) Maximum penetration	26.0	113.7
Temporal Grid Recovery (%)	47.6	

*   Preferred estimator

Table 13-2b. Tracer Recovery 1980

	5 Feb 1980		6 Feb 1980	
	Run #1	Run #2	Run #1	Run #2
Spatial Grid Recovery (%) Thickness estimator used for Z_o:				
Crickmore profile with				
*1) Abrupt decay (Eq 6B-7)	116.5	72.7	19.0	31.5
2) Crickmore estimator (Eq 6B-6)	98.6	65.9	13.0	22.2
Original profile with				
3) 80% cutoff	93.7	61.6	13.9	24.1
4) 90% cutoff	125.1	78.1	17.4	29.5
5) 1 grain/gm cutoff	116.8	76.0	18.7	32.3
6) ½ grain/gm cutoff	147.6	92.5	26.1	44.4
7) Abrupt decay (Eq 6B-7)	123.2	80.6	17.5	29.9
8) Maximum penetration	276.3	160.8	55.8	91.3
Temporal Grid Recovery (%)	90.6		55.6	

	21 Feb 1980		22 Feb 1980	
	Run #1	Run #2	Run #1	Run #2
Spatial Grid Recovery (%) Thickness estimator used for Z_o:				
Crickmore profile with				
*1) Abrupt decay (Eq 6B-7)	52.4	23.8	57.2	25.4
2) Crickmore estimator (Eq 6B-6)	47.0	19.6	43.5	21.7
Original profile with				
3) 80% cutoff	49.6	22.2	42.3	20.1
4) 90% cutoff	57.0	25.8	60.2	27.3
5) 1 grain/gm cutoff	45.2	19.3	48.6	19.6
6) ½ grain/gm cutoff	55.5	25.9	74.5	29.4
7) Abrupt decay (Eq 6B-7)	64.7	28.0	59.1	27.6
8) Maximum penetration	95.4	45.8	141.1	56.9
Temporal Grid Recovery (%)	107.7		87.7	

*   Preferred estimator

sampling (6B-8), and there is therefore only one estimate for temporal tracer recovery in Table 13-2. However, the longshore transport rates for both spatial and temporal sampling grids depend upon Z_o [see (6B-3),(6B-4)], so that there are different estimates of transport rate for each estimate of Z_o in Table 13-3.

It was not always feasible to extend our sampling all the way to the breakers. For such cases the most seaward value of sample concentration was assumed to

Table 13-3a. Longshore Sediment Transport Rates in Immersed Weight (Bedload I_b described in this chapter. Suspended load I_s is averaged from data listed in Table 11B-1, where values in parentheses use drift-bottle velocities.)

Transport Mode	Temporal Sampling		Spatial Sampling			
			Run 1		Run 2	
	Using x-variation in sand velocity V	Using average sand velocity V̄	Using x-variation in sand velocity V	Using average sand velocity V̄	Using x-variation in sand velocity V	Using average sand velocity V̄
11 March 1977						
Bedload Rate, I_b (nt/sec)						
Thickness estimator used for Z_o:						
Crickmore profile with						
*1) Abrupt decay (Eq 6B-7)			72	78		
2) Crickmore estimator (Eq 6B-6)			45	48		
Original profile with						
3) 80% cutoff			50	54		
4) 90% cutoff			82	88		
5) 1 grain/gm cutoff			36	39		
6) ½ grain/gm cutoff			54	59		
7) Abrupt decay (Eq 6B-7)			65	71		
8) Maximum penetration			86	201		
Suspended Load Rate, I_s (nt/sec)			74 (31)			
6 December 1978						
Bedload Rate, I_b (nt/sec)						
Thickness estimator used for Z_o:						
Crickmore profile with						
*1) Abrupt decay (Eq 6B-7)	296	297	Insufficient Data	169	201	205
2) Crickmore estimator (Eq 6B-6)	264	260		151	177	183
Original profile with						
3) 80% cutoff	246	222		141	167	172
4) 90% cutoff	286	246		163	194	198
5) 1 grain/gm cutoff	213	214		116	145	141
6) ½ grain/gm cutoff	280	278		160	191	194
7) Abrupt decay (Eq 6B-7)	335	333		193	225	235
8) Maximum penetration	503	518		319	389	387
Suspended Load Rate, I_s (nt/sec)	70 (73)		75 (80)		70 (74)	

* Preferred estimator

extend to the breaker zone. This may underestimate the transport, depending on the degree of cross-shore bimodality as indicated in Figure 13-4c. On 6 February 1980 our sampling extended to the breakers.

The temporal recovery rates for all NSTS experiments are sufficiently high to be useful measures of transport, although the temporal grid did not extend as far across the surf zone as desired on 6 December 1978. The spatial grids which

Table 13-3b. Longshore Sediment Transport Rates in Immersed Weight (Bedload I_b described in this chapter. Suspended load I_s is averaged from data listed in Table 11B-1, where values in parentheses use drift-bottle velocities.)

Transport Mode	Temporal Sampling		Spatial Sampling			
			Run 1		Run 2	
	Using x-variation in sand velocity V	Using average sand velocity $\bar{V}$	Using x-variation in sand velocity V	Using average sand velocity $\bar{V}$	Using x-variation in sand velocity V	Using average sand velocity $\bar{V}$
5 February 1980						
Bedload Rate, I_b (nt/sec)						
Thickness estimator used for Z_o:						
Crickmore profile with						
*1) Abrupt decay (Eq 6B-7)	405	354	210	192	59	69
2) Crickmore estimator (Eq 6B-6)	368	343	197	177	39	50
Original profile with						
3) 80% cutoff	349	316	170	153	38	48
4) 90% cutoff	436	381	235	215	52	61
5) 1 grain/gm cutoff	441	397	156	143	43	53
6) ½ grain/gm cutoff	518	455	169	155	51	60
7) Abrupt decay (Eq 6B-7)	457	414	258	234	50	62
8) Maximum penetration	894	737	441	413	213	232
Suspended Load Rate, I_s (nt/sec)	34		30		34	
6 February 1980						
Bedload Rate, I_b (nt/sec)						
Thickness estimator used for Z_o:						
Crickmore profile with						
*1) Abrupt decay (Eq 6B-7)	119	123	80	69	26	18
2) Crickmore estimator (Eq 6B-6)	103	101	104	88	16	11
Original profile with						
3) 80% cutoff	100	101	80	67	12	8
4) 90% cutoff	119	122	85	72	20	14
5) 1 grain/gm cutoff	133	139	78	66	18	13
6) ½ grain/gm cutoff	162	172	82	70	26	19
7) Abrupt decay (Eq 6B-7)	128	128	107	91	20	14
8) Maximum penetration	287	310	143	124	79	56
Suspended Load Rate, I_s (nt/sec)	6		7		6	

*   Preferred estimator

had recovery rates comparable in magnitude to the temporal recoveries were the first spatial grids on 5, 21, and 22 February 1980 and the second spatial grids on 5 February 1980 and 6 December 1978. The recovery rates exceeding 100% on 5 February 1980 emphasize the limits imposed by the methods of estimating tracer recovery. On 5 February 1980 the method of extrapolating the data to the break point overestimated recovery for almost all spatial grid samples, while the

Table 13-3c. Longshore Sediment Transport Rates in Immersed Weight (Bedload I_b described in this chapter. Suspended load I_s is averaged from data listed in Table 11B-1, where values in parentheses use drift-bottle velocities.)

Transport Mode	Temporal Sampling		Spatial Sampling			
			Run 1		Run 2	
	Using x-variation in sand velocity V	Using average sand velocity $\bar{V}$	Using x-variation in sand velocity V	Using average sand velocity $\bar{V}$	Using x-variation in sand velocity V	Using average sand velocity $\bar{V}$
21 February 1980						
Bedload Rate, I_b (nt/sec)						
Thickness estimator used for Z_o:						
Crickmore profile with						
*1) Abrupt decay (Eq 6B-7)	1090	1112	538	474	258	214
2) Crickmore estimator (Eq 6B-6)	1124	1135	437	403	181	151
Original profile with						
3) 80% cutoff	1084	1101	496	441	196	164
4) 90% cutoff	1248	1267	569	502	236	197
5) 1 grain/gm cutoff	1061	1064	463	430	258	213
6) ½ grain/gm cutoff	1370	1353	612	556	430	352
7) Abrupt decay (Eq 6B-7)	1517	1317	636	578	307	257
8) Maximum penetration	1909	1886	1040	892	684	565
Suspended Load Rate, I_s (nt/sec)	(149)		173 (147)		(154)	
22 February 1980						
Bedload Rate, I_b (nt/sec)						
Thickness estimator used for Z_o:						
Crickmore profile with						
*1) Abrupt decay (Eq 6B-7)	293	436	69	55	250	214
2) Crickmore estimator (Eq 6B-6)	264	370	89	67	355	302
Original profile with						
3) 80% cutoff	244	355	65	51	350	299
4) 90% cutoff	305	467	70	58	353	301
5) 1 grain/gm cutoff	284	354	65	36	156	141
6) ½ grain/gm cutoff	369	488	76	50	216	191
7) Abrupt decay (Eq 6B-7)	331	491	95	73	363	214
8) Maximum penetration	669	928	117	84	350	307
Suspended Load Rate, I_s (nt/sec)	124 (109)		109 (96)		113 (102)	

*   Preferred estimator

maximum-penetration estimates as expected, gave the maximum over estimations.

The five remaining spatial recovery rates are not sufficiently high to be considered good, and the transport rates from those grids are therefore not reliable. The recovery of tracer from the 11 March 1977 grid cannot be computed by (6B-5), since the measure of F, the number of dyed grains in a unit mass of injected sand, is unavailable. However, when Inman *et al.* (1980) estimated recovery from that grid (using a method of conversions between volumes and masses based on the assumption of spherical grains), they obtained a recovery of 65%. The grid on that day was far more extensive in terms of area covered and number of samples than any other grid ever used by us since then. Thus it would be surprising for the tracer recovery to be low on that day due to an inadequate number of samples.

The rates of transport of bed material obtained from (6B-4) are listed for the six days in Table 13-3. Two different methods were selected for computing transport velocity. [Refer to (6B-1) through (6B-3).] One of the methods uses all surf zone samples to obtain one average longshore velocity. This was the method used in the past by Inman, Komar and Bowen (1968), Komar (1969), and Komar and Inman (1970). The other method is to utilize the knowledge of cross-shore variation of longshore sand transport velocity, such as in Figure 13-4a. The suspended transport values listed are the average of the trough and crest values from Table 11B-1.

The cross-shore variation of the longshore transport rate is illustrated in Figure 13-4c for 6 February 1980. The longshore transport rate was found to be larger in the swash and near the breakers than in mid-surf for nearly all Santa Barbara tracer experiments.

The cross-shore distribution of longshore transport shown in Figure 13-4c differs from most theoretical models (Bijker, 1971; Komar, 1971; Thornton, 1972; Bailard, 1981). Maximum longshore transport rates are predicted somewhat shoreward of the breakers by Komar (at $x/X_b = 0.8$) and Bailard (at x/X_b between 0.8 and 0.9) and exactly at the breakers by Thornton (1973) who assumes that lateral mixing was zero. All of these models make assumptions that result in zero transport at the shoreline. The results in Figure 13-4c, indicating an increase in transport from mid-surf to the breakers, appear to agree with models such as Thornton's for the outer half of the surf zone. The fact that the swash data disagree with all of the aforementioned longshore current and sand transport models is not surprising. All these models make the rather unrealistic assumption of a stationary shoreline where all velocities are zero. Figure 13-4c shows that there is very active transport in this region, and it is also known now that orbital velocities and mean velocities actually increase shoreward (Guza and Thornton, 1985). The current and transport models also do not consider the effects of plunging breakers and overturning bores in the swash. These appear to have a significant effect on transport, as suggested in Figure 13-4c.

13.6 The Transport Modes

There has long been a question as to whether sand-tracer methods measure bedload transport only or total transport. In the surf zone there is an interaction and exchange between bedload and suspended load. Some bedload tracer may become suspended, travel longshore at a greater velocity, and result in an apparent increase in bedload transport velocity when reincorporated in the bed. Rip currents may produce the opposite effect in biasing estimates of transport velocity. When tracer is carried offshore, it is no longer sampled and generally results in a reduction in estimates of bedload velocity. The mass of grains in suspension is orders of magnitude smaller than the bedload mass. Therefore suspension processes should generally influence only a small proportion (on the order of one tenth of one percent) of the bedload mass. We believe that transport rates estimated from bed tracer sampling are principally estimates of bedload transport. Accordingly we consider the sum of the tracer-measured transport and the suspended transport measurements (Chapter 11B) to constitute total transport.

One reason that suspension of tracer has little effect on bedload sampling is that suspended load moves much faster than bedload. Longshore bedload velocities are of the order 1 cm/sec while longshore currents (and presumably suspended load velocities) are approximately 30 cm/sec. Under these conditions the suspended load, through exchange of tracer grains with the bed, should make a widely scattered and sparse contribution to the bed that is generally downdrift from the detailed sampling grids. The experiment at El Moreno using coarse and fine sand (Chapter 6B) supports this conclusion. The injected fine tracer had two areas of concentration. One was beneath the coarse sand where it was preferentially buried by the shear-sorting phenomenon (Inman *et al.*, 1966). A high concentration of tracer was also found at the surface nearly a kilometer downcoast from the sampling grid.

It is possible to estimate the maximum percent of time since injection which a tracer grain can spend in suspension and still be sampled in our grid of bedload samplers, namely,

$$\frac{y}{tv} \cdot 100\%$$

where y is the longshore distance between injection and sampling, t is the elapsed time between injection and sampling, and v is the longshore suspended velocity. This quantity was computed for the mean sample in each of the 16 NSTS sampling grids and was found to be less than 10% in all cases. An individual tracer grain can spend varying portions of time in suspension, saltation, bedload, or even buried beneath the bedload. But if it spends more than 10% of its time in the surf zone as suspension, then it will not be sampled by our grid of bedload samplers. Finally, tracer grains have not been found in the suspended load samples (Chapter 11B). This may be due to the shear-sorting burial phenomenon mentioned earlier or because of rapid motion of fine grains beyond the sampling region.

13.7 Evaluation of the Transport Relation

We have computed values of the transport coefficient K in (13-2) for both bedload transport and for total load. The total load is taken as the sum of the bedload transport and the water-column suspension measured by the methods detailed in Chapter 11B. The resulting values of K are listed in Table 13-4 (a-f) for all six days of sand-tracer experiments. The values for the two Torrey Pines experiments and the 21, 22 February 1980 experiments at Santa Barbara are generally between 0.3 and 0.6 for bedload thickness. The values for the 5 and 6 February 1980 experiments are much higher, however. The ranges of values of longshore transport I and the longshore transport potential $[C - S_{xy}]_b$ were used to plot boxes containing our estimates of K in Figure 13-5 for the tracer-measured bedload transport and in Figure 13-6 for the total transport.

Based on all of our sand-tracer experiments from Santa Barbara and Torrey Pines and on earlier experiments at El Moreno and Silver Strand, we believe the systematic range of K values we have found indicates that the transport coefficient is a variable, dependent upon the nature of the breaking waves and the slopes of the beach.

The most likely variable for parameterizing the effects of breaker type and beach slope is a form of the dimensionless surf similarity parameter of Battjes (1974), here given in the form of a reflection coefficient following Inman and Guza (1976)

$$c_{rb} = \frac{2g\,\tan^2\beta}{H_b\,\sigma^2} = \frac{L_\infty \tan^2\beta}{\pi H_b} \tag{13-4}$$

where β is the slope of the beach, $\sigma = 2\pi/T$ is the radian frequency of the incident waves, and L_∞ is the deepwater wavelength. A form of this relation was first used by Iribarren and Nogales (1949) for determining whether breaking would occur, and Carrier and Greenspan (1958) used the fully nonlinear shallow water equations to show that a pure standing wave is possible only if $c_{rb} \ge 1$. The relation has been used successfully for surf similarity purposes by Bowen *et al.* (1968), Galvin (1972), Guza and Inman (1975), and Munk and Wimbush (1969). The term c_{rb} is related to the Battjes (1974) form ξ of the surf similarity parameter by the relation

$$c_{rb} = \xi^2/\pi \tag{13-5}$$

Laboratory studies (Kamphuis and Readshaw, 1978) indicated a relation between K and a form of the surf similarity parameter. Expressing Kamphuis and Readshaw's equation 18 in terms of K and c_{rb}, the results of their experiments became

Table 13-4a. Values of the Coefficient K_b
and K_t, in the Longshore Transport Relation (13-2),
where Immersed-weight Bedload Transport Rate I_b
and the Suspended Load Rate I_s are from Table 13-3.

Data for 11 March 1977

Transport Coefficient	Temporal Sampling		Spatial Sampling			
			Run 1		Run 2	
	Using x-variation in sand velocity V	Using average sand velocity $\bar{V}$	Using x-variation in sand velocity V	Using average sand velocity $\bar{V}$	Using x-variation in sand velocity V	Using average sand velocity $\bar{V}$
Bedload, K_b						
Thickness estimator used for Z_o:						
Crickmore profile with						
*1) Abrupt decay (Eq 6B-7)			0.56	0.60		
2) Crickmore estimator						
(Eq 6B-6)			0.34	0.37		
Original profile with						
3) 80% cutoff			0.38	0.42		
4) 90% cutoff			0.63	0.68		
5) 1 grain/gm cutoff			0.28	0.31		
6) ½ grain/gm cutoff			0.42	0.46		
7) Abrupt decay (Eq 6B-7)			0.50	0.55		
8) Maximum penetration			1.43	1.55		
Total Load, K_t						
Thickness estimator used for Z_o:						
Crickmore profile with						
*1) Abrupt decay (Eq 6B-7)			0.73			
2) Crickmore estimator						
(Eq 6B-6)			0.51			
Original profile with						
3) 80% cutoff			0.55			
4) 90% cutoff			0.80			
5) 1 grain/gm cutoff			0.45			
6) ½ grain/gm cutoff			0.59			
7) Abrupt decay (Eq 6B-7)			0.67			
8) Maximum penetration			1.61			

*   Preferred estimator

Table 13-4b. Values of the Coefficient K_b
and K_t, in the Longshore Transport Relation (13-2),
where Immersed-weight Bedload Transport Rate I_b
and the Suspended Load Rate I_s are from Table 13-3.

Data for 6 December 1978

Transport Coefficient	Temporal Sampling		Spatial Sampling			
			Run 1		Run 2	
	Using x-variation in sand velocity V	Using average sand velocity $\overline{V}$	Using x-variation in sand velocity V	Using average sand velocity $\overline{V}$	Using x-variation in sand velocity V	Using average sand velocity $\overline{V}$
Bedload, K_b						
Thickness estimator used for Z_o:						
Crickmore profile with						
*1) Abrupt decay (Eq 6B-7)	0.48	0.48		0.26	0.33	0.33
2) Crickmore estimator (Eq 6B-6)	0.43	0.42	Insufficient Data	0.23	0.29	0.30
Original profile with						
3) 80% cutoff	0.40	0.36		0.22	0.27	0.28
4) 90% cutoff	0.46	0.40		0.25	0.32	0.32
5) 1 grain/gm cutoff	0.34	0.34		0.18	0.24	0.23
6) ½ grain/gm cutoff	0.45	0.45		0.25	0.31	0.32
7) Abrupt decay (Eq 6B-7)	0.54	0.54		0.30	0.37	0.38
8) Maximum penetration	0.81	0.84		0.49	0.63	0.63
Total Load, K_l						
Thickness estimator used for Z_o:						
Crickmore profile with						
*1) Abrupt decay (Eq 6B-7)	0.52			0.30	0.37	
2) Crickmore estimator (Eq 6B-6)	0.46			0.27	0.33	
Original profile with						
3) 80% cutoff	0.44			0.26	0.31	
4) 90% cutoff	0.50			0.29	0.35	
5) 1 grain/gm cutoff	0.38			0.22	0.25	
6) ½ grain/gm cutoff	0.49			0.28	0.28	
7) Abrupt decay (Eq 6B-7)	0.58			0.34	0.40	
8) Maximum penetration	0.85			0.53	0.67	

*   Preferred estimator

Table 13-4c. Values of the Coefficient K_b
and K_t, in the Longshore Transport Relation (13-2),
where Immersed-weight Bedload Transport Rate I_b
and the Suspended Load Rate I_s are from Table 13-3.

Data for 5 February 1980

Transport Coefficient	Temporal Sampling		Spatial Sampling			
			Run 1		Run 2	
	Using x-variation in sand velocity V	Using average sand velocity $\overline{V}$	Using x-variation in sand velocity V	Using average sand velocity $\overline{V}$	Using x-variation in sand velocity V	Using average sand velocity $\overline{V}$
Bedload, K_b						
Thickness estimator used for Z_o:						
Crickmore profile with						
*1) Abrupt decay (Eq 6B-7)	1.87	1.63	0.99	0.91	0.27	0.32
2) Crickmore estimator (Eq 6B-6)	1.69	1.58	0.93	0.84	0.18	0.23
Original profile with						
3) 80% cutoff	1.61	1.46	0.80	0.72	0.17	0.22
4) 90% cutoff	2.01	1.76	1.11	1.02	0.24	0.28
5) 1 grain/gm cutoff	2.03	1.83	0.74	0.68	0.20	0.24
6) ½ grain/gm cutoff	2.38	2.10	0.80	0.73	0.23	0.28
7) Abrupt decay (Eq 6B-7)	2.11	1.91	1.22	1.11	0.23	0.28
8) Maximum penetration	4.12	3.39	2.09	1.96	0.97	1.06
Total Load, K_t						
Thickness estimator used for Z_o:						
Crickmore profile with						
*1) Abrupt decay (Eq 6B-7)	2.01	1.77	1.12	1.03	0.41	0.45
2) Crickmore estimator (Eq 6B-6)	1.83	1.72	1.06	0.97	0.32	0.37
Original profile with						
3) 80% cutoff	1.75	1.60	0.93	0.85	0.31	0.36
4) 90% cutoff	2.15	1.90	1.24	1.14	0.37	0.42
5) 1 grain/gm cutoff	2.17	1.97	0.86	0.80	0.33	0.38
6) ½ grain/gm cutoff	2.52	2.24	0.93	0.86	0.37	0.41
7) Abrupt decay (Eq 6B-7)	2.25	2.05	1.35	1.23	0.37	0.42
8) Maximum penetration	4.26	3.54	2.21	2.08	1.11	1.20

*   Preferred estimator

Table 13-4d. Values of the Coefficient K_b
and K_t, in the Longshore Transport Relation (13-2),
where Immersed-weight Bedload Transport Rate I_b
and the Suspended Load Rate I_s are from Table 13-3.

Data for 6 February 1980

Transport Coefficient	Temporal Sampling		Spatial Sampling			
			Run 1		Run 2	
	Using x-variation in sand velocity V	Using average sand velocity $\bar{V}$	Using x-variation in sand velocity V	Using average sand velocity $\bar{V}$	Using x-variation in sand velocity V	Using average sand velocity $\bar{V}$
Bedload, K_b						
Thickness estimator used for Z_o:						
Crickmore profile with						
*1) Abrupt decay (Eq 6B-7)	1.30	1.34	0.86	0.74	0.28	0.20
2) Crickmore estimator						
(Eq 6B-6)	1.12	1.10	1.12	0.95	0.18	0.12
Original profile with						
3) 80% cutoff	1.09	1.10	0.86	0.73	0.13	0.09
4) 90% cutoff	1.29	1.32	0.91	0.78	0.22	0.16
5) 1 grain/gm cutoff	1.45	1.52	0.84	0.71	0.20	0.14
6) ½ grain/gm cutoff	1.76	1.87	0.89	0.75	0.29	0.20
7) Abrupt decay (Eq 6B-7)	1.39	1.39	1.16	0.98	0.22	0.15
8) Maximum penetration	3.12	3.37	1.54	1.34	0.86	0.61
Total Load, K_l						
Thickness estimator used for Z_o:						
Crickmore profile with						
*1) Abrupt decay (Eq 6B-7)	1.38	1.42	0.96	0.83	0.36	0.28
2) Crickmore estimator						
(Eq 6B-6)	1.20	1.18	1.22	1.04	0.26	0.20
Original profile with						
3) 80% cutoff	1.17	1.18	0.96	0.82	0.21	0.17
4) 90% cutoff	1.37	1.41	1.01	0.87	0.30	0.24
5) 1 grain/gm cutoff	1.53	1.60	0.93	0.81	0.28	0.22
6) ½ grain/gm cutoff	1.84	1.96	0.98	0.85	0.37	0.28
7) Abrupt decay (Eq 6B-7)	1.47	1.47	1.25	1.08	0.30	0.23
8) Maximum penetration	3.20	3.45	1.69	1.43	0.94	0.69

*   Preferred estimator

Table 13-4e. Values of the Coefficient K_b
and K_t, in the Longshore Transport Relation (13-2),
where Immersed-weight Bedload Transport Rate I_b
and the Suspended Load Rate I_s are from Table 13-3.

Data for 21 February 1980

| Transport Coefficient | Temporal Sampling | | Spatial Sampling | | | |
| | | | Run 1 | | Run 2 | |
	Using x-variation in sand velocity V	Using average sand velocity $\overline{V}$	Using x-variation in sand velocity V	Using average sand velocity $\overline{V}$	Using x-variation in sand velocity V	Using average sand velocity $\overline{V}$
Bedload, K_b						
Thickness estimator used for Z_o:						
Crickmore profile with						
*1) Abrupt decay (Eq 6B-7)	0.67	0.69	0.37	0.33	0.15	0.12
2) Crickmore estimator						
(Eq 6B-6)	0.69	0.70	0.30	0.28	0.11	0.09
Original profile with						
3) 80% cutoff	0.67	0.68	0.34	0.30	0.11	0.10
4) 90% cutoff	0.77	0.78	0.39	0.35	0.14	0.11
5) 1 grain/gm cutoff	0.65	0.66	0.32	0.30	0.15	0.12
6) ½ grain/gm cutoff	0.85	0.83	0.42	0.38	0.25	0.20
7) Abrupt decay (Eq 6B-7)	0.94	0.81	0.44	0.40	0.18	0.15
8) Maximum penetration	1.18	1.16	0.71	0.61	0.40	0.33
Total Load, K_l						
Thickness estimator used for Z_o:						
Crickmore profile with						
*1) Abrupt decay (Eq 6B-7)	0.76	0.78	0.47	0.43	0.24	0.21
2) Crickmore estimator						
(Eq 6B-6)	0.79	0.79	0.40	0.38	0.19	0.18
Original profile with						
3) 80% cutoff	0.76	0.77	0.44	0.40	0.20	0.18
4) 90% cutoff	0.86	0.87	0.49	0.45	0.23	0.20
5) 1 grain/gm cutoff	0.75	0.75	0.42	0.40	0.24	0.21
6) ½ grain/gm cutoff	0.94	0.93	0.52	0.48	0.34	0.29
7) Abrupt decay (Eq 6B-7)	1.03	0.90	0.54	0.50	0.27	0.24
8) Maximum penetration	1.27	1.25	0.82	0.71	0.49	0.42

*   Preferred estimator

Table 13-4f. Values of the Coefficient K_b
and K_t, in the Longshore Transport Relation (13-2),
where Immersed-weight Bedload Transport Rate I_b
and the Suspended Load Rate I_s are from Table 13-3.

Data for 22 February 1980

| Transport Coefficient | Temporal Sampling | | Spatial Sampling | | | |
| | | | Run 1 | | Run 2 | |
	Using x-variation in sand velocity V	Using average sand velocity $\bar{V}$	Using x-variation in sand velocity V	Using average sand velocity $\bar{V}$	Using x-variation in sand velocity V	Using average sand velocity $\bar{V}$
Bedload, K_b						
Thickness estimator used for Z_o:						
Crickmore profile with						
*1) Abrupt decay (Eq 6B-7)	0.43	0.64	0.10	0.08	0.36	0.31
2) Crickmore estimator (Eq 6B-6)	0.39	0.58	0.13	0.10	0.51	0.43
Original profile with						
3) 80% cutoff	0.36	0.52	0.10	0.08	0.50	0.43
4) 90% cutoff	0.45	0.69	0.10	0.09	0.51	0.43
5) 1 grain/gm cutoff	0.42	0.52	0.10	0.05	0.22	0.20
6) ½ grain/gm cutoff	0.54	0.72	0.11	0.07	0.31	0.27
7) Abrupt decay (Eq 6B-7)	0.49	0.72	0.14	0.12	0.52	0.44
8) Maximum penetration	0.98	1.36	0.18	0.13	0.50	0.44
Total Load, K_t						
Thickness estimator used for Z_o:						
Crickmore profile with						
*1) Abrupt decay (Eq 6B-7)	0.58	0.79	0.24	0.22	0.49	0.44
2) Crickmore estimator (Eq 6B-6)	0.54	0.73	0.27	0.23	0.64	0.57
Original profile with						
3) 80% cutoff	0.51	0.67	0.23	0.21	0.64	0.56
4) 90% cutoff	0.60	0.83	0.24	0.22	0.64	0.57
5) 1 grain/gm cutoff	0.56	0.67	0.23	0.19	0.36	0.34
6) ½ grain/gm cutoff	0.69	0.86	0.25	0.21	0.44	0.41
7) Abrupt decay (Eq 6B-7)	0.63	0.87	0.27	0.24	0.65	0.58
8) Maximum penetration	1.13	1.51	0.31	0.26	0.64	0.57

*　Preferred estimator

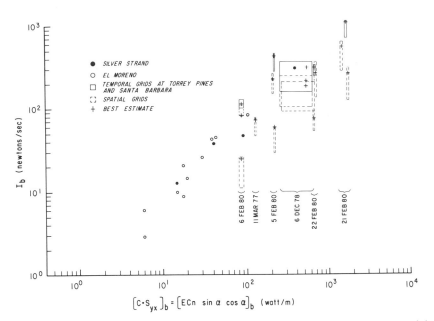

Figure 13-5. Estimates of bedload transport versus longshore sand transport potential. Best estimates (Method 1, in Tables 2-4) are shown as crosses "+". The slope of a line drawn through the origin and the "best estimate" is an estimate of bedload transport coefficient K_b.

$$K = 0.74\sqrt{c_{rb}} \qquad\qquad (13\text{-}6)$$

for

$$0.48 < c_{rb} < 0.62$$

We computed values of c_{rb} from the peak spectral frequency on each of our six transport days and compared them to the average values of K for each day, where the K values were restricted to the methods of computing transport thickness that were the most consistent (i.e., methods 1, 2, and 3 in the tables). We fit K and c_{rb} to the model

$$K = mc_{rb}^n + b \qquad\qquad (13\text{-}7)$$

by least-squares methods for values of n of 2, 1, and 1/2. The smallest standard deviation was found for the equation

$$K_t = 2.16\sqrt{c_{rb}} \text{ for } 0.02 < c_{rb} < 0.42 \qquad\qquad (13\text{-}8)$$

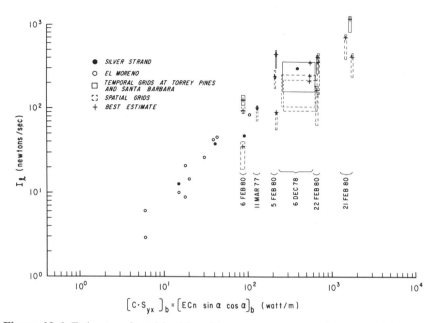

Figure 13-6. Estimates of total load (sand-tracer measurements plus suspension) versus sediment transport potential. Best estimates (Method 1, in Tables 2-4) are shown as crosses "+". The data points for Silver Strand (only) are all underestimates of total load, since suspension is not included in those data.

The K-intercept value b in (13-7) was found to be -0.01 and thus may be considered zero.

Linear correlation analysis for the three values of n attempted indicated best fit for $n = 1/2$ with a correlation coefficient R of 0.917, significant at the 99% level.

The increase of the transport efficiency K with increasing values of the surf similarly parameter c_{rb}, implies that the transport is more efficient for plunging waves on steep beaches with narrow surf zones. This is probably because plunging waves produce more bottom stress and turbulence, and steeper beaches result in a higher concentration of bottom dissipation per unit area of surf zone. It has been observed that plunging waves have a much greater influence than spilling waves on the sand bed, presumably due to large vertical velocities induced by the plunging process. Our measures of transport thickness suggest this, since Z_o is much greater under plunging breakers than in mid-surf.

Another observation about the K versus c_{rb} relation is that the day with pronounced cusps, 5 February 1980, has the largest values of K, the day with remnant cusps, 6 February 1980, has the next highest K, and the other days, during which no cusps were present, had noticeably smaller values of K. It is likely that some of the reflected energy drives cusp-induced circulation (Inman and Guza, 1982) which may be responsible for some longshore sediment transport. Finally, it should be noted that this data covers only spilling and plunging breakers, but not surging. It is possible, and indeed logical, that at very high values of c_{rb} [>0.84, Inman and Guza (1976)] the values of K will decrease as c_{rb} increases.

In summary, the longshore transport coefficient K appears to be a variable which is related to easily measurable parameters such as a form of the surf similarity parameters c_{rb} and ξ.

13.8 References

Bailard, J. A., 1981, An energetics total load sediment transport model for a plane sloping beach, *Journal of Geophysical Research*, 86: 10938-10954.

Battjes, J. A., 1974, Surf similarity, *Proceedings*, Fourteenth Coastal Engineering Conference, June 24-28, 1974, Copenhagen, Denmark, American Society of Civil Engineers, New York: 466-479.

Bijker, E. W., 1971, Longshore transport computations, *Journal of Waterways, Harbors and Coastal Engineering Division*, Proceedings of American Society of Civil Engineers, New York, WW4: 687-701.

Bowen, A. J., D. L. Inman and V. P. Simmons, 1968, Wave set-down and set-up, *Journal of Geophysical Research*: 73(8): 2569-2577.

Carrier, G. F. and H. P. Greenspan, 1958, Water waves of finite amplitude on a sloping beach, *Journal of Fluid Mechanics*, 4: 97-109.

Eaton, R. O., 1951, Littoral processes on sandy coasts, *Proceedings*, Second Conference on Coastal Engineering, November, 1951, Houston, Texas, Council on Wave Research, J. W. Johnson, ed., Berkeley, California: 140-154.

Galvin, C. J., Jr., 1972, Wave breaking in shallow water, In: *Waves on Beaches and Resulting Sediment Transport*, R. E. Meyer (ed): 413-456, New York and London, 462 pp.

Guza, R. T. and D. L. Inman, 1975, Edge waves and beach cusps, *Journal of Geophysical Research*, 80 (21): 2997-3012.

Guza, R. T. and E. B. Thornton, 1985, Velocity moments in the nearshore, *Journal Waterways, Port, Coastal and Ocean Engineering*, American Society of Civil Engineers, New York, 111(2): 235-256.

Inman, D. L., 1978, Status of surf zone sediment transport relations, *Proceedings*, Workshop on Coastal Sediment Transport with Emphasis on the National Sediment Transport Study, University of Delaware, Sea Grant Report DEL-SG-16-78: 9-20.

Inman, D. L. and R. A. Bagnold, 1963, Littoral processes, In: *The Sea: Ideas and Observations*, Interscience Publishers, New York, 3: 529-553.

Inman, D. L. and T. K. Chamberlain, 1959, Tracing beach sand movement with irradiated quartz, *Journal of Geophysical Research*, 64 (1): 41-47.

Inman, D. L. and R. T. Guza, 1976, Application of nearshore processes to the design of beaches, *Abstracts*, Fifteenth Coastal Engineering Conference, July 11-17, 1976, Honolulu, Hawaii, American Society of Civil Engineers, New York: 526-529.

_____. 1982, The origin of swash cusps on beaches, *Marine Geology*, 49 (9): 133-148.

Inman, D. L., G. C. Ewing and J. B. Corliss, 1966, Coastal sand dunes of Guerrero Negro, Baja, California, Mexico, *Geological Society of America*, Bulletin, 77(8): 787-802.

Inman, D. L., P. D. Komar and A. J. Bowen, 1968, Longshore transport of sand, *Proceedings*, Eleventh Conference on Coastal Engineering, London, 1968, American Society of Civil Engineers, New York, 1: 298-306.

Inman, D. L., J. A. Zampol, T. E. White, D. M. Hanes, B. W. Waldorf and K. A. Kastens, 1980, Field measurements of sand motion in the surf zone, *Proceedings*, Seventeenth Coastal Engineering Conference, March 23-28, 1980, Sydney, Australia, American Society of Civil Engineers, New York, 2: 1215-1234.

Iribarren, C. R. and C. Nogales, 1949, Protection des ports, Section II, Comm. 4, XVIIth International Naval Congress, Lisbon: 31-80.

Kamphuis, J. W. and J. S. Readshaw, 1978, A model study of alongshore sediment transport rate, *Proceedings*, Sixteenth Coastal Engineering Conference, August 27-September 3, 1978, Hamburg, Germany, American Society of Civil Engineers, New York: 1656-1674.

King, C. A. M., 1951, Depth of disturbance of sand on sea beaches by waves, *Journal of Sedimentary Petrology*, 21: 131-140.

Komar, P. D., 1969, The longshore transport of sand on beaches, Ph.D. thesis in Oceanography, University of California, San Diego, 143 pp.

_____. 1971, The mechanics of sand transport on beaches, *Journal of Geophysical Research*, 76: 713-721.

Komar, P. D. and D. L. Inman, 1970, Longshore sand transport on beaches, *Journal of Geophysical Research*, 75: 5914-5927.

Munk, W. and M. Wimbush, 1969, A rule of thumb for wave breaking over sloping beaches, *Oceanography*, 9: 56-59.

Pawka, S. S., D. L. Inman and R. T. Guza, 1983, Radiation stress estimator, *Journal of Physical Oceanography*, 13(9): 1698-1708.

Thornton, E. B., 1972, Distribution of sediment transport across the surf zone, *Proceedings*, Thirteenth Coastal Engineering Conference, July 10-14, 1972, Vancouver, B.C., American Society of Civil Engineering, New York: 1049-1068.

Chapter 14

MEASURING LONGSHORE TRANSPORT
WITH TRAPS

Robert G. Dean

University of Florida

14.1 Introduction

The Nearshore Sediment Transport Study (NSTS) included two total trap experiments conducted to evaluate the longshore sediment transport relationship $I_\ell = KP_{\ell s}$ in which I_ℓ is the total longshore immersed weight sediment transport rate, $P_{\ell s}$ is the longshore component of wave energy flux and K is a proportionality factor which could depend on a number of wave and sediment parameters. The locations of the two experiments were at Santa Barbara, California and Rudee Inlet, Virginia. The trap characteristics for each of these two sites are quite different. The Santa Barbara trap consists of the spit formed in the lee of the Santa Barbara breakwater. At Rudee Inlet, the trap is inside of a weir-type jetty with a crest elevation at about mean sea level.

This chapter examines the basis for total trap experiments, discusses the types of possible errors in determining the total transport coefficient, K, describes the susceptibility of the two experiments conducted to such errors, and presents the results from the two experiments.

14.2 Governing Relationships

As a preparation for discussing the two total trap experiments and the results obtained, it is useful to provide a framework by presenting the governing equations.

14.2.1 Conservation of Sediment

The conservation of sediment can be expressed in vector form as

$$\frac{\partial h}{\partial t} - \vec{\nabla} \cdot \vec{q_s} = 0 \tag{14-1}$$

in which h is the water depth referenced to some fixed horizontal plane, t is time, $\vec{\nabla}$ is the horizontal divergence vector operator and $\vec{q_s}$ is the sediment transport vector in terms of bulk volume. Equation (14-1) is exact and will be employed in interpreting the measurements of sediment volume. Expanding

313

(14-1) in Cartesian form

$$\frac{\partial h}{\partial t} = \frac{\partial q_x}{\partial x} + \frac{\partial q_y}{\partial y} \tag{14-2}$$

where, for purposes here, x and y are taken in the *local* offshore and longshore directions, respectively. Integrating across the entire zone of active sediment transport,

$$\int_{x_1}^{x_2} \frac{\partial h}{\partial t} \, dx = \int_{x_1}^{x_2} \frac{\partial q_x}{\partial x} \, dx + \int_{x_1}^{x_2} \frac{\partial q_y}{\partial y} \, dx \tag{14-3}$$

The first term in (14-3) represents the time rate of change of water cross-sectional area, A_w, in the profile, and is the complement of the time rate of change of the sand volume, A, in the profile, i.e., $(\partial A / \partial t = -\partial A_w / \partial t)$ the second term is $(q_x)_2 - (q_x)_1$ each term of which is zero since x_1 and x_2 represent the limits of cross-shore sediment transport. The last term is the gradient $\partial Q / \partial y$ of the total longshore sediment transport. Equation (14-3) can now be written

$$\frac{\partial A}{\partial t} = -\frac{\partial Q}{\partial y} \tag{14-4}$$

Integrating from an alongshore position y_1 representing the location where the wave measurements are conducted to the downdrift limit, y_2, of the trap location (i.e., $Q(y_2) \equiv 0$),

$$\frac{\partial}{\partial t} \int_{y_1}^{y_2} A \, dy = -\int_{y_1}^{y_2} \frac{\partial Q}{\partial y} \, dy = Q(y_1) - Q(y_2) = Q(y_1) \tag{14-5}$$

The term on the left-hand side represents the time rate of sediment volume change between y_1 and y_2 so that (14-5) becomes

$$\frac{\partial V}{\partial t} = Q(y_1) \tag{14-6}$$

Finally, now integrating over the intersurvey period between two successive surveys.

$$\int_{t_1}^{t_2} \frac{\partial V}{\partial t} \, dt = \int_{t_1}^{t_2} Q(y_1) \, dt \tag{14-7}$$

or

$$V(t_2) - V(t_1) = \Delta V = \int_{t_1}^{t_2} Q(y_1) \, dt \tag{14-8}$$

where it is stressed that ΔV represents all the volume passing the reference point, y_1, in the time interval from t_1 to t_2.

14.2.2 Sediment Transport Equation

The equation for sediment transport which will be evaluated is

$$I_\ell = KP_{\ell s} \tag{14-9}$$

in which I_ℓ is the immersed weight sediment transport rate and is related to the bulk volumetric sediment transport rate, Q, by

$$I_\ell = (\rho_s - \rho_w)g\,(1-p)\,Q \tag{14-10}$$

where ρ_s and ρ_w are the mass densities of sediment and water, respectively, g is the gravitational constant and p is the sediment porosity. Equations (14-9) and (14-10) can be combined to

$$Q = K'P_{\ell s} \tag{14-11}$$

in which K' is a dimensional parameter defined by

$$K' \equiv \frac{K}{(\rho_s - \rho_w)g\,(1-p)} \tag{14-12}$$

14.2.3 Determination of the Sediment Transport Factor, K'

The sediment transport factor is determined from the experiments by combining (14-8) and (14-11),

$$\Delta V = \int_{t_1}^{t_2} K'P_{\ell s}(t, y_1)\,dt \tag{14-13}$$

or if K' is independent of time

$$K' = \frac{\Delta V}{\int_{t_1}^{t_2} P_{\ell s}(t, y_1)\,dt} \tag{14-14}$$

The application of (14-14) can be illustrated by reference to the Santa Barbara site, Figure 14-1. The east slope array was used to establish the wave characteristics and was located approximately twenty percent of the distance from the east to the west end of Leadbetter Beach. Therefore, the volumetric changes that were included in the trap included the spit, which represents the downdrift limit of transport, the south face of the breakwater and approximately 200 meters of Leadbetter Beach to a location directly onshore of the eastern slope array.

14.3 Relative Merits of Total Trap Experiments

There are inherent advantages and disadvantages of total trap experiments in which the transport is accumulated in a trap over a period of time, then measured and correlated with the appropriate wave characteristics integrated

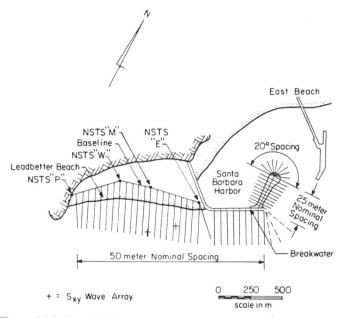

Figure 14-1. Santa Barbara survey plan and location of S_{xy} wave gages.

over the same period of time. An alternate approach to investigating longshore sediment transport is through tracer techniques and/or measurement of bedload transport and suspended sediment concentration and the associated longshore current which have the potential of providing more detail but greater possibility of substantial errors.

The primary advantage of a total trap experiment is the potential of accumulating large, and therefore easy to measure, quantities of sediment. Moreover, it is not necessary to be concerned about the partitioning of the transport between bedload and suspended load. If the trap is complete, the transport which occurs during very substantial wave events can be included, which would be quite difficult with other methods. Major potential drawbacks result from the time and spatial averaging or smoothing of the data.

The advantages of other types of experiments include the possibility of examining the transport over short time spans while the wave characteristics are essentially constant and examining the distribution of sediment transport across the surf zone. In designing experiments to establish the sediment transport distribution, it is advantageous to avoid tidal smearing by conducting the experiments at neap tide conditions and/or at peak or low tide when changes are minimal. Primary disadvantages with tracer experiments include the need to quantify the depth of active motion of bedload which is usually approximated by implanting a plug of tracer material, then determining the depth to which this tracer has been disturbed. Alternatively, cores can be taken throughout the area in which tracer is present to determine the depth of tracer mixing. However, the depth of active motion is not a definite value; but rather in nature the *probability* of motion decreases with distance into the sand bed. Thus it would appear that the depth to which motion had actually occurred during an experiment would depend strongly on the *duration* of the experiment, with the apparent depth being too shallow for short experiments and too deep for experiments of long duration. Tracer experiments must also be carried out during periods of relative stable beach conditions. If the beach should be in a transient phase of erosion or recovery, there is the possibility that portions of the tracer could be completely lost due to burial or yield a false indication of depth of motion through local erosion.

In summary, tracer and other short-term methods provide the opportunity to examine the details of the process such as distribution with time and across the surf zone whereas total trap experiments yield much greater confidence in total volumes transported and ultimately the sediment transport coefficient, K.

14.4 Possible Extraneous Effects or Sources of Errors in Total Trap Experiments

In assessing the results of the field experiments conducted at Santa Barbara and Rudee Inlet, it is useful to first discuss the types of errors that can occur in the data. In the following paragraphs, three types of errors or extraneous effects will be considered. These include:

(1) errors in sediment volumes,
(2) errors in wave data, and
(3) errors associated with the adopted sediment transport relationship.

14.4.1 Errors in Sediment Volumes

The most obvious source of error in measured sediment volume is caused by the measurement (implied in measure) of the changes in volumes. An analysis of such errors has been presented in Chapter 3B with special reference to the Santa Barbara experiments and it has been found that relative to the changes measured, such errors are small. Additionally, at both sites, repeated measurements of volume were carried out and since such errors are not cumulative, any errors associated with summed volumes over all surveys should be much less.

A less obvious source of error in total trap experiments is due to the character of the trap itself. This effect is best demonstrated by considering an offshore breakwater with normally incident waves, see Figure 14-2. Due to wave diffraction, even though there is no longshore sediment transport attributable to wave directionality, sediment will accumulate behind the breakwater, i.e., the physical characteristics of the trap are such that it causes a

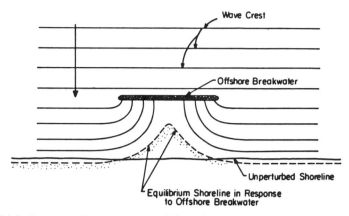

Figure 14-2. Transport bias caused by offshore breakwater illustrated by deposition in lee of breakwater.

trapping bias. Although the sediment volume accumulated behind the offshore breakwater will approach an equilibrium, if the sediment is removed, say by a dredge, additional sand will be transported to and deposited behind the breakwater. In order to account for any bias, it is necessary to measure accurately the wave characteristics relative to the *local* shoreline orientation and the sediment volume changes occurring in the sand trap and between the sand trap and the location of wave measurements, i.e., in accordance with (14-14). At any given time and location, the sediment transport relationship should be in accordance with (14-14) or the associated differential form; however, if local or temporal changes in shoreline alignment occur, these must be accounted for in the correlation to determine K. Moreover, if the trap effect is such that the depth contours inside the surf zone are no longer parallel, difficulties arise in specifying an appropriate shoreline orientation which is necessary for the application of (14-14).

A second type of bias that can occur is caused by the *elevation* of a template structure such as a weir jetty over which the sediment flows into the trap. As sediment is removed from the trap by dredging, the updrift (south) beach at Rudee Inlet will be drawn down, and the flow will continue to be induced into the trap. As in the case for the previous example, the correct analysis method is to measure the wave characteristics at a point (including wave direction relative to the beach alignment) and to measure the sediment volume changes occurring in the sand trap and in the segment between the sand trap and the wave measurement location. The correlation of the change in volume with the integrated longshore energy flux should yield an unbiased estimate of the sediment transport parameter, K.

Removal of material from a trap (usually by dredging), can introduce noise in the sediment volume data unless the volumes removed are quantified accurately. The optimum situation is one in which a survey is conducted immediately before and after dredging thereby yielding a volume accumulated before dredging and a valid initial trap volume prior to the next accumulation phase. This optimum plan requires considerable coordination with those responsible for the dredging and in some cases, one is required to accept the somewhat less accurate estimates of those performing the dredging.

14.4.2 Errors in Wave Data

The wave data for the two total trap experiments were measured using slope arrays (Chapter 2). At Santa Barbara, two slope arrays were located as shown in Figure 14-1, whereas for the Rudee Inlet experiment, only one slope array was available (Figure 14-3). Two possible sources of error will be discussed: water surface elevation and wave direction, both important parameters to the computation of $P_{\ell s}$. As described in Chapter 2, each slope array consists of four sensitive pressure gages. These gages allow pressure intercomparison, the results of which indicate that the pressure fluctuations were measured reasonably accurately. The pressure fluctuations were transformed to water level fluctuations through the linear pressure response with a cut-off for periods less than 4 sec. The wave climate at Santa Barbara is such that this cut-off probably does not seriously affect the data. The shorter periods are of greater relative importance at the Rudee Inlet site.

In the establishment of wave direction from a slope array or other directional gage, it is essential to determine, through field measurements, the orientation of the system. For the slope arrays, the orientation was established by means of a diver-held digital compass mounted on a non-magnetic extension to provide distance from the steel frame of the slope array. The compass should be reasonably accurate; Seymour (personal communication) has estimated an orientation accuracy of ± 2 degrees. To evaluate the error in $P_{\ell s}$ associated with an error in slope array orientation we note that, for the idealized case of straight and parallel bottom contours and no energy dissipation, S_{xy} does not vary up to the breaker line and S_{xy} and $P_{\ell s}$ are related by

$$(P_{\ell s})_b = C_b (S_{xy})_b = C_b (S_{xy})_g$$

$$= C_b \frac{E_g}{2} \left[\frac{C_G}{C} \right]_g \sin 2\theta_g \qquad (14\text{-}15)$$

where C and C_G denote wave celerity and group velocity, respectively, and the subscripts b and g denote breaker line and gage location, respectively. The relative error in $P_{\ell s}$ due to an error in the gage orientation, $\Delta\theta_g$, is then

$$\frac{\Delta P_{\ell s}}{P_{\ell s}} = \frac{1}{P_{\ell s}} \frac{\partial P_{\ell s}}{\partial \theta_g} \qquad (14\text{-}16)$$

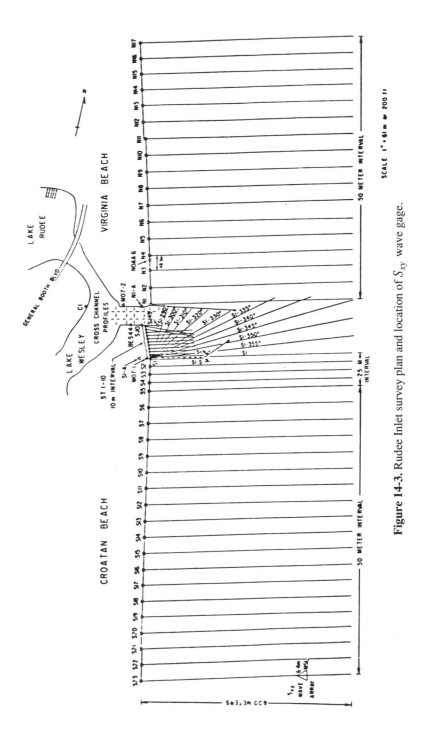

Figure 14-3. Rudee Inlet survey plan and location of S_{xy} wave gage.

$$\frac{\Delta P_{\ell s}}{P_{\ell s}} = \frac{2\Delta\theta_g}{\tan 2\theta_g} \tag{14-17}$$

As an example, if the error in gage orientation is 2 degrees and the wave direction at the gage relative to the shoreline is 20 degrees, the percentage error in $P_{\ell s}$ is 8 percent.

Relative to the question of accuracy of wave data, Seymour (1983) has reported that the S_{xy} values of the east gage at Santa Barbara were approximately one-half the values determined from the west gage. The reasons for this difference are not completely understood, but may be due to the greater effects of wave diffraction from Santa Barbara Point on the westerly gage.

14.4.3 Possible Errors in Adopted Sediment Transport Relationship

The sediment transport relationship evaluated in the NSTS total trap experiment is expressed as (14-9)

$$I_\ell = KP_{\ell s}$$

and, in the form utilized as (14-14)

$$K' = \frac{\Delta V}{\int_{t_1}^{t_2} P_{\ell s}(t, y_1)\, dt} \tag{14-18}$$

The transport relationship (14-9) may not allow for all physical phenomena. For example, the transport might depend on the relative proportions of onshore directed wave energy (as an agitation mechanism) and the alongshore directed wave energy. The presence of an offshore bar could also be a factor. In such cases, one would expect the values of K as determined for the various intersurvey periods to vary with the level and character of wave energy, beach morphology, etc. As noted previously, total trap experiments rely on averages (14-18) over intersurvey periods, which represents one limitation of this type of experiment.

14.5 Brief Review of the Two Trap Experiments

This section presents the characteristics of the two total trap experiments.

14.5.1 Santa Barbara, California

The Santa Barbara experiment was carried out over the period October 13, 1979 through December 17, 1980, a duration of fourteen months. A total of eight complete surveys were conducted documenting a total sediment accumulation of 288,600 m^3. Three of the seven intersurvey volumes were affected by dredging conducted during the intersurvey periods; however, for one of these periods, the amount removed by dredging was relatively minor. The waves at Santa Barbara were measured by two slope arrays located in approximate water depths of 9 m. The survey plan for this site consisted of 59 lines for which the beach profiles and soundings were colinear. An additional 5 beach profiles and 3 sounding lines were measured. The survey plan has been presented previously as Figure 14-1. As described in Chapter 3B, three fathometer passes were conducted along each line, to allow averaging of the results. The trap characteristics are described in §14.2.3.

Three sources of tidal information were utilized at Santa Barbara to correct the sounding data. One was a NOAA tide gage mounted on Stearn's Wharf to the east of the Santa Barbara breakwater spit. The second source was a graduated tide board inside and at the base of the breakwater. Finally, during soundings inside the spit, frequent readings were taken on the relatively still water level intersection with the sloping sand surface.

Two distinguishing features of the Santa Barbara location are:

(1) the relatively constant west to east direction of longshore sediment transport, and

(2) the relatively large amounts of longshore sediment transport. Both of these features are conducive to a quality trap experiment.

14.5.2 Rudee Inlet, Virginia

The general physiography of the Rudee Inlet site was described in Chapter 1C. The field program extended between June 26, 1981 and November 4, 1981, a duration of approximately four and one-half months. During this period, a total of six complete surveys were commenced; however, one of the surveys was aborted due to weather conditions. The surveys documented a gross sediment volumetric change of approximately 56,700 m^3, substantially less than at the Santa Barbara experiment. Wave and current action causes sediment to be carried over the southern weir jetty where deposition occurs in the sand trap, Figure 14-3. The volumetric capacity of this trap is fairly small, on the order of 15,000 m^3. As will be discussed in greater detail later, it was found that the area to the south of Rudee Inlet functions as a storage area, with the sand volume alternately increasing and decreasing, depending on wave direction. On occasions when the sand trap was filled to capacity, sediment was carried past the trap and deposited in the inlet throat. Sediment also was deposited near the entrance to Rudee Inlet, due to jetting action of the ebb tidal currents or by transport along the outside of the south jetty.

The wave characteristics were obtained from a slope array in 6.4 m of water approximately 1000 m south of the south jetty and 500 m offshore. Tidal data were obtained from a tide gage located at the Marina immediately inside Lake Rudee or from readings at a graduated tide staff near the Pump House, just inside Lake Wesley, see Figure 14-3 which also presents the survey plan at Rudee Inlet. The outer depths of offshore lines from sequential surveys were compared for closure and, if necessary, the tidal values used in correcting the sounding data were adjusted. Volume computations were carried out to a distance of 400 m from the baseline, i.e., to depths of 5 to 6 m. On the outer beach, a total of 40 beach profiles and colinear sounding lines were measured. An additional 10 wading profiles and 10 sounding lines which were not colinear defined the volume changes inside the entrance including the sand trap. Finally, volume changes within the throat of the entrance channel were determined by a series of cross channel profiles (sounding lines). As was the case for Santa Barbara, each sounding line was surveyed three times to improve the accuracy of the measurements through averaging. The trap dredging events were of short duration and the volumes were generally established by before and after surveys.

Distinguishing characteristics of the Rudee Inlet experiment were:

(1) a relatively small trap that was easily filled to capacity,
(2) a weir jetty which, due to its profile at approximately mean sea level (MSL), admitted sand to the trap even during conditions when the dominant wave direction caused general sand transport away from the sand trap, and
(3) somewhat smaller magnitudes of sediment transport than at Santa Barbara.

14.6 Analysis Procedures

Because the analysis procedures differed to some degree for the Santa Barbara and Rudee Inlet sites, the methods for these two sites will be presented separately.

14.6.1 Santa Barbara

14.6.1.1 Wave Data

Two types of wave data from the east slope array were used for correlation with sediment volume changes representing transport rates. First, values of $P_{\ell s}$ were calculated through a singular wave analysis in which the wave characteristics of the slope array were characterized by a wave height, period and direction and using linear wave theory, this singular wave was refracted and shoaled to the breaker line where $P_{\ell s}$ was calculated. Breaking was based on a depth limited wave with a proportionality factor (wave height/depth) of 0.78. The representative wave height was obtained by calculating the spectrum based on the pressure transfer function from linear wave theory. The rms wave height and an energy weighted frequency were obtained from the spectrum. The wave direction at the gage was obtained from the cospectra of the water surface displacement and sea surface slope. Secondly, S_{xy} values were obtained through correlations of the sea surface slope and the sea surface displacement as described by Higgins, Seymour and Pawka (1981). As noted previously, Seymour (1983) found that the west gage yielded S_{xy} values which were approximately twice those based on the east gage. In all of the results to be presented here, the wave characteristics were based on data from the east gage.

14.6.1.2 Sediment Data

Two types of analyses yielding two distinct types of results were performed on the sediment data.

14.6.1.3 Total Transport

The purpose of the first analysis was to establish the total transport in the region including the spit, along the outer face of the breakwater and along approximatley 200 m of Leadbetter Beach, Figure 14-1. For this purpose the intersurvey volume changes were calculated out to what was judged to be an optimum depth which minimized the combination of errors which can result from: (a) not including a sufficient distance offshore so as to incorporate the complete active zone and (b) including such a very large offshore distance that a small bias in elevation can contribute significant errors. This offshore limit of computations was taken as 5 m and is believed to be *near-optimal*. The results of these intersurvey volume changes have been presented in Table 3B-1.

14.6.1.4 Cross-shore Distribution of Longshore Sediment Transport

Although the initial planning for the Santa Barbara field program envisioned it only as a total trap experiment, during the conduct of the program it was noted that due to changes in wave directions, there were considerable readjustments in the alignment of the nearshore contours in the Leadbetter Beach compartment. The analysis of these orientation changes to determine an estimate of the cross-shore distribution of longshore sediment transport is described below.

Consider the idealized beach presented in Figure 14-4 with the alongshore coordinate origin now centered in the compartment. Any change in the location of a particular contour can be represented as the sum of an even and an odd component, i.e., if $x_c(y)$ represents the distance to a particular contour and $\Delta x_c(y)$ represents the change occurring over an intersurvey period, then the even $(\Delta x_c)_e$ and odd $(\Delta x_c)_o$ components are

$$\left[\Delta x_c(y)\right]_e = \frac{1}{2}\left[\Delta x_c(y) + \Delta x_c(-y)\right]$$

$$\left[\Delta x_c(y)\right]_o = \frac{1}{2}\left[\Delta x_c(y) - \Delta x_c(-y)\right] \tag{14-19}$$

We note from (14-19) that

$$\Delta x_c(y) = \left[\Delta x_c(y)\right]_e + \left[\Delta x_c(y)\right]_o \tag{14-20}$$

Referring to Figure 14-4, it is reasonable to assume that the odd component of the change is due to a change in wave direction and therefore due to longshore sediment transport. As an example of the concept, if the pocket beach has no losses and if the contours shallower than (say) 3 m rotate while the deeper contours do not change orientation, then it is clear that all of the longshore

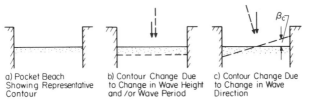

a) Pocket Beach
Showing Representative
Contour

b) Contour Change Due
to Change in Wave Height
and /or Wave Period

c) Contour Change Due
to Change in Wave
Direction

Figure 14-4. Representative contour displacement signatures for an idealized pocket beach due to cross-shore (sketch b) and longshore (sketch c) sediment transport.

sediment transport is at depths shallower than 3 m. This concept is formalized below to develop a procedure for extracting the cross-shore distribution of longshore sediment transport.

The three-dimensional equation of sediment conservation is

$$\frac{\partial h}{\partial t} = \frac{\partial q_x}{\partial x} + \frac{\partial q_y}{\partial y} \tag{14-21}$$

Separating the depth change effect into an even component, $(\partial h/\partial t)_e$, due to cross-shore sediment transport and an odd component, $(\partial h/\partial t)_o$, due to alongshore sediment transport, then by assumption

$$\frac{\partial q_x}{\partial x} = \left(\frac{\partial h}{\partial t}\right)_e$$

$$\frac{\partial q_y}{\partial y} = \left(\frac{\partial h}{\partial t}\right)_o \tag{14-22}$$

and it is the odd component that will be of primary interest here. Since in terms of representing the distribution it is most orderly to determine the sediment transport along a given contour *relative to the water level*, it is useful to transform the odd component of the depth changes to changes in contour position by the one-dimensional transformation

$$\frac{\partial h}{\partial t} = \frac{\partial h}{\partial x}\frac{\partial x}{\partial t} \tag{14-23}$$

or

$$\left(\frac{\partial x_c}{\partial t}\right)_o = \frac{1}{m_c}\left(\frac{\partial h}{\partial t}\right) = \frac{1}{m_c}\frac{\partial q_y}{\partial y} \tag{14-24}$$

where m_c is the beach slope at a contour of interest, h_c.

Integrating (14-24) along a contour from the end of the beach compartment where q_y is zero to some arbitrary y

$$q_y\left[h_c, y\right] = -m_c \int_y^{\ell} \left[\frac{\partial x_c}{\partial t}\right] dy \qquad (14\text{-}25)$$

which relates the local longshore sediment transport rate along a contour to the integral of the time rate of change of that contour h_c. One rather critical requirement of this method is that the system should not have reached planform equilibrium. If equilibrium with the new wave direction has been achieved for some contours, then the apparent transport rate along those contours simply will be proportional to the beach slope at that contour, cf. (14-25). Another requirement is that the system be relatively free of losses or additions of sediment, at least for the contours being investigated.

It was found that the odd components of the contour changes were fairly linear. Denoting the slope of the *planform* changes as $\tan\beta_c$, the local longshore sediment transport can be expressed in proportionality form as

$$|q_y(h_c)| \; \alpha \; m_c \; \tan\beta_c \qquad (14\text{-}26)$$

in which α denotes proportionality.

The previous paragraphs have described the concept of relating the longshore sediment transport distribution to the odd component of the Leadbetter Beach contour changes. In application, this concept was complicated by tidal smearing and varying wave height such that the width of the active zone changed with time. To test various candidate longshore transport distributions, $q_y'(x)$, these distributions as a function of offshore distance were first transformed to a function of depth $q_y''(h)$ by the relationship

$$q_y''(h) = \frac{1}{m_c} q_y'(x) \qquad (14\text{-}27)$$

where the distribution, q''_y, is normalized to unit area and is scaled in accordance with the breaking depth. In application, the transport along a given contour, h_c at time, t, is given by

$$q_y(h_c, t) = P_{\ell s}(t) \, q_y'' \left[h + \eta_T(t)\right] \qquad (14\text{-}28)$$

where $\eta_T(t)$ represents the tide. This procedure is illustrated schematically in Figure 14-5. The reader is referred to Berek and Dean (1982) for greater detail. Results from these calculations will be presented in a following section.

14.6.2 Rudee Inlet

14.6.2.1 Wave Data

The wave data from the slope array at Rudee Inlet were analyzed by Castel and Seymour (1982) and results were provided from which the cumulative P_{ls} data could be extracted. It is noted that earlier comparisons of the Castel-Seymour predictions of P_{ls} with those which we carried out at Santa Barbara yielded an average ratio of 1/5.6 (= Castel-Seymour/our calculations). This same factor was applied to their predictions at Rudee Inlet to establish the P_{ls} correlated with documented sediment impoundment.

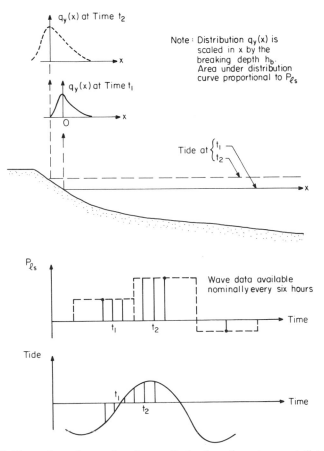

Figure 14-5. Illustration of procedure for predicting longshore transport distribution accounting for varying wave conditions and fluctuating tides.

The correlation of the $P_{\ell s}$ results for the intersurvey periods of interest with the intersurvey volumes will be presented later.

14.6.2.2 Sediment Data

The intersurvey volume changes at Rudee Inlet were based on the 23 profiles comprising the South (Croatan) Beach and those profiles encompassing the sand trap and interior of Rudee Inlet. Referring to Figure 14-3, the only region surveyed which was not included in the volumes for the sand trap was the North Beach as defined by profiles N1 through N17.

Due in part to the relatively small volume associated with the sand trap area at Rudee Inlet, there was greater interest in frequent dredging and this represented a complicating factor in the scheduling of surveys to ensure that the results would represent the minimum contamination due to dredging. However, five intensive surveys, supplemented by several additional surveys of the immediate trap area provided four intersurvey data sets that are relatively free from dredging effects. The total number of intersurvey periods from which K values could be extracted was reduced to three since the wave gage was inoperable for a substantial duration during the third intersurvey period.

14.7 Results

The results will be presented separately for the two field sites.

14.7.1 Santa Barbara, California

The total trap data for the seven intersurvey periods documented at Santa Barbara, California are summarized in Table 14-1.

14.7.1.1 I_ℓ versus $P_{\ell s}$

The data for the seven intersurvey periods were plotted as I_ℓ versus $P_{\ell s}$ as presented by the solid circles in Figure 14-6. Examination of Table 14-1 and Figure 14-6 demonstrates that the average sediment transport during some intersurvey periods was much greater than during others. For example, during the third intersurvey period, the immersed weight sediment transport rate was almost 300 N/s whereas during the following intersurvey period, the rate was less by an order of magnitude. Also as evident from Table 14-1 with the exception of the low outlier, the individual K values range from 0.84 to 1.63. This outlier was excluded in the analysis results to follow. Two types of regressions were applied to the data in Figure 14-6 to determine overall K values. Values of K were determined which provide best least squares fits between: (a) I_ℓ and $P_{\ell s}$ and (b) $\log I_\ell$ and $\log P_{\ell s}$. The corresponding K values were 0.93 and 1.23, respectively. These values are generally higher than found in previous studies and it is possible that the relative fine sand at Santa Barbara ($D \approx 0.22$ mm) is responsible for the higher K values. It is noted that a previous study of Santa Barbara by Galvin (1969) had combined sediment accumulation

Table 14-1. Summary of Santa Barbara Field Results

Intersurvey Period	No. of Days	Dredging Event	Total Volume Change (m^3)	Immersed Weight Transport Rate I_ℓ (N/s)	Net Longshore Component of Wave Energy Flux at Breaking $P_{\ell s}$ (N/s)	$K = I_\ell / P_{\ell s}$	Net Onshore Flux of Longshore Component of Momentum S_{xy} (N/m)	$K_* = I_\ell / S_{xy}$ (m/s)
Oct. 13, 1979-Nov. 30, 1979	48	No	32,820	85.3	52.2	1.63	27.8	3.06
Dec. 1, 1979-Jan. 20, 1980	31	Yes, Major	65,070	159.1	101.4	1.57	45.4	3.50
Jan. 21, 1980-Feb. 25, 1980	35	Yes, Minor	82,810	295.0	352.4	0.84	119.6	2.47
April 11, 1980-June 3, 1980	53	No	10,290	24.2	76.6	0.32	37.9	0.64
June 4, 1980-Aug. 25, 1980	82	No	22,220	33.8	31.7	1.07	17.6	1.91
Aug. 26, 1980-Oct. 23, 1980	57	No	38,760	84.8	63.8	1.33	32.6	2.60
Oct. 24, 1980-Dec. 17, 1980	54	Yes, Major	35,640	84.6	64.4	1.31	34.2	2.47

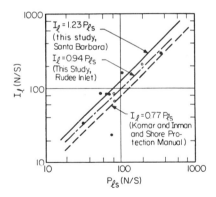

Figure 14-6. Data from Santa Barbara (●) and Rudee Inlet (○) field experiments. I_ℓ versus $P_{\ell s}$, present and past correlations.

values developed by J. W. Johnson with wave characteristics deduced by Galvin from wave hindcasts and a wave direction which yielded the maximum $P_{\ell s}$ (i.e., minimum K); the resulting mean value of K was 1.60.

Figure 14-7 presents the K value (1.23) from this study with those from earlier investigations and it appears that there is a trend of decreasing K with increasing sediment diameter.

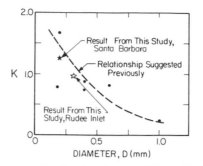

Figure 14-7. Plot of K versus D. Results of present [Santa Barbara (*) and Rudee Inlet (open star)] and previous studies. (Modified from Dean, 1978)

Based on the cumulative sediment volumes and P_{ls} values over the various intersurvey periods, the values of K are 1.17 and 1.07 without and with the outlier intersurvey period included.

14.7.1.2 I_l versus S_{xy}

A second correlation was carried out to determine K_*, where

$$I_l = K_* S_{xy} \tag{14-29}$$

One disadvantage of this form is that K_* is dimensional, whereas K (14-1) is dimensionless. The I_l versus S_{xy} values are summarized in Table 14-1 and plotted in Figure 14-8. Again, excluding the one low outlier value, the best-fit linear and logarithmic K_* values are 2.60 and 2.63 m/s, respectively. The K_*

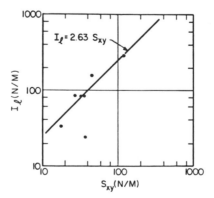

Figure 14-8. Data from Santa Barbara field experiment. I_l versus S_{xy} and best least squares fit.

values based on cumulative volumes and S_{xy} values are 2.67 and 2.40 m/s without and with the one outlier intersurvey period included.

14.7.2 Longshore Sediment Transport Distribution, $q_y(h)$

The results from two intersurvey periods appeared most suitable for analysis of q_y. The odd components of the contour changes are presented in Figure 14-9. The relative longshore sediment transport, $q_y(h)/q_y(0)$ was inferred from these data in accordance with (14-26) and the results are presented as the solid lines in Figure 14-10(a) and 14-10(b). These results were compared with the previous analytical distributions by Komar (1977) and Tsuchiya (1982) and the empirical distribution determined by Fulford (1982), see Figure 14-11. The results are presented in Figures 14-10(a) and 14-10(b). The reader is referred to Berek and Dean (1982) for greater details of this method. It is seen that during both intersurvey periods, changes occurred out to approximately the 3 m contour; however, during the latter intersurvey period, the field distribution decreases much more rapidly than for the earlier period. These results appear to be so limited as to justify only qualitative interpretation. It is noteworthy that while the candidate transport distributions tested differ substantially (Figure 14-11), the resulting transports based on calculated contour changes presented in Figures 14-10(a) and 14-10(b) are quite similar. Tidal smearing is responsible for the reduction of distinguishing differences.

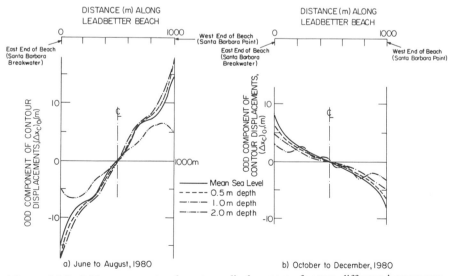

Figure 14-9. Odd components of contour displacements for two different intersurvey periods. Santa Barbara field experiment.

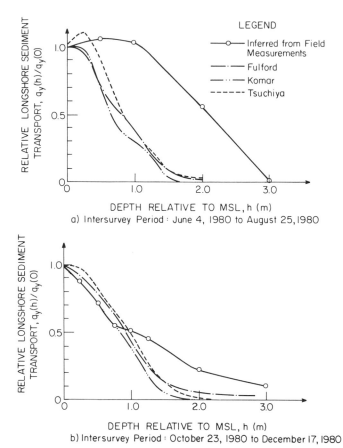

Figure 14-10. Comparison of predicted relative longshore sediment transport distribution with the distribution inferred from field measurements.

Finally, the method developed appears to provide an additional and nonintrusive approach to measuring longshore sediment transport distribution. In any future applications of this contour rotation method to establish longshore sediment transport distribution, an attempt should be made to select a site at which the vertical extent of substantial motion is significantly greater than the tidal range responsible for smearing over this vertical dimension.

14.7.3 Rudee Inlet

14.7.3.1 General

The analysis of the Rudee Inlet sediment data yielded an interesting and unanticipated character of the sediment transport and deposition/erosion system south of Rudee Inlet. The seasonal wave characteristics are such that the net longshore transport is toward the north during the spring and summer and toward the south during the fall and winter. The gross sediment transport appears to be large relative to the net transport. The beach and offshore area south of Rudee Inlet function as a large storage area responding to changes in seasonal influx and efflux of sediments while the weir jetty represents a small leak in the system allowing sediment to flow into the sediment trap during waves incident from a wide range of directions. This was evident by accumulation in the trap during periods of waves both from the northeast and southeast. Also the previously-mentioned draw-down (erosion) of the updrift beach and strong northerly flowing currents across the weir during flood tide support this description. This behavior is illustrated quantitatively in Table 14-2 which summarizes the results from the five major surveys conducted in the Rudee Inlet field program. Specifically, it is seen that during the last intersurvey period (September 22, 1981 to November 4, 1981), the cumulative net $P_{\ell s}$ was toward the south, yet sediment accumulation occurred in the sand trap and inlet throat, and the net change including the effect of the south beach was erosion. The sediment system described above raises the interesting question of whether or not the weir jetty and sand bypass system transfers the net longshore sediment transport as would be most equitable for the updrift and downdrift beaches. Unfortunately, the data available are not adequate to resolve this question.

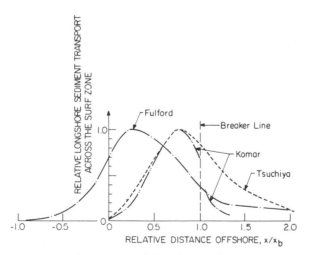

Figure 14-11. Cross-shore distributions of longshore sediment transport tested in this study.

Table 14-2. Summary of Rudee Inlet Field Results

Intersurvey Period	No. of Days	Total Volume Change (m^3) *	Net Immersed Weight Transport Rate I_ℓ (N/s)	Net Longshore Component of Wave Energy Flux at Breaking $P_{\ell s}$ (N/s)	$K = I_\ell / P_{\ell s}$
June 22, 1981-July 7, 1981	15	+23,600	203.1	185.7	1.09
July 8, 1981-July 27, 1981	19	+12,760	86.7	80.9	1.07
July 28, 1981-Sept. 27, 1981	62	-10,180	-21.2	Sixteen Days of Inoperative Wave Gage	-----
Sept. 28, 1981-Nov. 4, 1981	38	-20,370	-69.2	-82.4	0.84

*Note: Approximately 80% of the volume changes documented occurred on the beach south of Rudee Inlet.

14.7.3.2 I_ℓ versus $P_{\ell s}$

The only analysis performed for the Rudee Inlet data was a correlation of I_ℓ versus $P_{\ell s}$. These results are presented in Table 14-2 and as the open circles in Figure 14-6. Although five complete surveys were conducted, thereby documenting four intersurvey periods from which it was potentially possible to determine four values of K. During the July 27 to September 27, 1981 intersurvey period, the wave gage array was inoperable for approximately 25% of this period. Thus only three K values could be computed. During each of these three intersurvey periods, the wave array was inoperable a maximum of three days. The three values of K range from 0.84 to 1.09 with a least squares value (logI_ℓ versus log$P_{\ell s}$) determined as 0.94.

This value has been presented as the open star in the K versus D plot in Figure 14-7 with a representative $D \approx 0.3$ mm for the south beach at Rudee Inlet. Reasonable agreement with the trend is evident.

14.8 Summary

The two total trap experiments conducted at Santa Barbara, California and Rudee Inlet, Virginia have yielded a total of 10 intersurvey volumes and associated wave characteristics. Although the resulting sediment transport coefficients ($K \equiv I_\ell / P_{\ell s}$) are in general agreement with those of earlier investigators, the results obtained here appear to exhibit a dependency on sediment size which is in general agreement with earlier results (Figure 14-7).

A new and potentially useful method was developed for establishing the *distribution* of longshore sediment transport across the surf zone, an important engineering characteristic. Ideally the method should be applied to pocket beaches which have only small losses or additions with a relatively small tidal range and located such that substantial changes in wave direction occur reasonably frequently. Although the Santa Barbara data were not ideally suited for the purpose of extracting longshore transport distribution due to the relative large tidal range causing tidal smearing, they did yield qualitative results and served to illustrate the method.

Compared to other field methods of determining longshore sediment transport, total trap experiments provide a major advantage in accuracy of sediment volumes transported. Traps are not ideally suited for resolving spatial and temporal distributions of sediment transport.

14.9 Acknowledgements

Many individuals have contributed differently and significantly to the results presented herein. Chris Gable effectively contributed to both field experiments through his organization and hard work as did Bob Catalano of Ocean Surveys, Inc. Gene Berek participated in the Santa Barbara field work and carried out the resulting wave and sediment volume analysis. Kevin Bodge provided overall field direction and documentation at Rudee Inlet. Richard Seymour and David Duane, through their interest, encouragement, support and, by no means the least, patience were effective in the completion of this effort as presented here. To those many additional individuals who contributed, you are not mentioned only due to lack of space, not due to lack of appreciation.

14.10 References

Berek, E. P. and R. G. Dean, 1982, Field investigation of longshore transport distribution, *Proceedings*, Eighteenth Coastal Engineering Conference, November 14-19, 1982, Cape Town, Republic of South Africa, American Society of Civil Engineers, New York: 1620-1639.

Castel, D. and R. J. Seymour, 1982, Longshore sand transport report, February 1978 through December 1981, Report to U. S. Army Corps of Engineers and California Department of Boating and Waterways. IMR Ref. No. 86-2, 216 pp. (Reprinted March 1986)

Dean, R. G., 1978, Review of sediment transport relationship and the data base, *Proceedings*, Workshop on Coastal Sediment Transport, University of Delaware Sea Grant Program: 25-39.

Fulford, E., 1982, Distribution of longshore sediment transport across the surf zone, M.S. thesis, Department of Civil Engineering, University of Delaware, Newark, DE, September.

Galvin, C. J., 1969, Comparison of Johnson's littoral drift data for Santa Barbara with the empirical relation of CERC TR 4, Memorandum for Record, Coastal Engineering Research Center, February.

Higgins, A. L., R. J. Seymour and S. S. Pawka, 1981, A compact representation of ocean wave directionality, *Applied Ocean Research*, 3(3): 105-112.

Johnson, J. W., 1953, Sand transport by littoral currents. *Proceedings*, Fifth Hydraulics Conference, State University of Iowa Studies in Engineering, Bulletin 34: 89-109.

Komar, P. D., 1977, Beach sand transport; distribution and total drift, *Journal of the Waterways, Port, Coastal and Ocean Division*, American Society of Civil Engineers, 103(WW2): 225-239.

Komar, P. D. and D. L. Inman, 1970, Longshore sand transport on beaches, *Journal of Geophysical Research*, 75(30), October 20: 5914-5927.

Seymour, R. J., 1983, Personal Communication, letter dated January 10, 1983.

Tsuchiya, Y., 1982, The rate of longshore sediment transport and beach erosion control, *Proceedings*, Eighteenth Coastal Engineering Conference, November 14-19, 1982, Cape Town, Republic of South Africa, American Society of Civil Engineers, New York: 1326-1334.

Chapter 15

MODELS FOR SURF ZONE DYNAMICS

Edward B. Thornton
Oceanography Department
Naval Postgraduate School
Monterey, California
and
R. T. Guza
Center for Coastal Studies
Scripps Institution of Oceanography
La Jolla, California

15.1 Introduction

Longshore current models have been tested using NSTS measurements at Leadbetter Beach, Santa Barbara. The NSTS provided data on nearshore currents for a variety of wave conditions including narrow banded (in frequency and direction) swell waves of small and moderate height and wide banded waves during local storms. The theoretical models include a two-dimensional, finite element, monochromatic wave driven model (Wu *et al.*, 1985) and one-dimensional (i.e., no longshore variations) random wave driven model (Thornton and Guza, 1986). These models assume the waves are narrow banded in frequency and direction and use a single frequency and direction to describe the input wave field. In keeping with the narrow band wave assumption, data from four days with narrow band swell (3-6 February) were used for comparisons. Empirical orthogonal eigenfunctions were used to find longshore current patterns during all directional wave conditions (Guza *et al.*, 1986). This chapter summarizes these papers and presents additional comparisons to a simple one-dimensional monochromatic wave model (Longuet-Higgins, 1970).

The Torrey Pines experiment proved to be excellent for testing wave transformation models (Chapter 8), because the waves approached nearly normal to the almost straight and parallel contours. But because of the small angle of approach and often bimodal directional spectra due to the location of offshore islands, the Torrey Pines data are not ideal for testing longshore current models. On the other hand, data acquired at Leadbetter Beach are suitable and are described in the following section. Then in §15.3.1-15.3.4, the various longshore current models are described and compared to the data.

15.2 Longshore Currents, Waves and Radiation Stress Measurements at Leadbetter Beach

As was described in Chapter 2, the wave climate varied dramatically during the experiment. Measurements commenced on the 30th of January. During the first week, the local winds were generally light and the incident narrow band swell waves were almost entirely derived from distant sources. In the late afternoon of the 6th of February the barometric pressure started to drop, signaling the first of a three week series of storms. Strong winds were associated with these storms creating a wind driven sea inside the Channel Islands from a range of directions. Swell from distant sources was also often present, resulting in bimodal wave spectra in both frequency and direction.

Leadbetter Beach is highly sheltered from waves generated in the open ocean because of offshore islands and its east-west orientation on a coastline predominantly oriented north-south. There are two directional sectors, from west and southeast, with substantial fetches. Open ocean swell pass through the narrow west window between Point Conception and the Channel Islands (±9 degrees centered on 249 degrees). The southeast window (123-143 degrees) has a restricted fetch (150 km) from which local wind driven waves were generated during the storms. Oltman-Shay and Guza (1984) used a high resolution, data-adaptive technique to show that the directional spectra at the slope array are consistent with these topographic and refractive constraints.

Longshore currents, wave heights, and energy and radiation stress spectra were calculated using 68.2 minute records. Records were chosen on the basis of a subjective judgment that there were enough functioning current meters to define the cross-shore structure of the longshore currents. Data from individual current meters were rejected only if the fouling, bending or other experimental malfunction was severe enough to grossly distort the mean currents compared with nearby sensors. Vector plots of the mean currents for 17.1 minute data sections were used for this purpose, and an example is shown in Figure 15-1; the locations of the current meters and pressure sensors, with the exception of the offshore slope arrays and swash *spiders*, are indicated. A total of 64 runs unevenly distributed over 17 days between January 30 to February 15 were selected. The minimum, maximum, and average number of functional current meters during the runs selected were 8, 23, and 17.5.

Independent estimates of $S_{xy}(f)$ were obtained from the slope array (Chapter 2) and the two deepest current meters on the cross-shore instrument transect (EMD and EM1 in Figure 4-2). Estimates of total radiation stress, S_{xy}^T, from the slope array and the two current meters, for each of the 64 runs are shown in Figure 15-2. The contours are nearly plane and parallel between EMD and EM1, so S_{xy} should be a conserved quantity, but the EM1 estimates are

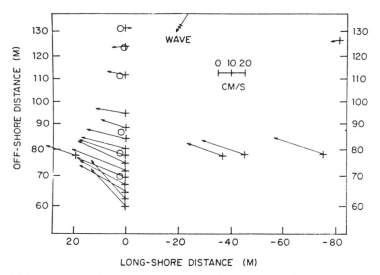

Figure 15-1. Mean (17 min) current vectors showing locations of current (+) and pressure (O) sensors on 4 February. Shoreline is at approximately 50 m relative to an arbitrary datum. (Wu *et al.*, 1985)

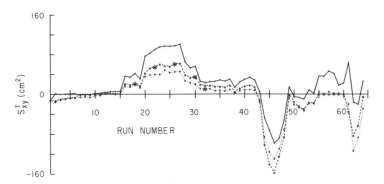

Figure 15-2. Total radiation stress (S_{xy}^T) versus run number; slope array is solid line, EMD dashed line, EM1 dotted line, asterisks are refracted from the slope array on narrow banded days 2-6 February. (Guza *et al.*, 1986)

lower. There does not appear to be an orientation problem with EM1, because its S_{xy}^T estimates are lower than that EMD for waves coming from both quadrants. However, it was found that the S_{xy}^T estimates from EM1 were nearly identical with those from EMD if the energy spectrum at location 1 is estimated with a pressure sensor located there, and the directional spreading estimated from EM1. This suggests a gain problem with EM1. A reasonable estimate of S_{xy} is the average of EMD and this corrected EM1. Unless otherwise noted, the

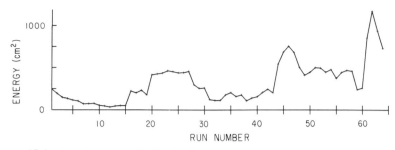

Figure 15-3. Total energy $(0.04 < f < 0.4 \, \text{Hz})$ of sea surface elevation versus run number. (Guza *et al.*, 1986)

S_{xy} values in this section refer to the average of EMD and corrected EM1. In Figure 15-2, the ρg factor in S_{xy} (10A-14) is not included, similar to the common convention of dropping that factor in energy spectra.

The total energy (E^T) of the sea surface elevation, averaged using EMD and the pressure sensor at EM1 is shown in Figure 15-3. Comparison of E^T (Figure 15-3) and S_{xy}^T (Figure 15-2) shows that the most energetic waves (run 62) do not correspond to the maximum S_{xy}^T (run 46). This is due to substantial amounts of energy coming from both windows (which are in different directional quadrants) in run 62. Figure 15-4 compares the radiation stress spectra for runs 46 and 62. It is difficult to surmise what single value of the breaking wave angle would be estimated by an observer (the method used in many longshore current studies) for run 62.

A measure of the relative amounts of $S_{xy}(f)$ coming from up and down coast quadrants is given by

$$ r = \frac{\left| \sum_{j} S_{xy}(f) \right|}{\sum_{j} \left| S_{xy}(f) \right|} \tag{15.2-1} $$

which is plotted in Figure 15-5. The values of r range from nearly unity (all waves in 1 quadrant) to zero (equal radiation stress in opposite quadrants). Figure 15-6 shows three cases of small S_{xy}^T, but widely varying r. Run 8 has only 15 percent of the energy of runs 56 and 58, yet has a larger S_{xy}^T. S_{xy}^T is the small difference of large numbers. Figures 15-2 through 15-6 illustrate that S_{xy}^T is a strong function of both the total energy and the directional distribution of that energy.

The observed longshore currents (V) for each of the runs in Figures 15-4 and 15-6 are shown in Figure 15-7. The largest S_{xy}^T (run 46) has the strongest currents (note scale variation in plots). Runs 8, 56, and 58 have comparable

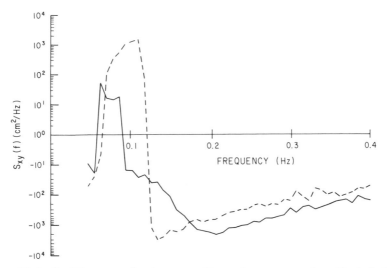

Figure 15-4. $S_{xy}(f)$ spectra for the runs with maximum energy (run 62, dashed line, $E^T = 1176$ cm^2, $S_{xy}^T = -112$ cm^2) and maximum S_{xy}^T (run 46, solid line, $E^T = 759$ cm^2, $S_{xy}^T = -171$ cm^2). (Guza *et al.*, 1986)

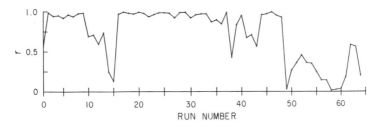

Figure 15-5. A measure of the relative amount of S_{xy} in different quadrants (r, 15.2-1) versus run number. (Guza *et al.*, 1986)

(weak) mean currents, consistent with the small S_{xy}^T values. Note that much of the variability in the weak current cases may simply be the uncertainty of sensor offsets, roughly ±5 cm/sec. Apparently the strong bidirectionality of the incident waves in runs 56 and 58 does not lead to significant current reversals. Run 25 is associated with a very narrow-banded $S_{xy}(f)$ and energy spectra ($r = 0.996$) but has a V pattern very similar (but opposite in sign) to a biquadrant run 62 ($r = -0.59$). Figure 15-7 suggests that the structure of $S_{xy}(f)$ is of secondary importance to longshore currents. It will be shown in §15.4 that most of the variation in $V(x)$, between runs, can be explained in terms of S_{xy}^T. We note that the average current (over all sensors and all runs) is -1.3 cm/sec, consistent with the small average S_{xy}^T value of -5.5 cm^2.

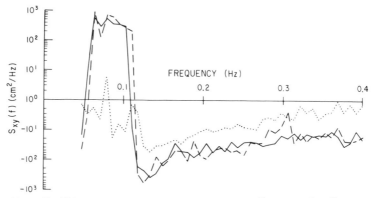

Figure 15-6. $S_{xy}(f)$ for runs 8 (dotted line, $r = 0.97$, $E^T = 71$ cm^2, $S_{xy}^T = 7$ cm^2), 56 (dashed line, $r = 0.13$, $E^T = 445$ cm^2, $S_{xy}^T = 6$ cm^2), and 58 (solid line, $r = 0.04$, $E^T = 460$ cm^2, $S_{xy}^T = 1.6$ cm^2). (Guza *et al.*, 1986)

15.3.1 Narrow-Banded Wave Models-Data

A basic assumption of existing longshore current models is that the incident waves are narrow-banded in frequency and direction. Only the narrow-banded days occurring at the beginning of the experiment from 3-6 February are appropriate for the narrow-band wave model comparisons. The waves during this time were derived from a distant North Pacific storm so that wave incidence was restricted to the very narrow window from the west approaching Leadbetter Beach at relatively large oblique angles to the local bottom contours.

The incident waves during this time exhibited an almost classical dispersive pattern of swell from a distant storm. Weak forerunners, having a peak frequency of 0.05 Hz, appeared on the morning of February 1. The peak frequency gradually increased to 0.09 Hz on the sixth. The H_{rms} increased to a maximum on the fourth and then decreased (see Table 15-1). Plunging breakers were most often observed. Although the waves were small in height, the large incident wave angles resulted in moderately strong longshore currents.

The boundary conditions for wave height and direction for the two dimensional model (Wu *et al.*, 1985) were determined using the slope array in 9 m depth (Chapter 2). A significant wave angle at each frequency is defined (Higgins *et al.*, 1981)

$$\alpha(f) = \frac{1}{2}\sin^{-1}\left[\frac{S_{xy}(f)}{E(f)n(f)}\right] \qquad (15.3-1)$$

where E and S_{xy} are the measured energy and radiation stress, $n = Cg/C$ is calculated from the known water depths. An example of energy, radiation stress, and significant wave angle spectra measured at the slope array is shown in Figure 15-8. The modeling frequency, f_p, is chosen as the value corresponding to the peak of the narrow-banded energy spectrum. Due to the

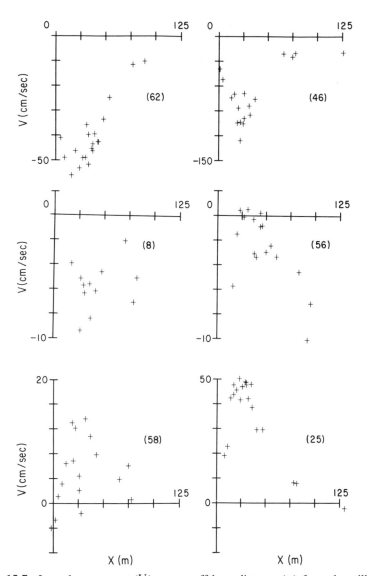

Figure 15-7. Longshore current (V) versus offshore distance (x) from the still water line. Run numbers are indicated on each panel in parentheses. Note that the velocity scale changes from panel to panel. (Guza *et al.*, 1986)

Table 15-1. Wave and Beach Conditions

February	3	4	5	6	Average
At S_{xy} array					
h (m)	9.1	9.1	9.0	8.8	
f_p (Hz)	.070	.070	.078	.090	
T_p (sec)	14.3	14.3	12.8	11.1	
[1] $\hat{\alpha}(°)$	16.6	18.4	17.8	17.8	
H_{rms}(m)	.49	.52	.41	.26	
S_{xy} (N/m)	-78.4	-88.9	-53.7	-21.4	
At 4 m					
h (m)	3.8	3.8	3.6	3.5	
[2] H_{rms}(m)	.55	.56	.45	.26	
[3] $\hat{\alpha}$ (°)	7.8	9.0	8.4	8.3	
Surf Zone					
Breaker Type	Plunge	Plunge	Plunge/ Spill	Plunge/ Spill	
[4] γ_b	.40	.35	.39	.31	.35
[5] γ	.48	.45	.43	.34	.43
[6] $V_{max}/|u_m|$	.8	1.0	1.1	.8	.9
[7] bottom slope	.044	.038	.035	.033	.042
foreshore slope	.053	.040	.052	.064	.058

(1)   Defined in (15.3-2).

(2)   Measured in 4 m depth.

(3)   Result of refracting the peak frequency wave from 9 m to 4 m depth; see text.

(4) $\gamma_b = H_b/h_b$, where H_b is measured as maximum H_{rms} in cross-shore distribution.

(5)   Defined as the mean slope of H_{rms}/h in the inner surf zone (saturation region).

(6) $|u_m|$ calculated as defined in (15.3-10).

(7)   Mean bottom slope measured between shoreline and mean breaker line.

decreased refraction of waves towards higher frequencies there is a slow increase in significant wave angle across the energetic band ($0.06 < f < 0.12$ Hz) of the spectrum. To insure that the model wave forcing equals the measured total S_{xy}, a single weighted wave angle, $\hat{\alpha}$, is defined using the summed components of $S_{xy}(f)$ and $E(f)$ over the sea-swell band of frequencies (0.05-0.5 Hz)

$$\hat{\alpha} = \frac{1}{2}\sin^{-1}\left[\frac{S_{xy}^T}{E^T n(f_p)}\right] \tag{15.3-2}$$

where the superscript T refers to summed quantities. The wave heights and angles are summarized in Table 15-1.

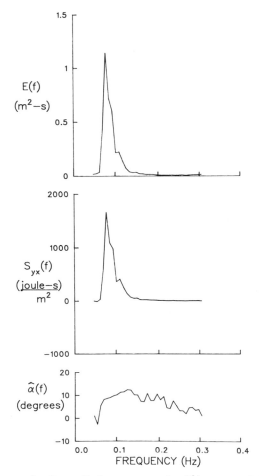

Figure 15-8. Energy density, radiation stress and $\alpha(f)$ spectra, 5 February 1980, h = 9.0 m. (Thornton and Guza, 1986)

To apply a one-dimensional longshore current model, not only must the beach contours be essentially straight and parallel but the beach orientation must be accurately determined. A misalignment of the beach orientation can lead to substantial error in the radiation stress calculations (Chapter 10). The beach orientation was determined by fitting straight lines (in the least-square-sense) to the +50, 0 MSL, −50, and −100 cm depth contours across measured profile lines and determining the average orientation for the four contours. An unresolved question is what is the approximate alongshore scale of longshore currents. However, the average alongshore orientations measured over 200 m (five profile lines at y = 100, 50, 10, −60, −100 m) and over 110 m (three profile lines at y = 50, 10, −60 m) were essentially the same. The beach orientation used in this analysis was averaged over 110 m for the 16 days from 29 January to 13

February. The average beach orientation varied little day to day during this time period (standard deviation = 0.43 degrees). All angles given in Table 15-1 are relative to the coordinate system determined in this manner.

It is pointed out that the two-dimensional nearshore current analysis of §15.3.2 uses the original NSTS coordinate system, which is rotated 1.2 degrees counterclockwise relative to the coordinate system described above. The results of a two-dimensional analysis should not be dependent on the coordinate system employed.

The offshore bathymetry measured on 22 January and the daily beach surveys are combined to give the total bathymetry. The bottom contours are nearly straight and parallel inside the 4 m contour, but are non-parallel seaward (Figure 1B-2). Therefore, one-dimensional models will be used to describe the currents inside 4 m depth. The analytical longshore current models assume a plane sloping beach, and the surf zone profiles were nearly planar during 2-6 February (Figure 15-9). A mean beach slope, $\tan\beta$, representing the plane sloping beach was determined by a least-square linear fit to the bottom profile across the surf zone measured between the mean breaker-line and the intersection of the mean water line with the beach. The breaker-line is defined here as the location of maximum measured H_{rms}. A foreshore slope was also determined and is always steeper than $\tan\beta$ (Table 15-1).

The one-dimensional models require boundary conditions specified in 4 m depth. Therefore, the waves measured at the slope array in 9 m were refracted to the approximate 4 m depth as input to the one-dimensional models. The wave angles in 4 m were determined in two ways. The first, and more complex, was to measure the directional spectra at the pressure sensor array using maximum likelihood estimator techniques (Pawka, personal communication); the directional spectra measured in 9 m were then refracted to the 4 m depth, $S_{xy}(f)$ calculated, and a mean angle determined using (15.3-1). The second method was to simply treat the waves as monochromatic and unidirectional and refract the peak frequency wave to 4 m depth, starting with the mean wave angle in 9 m defined using (15.3-2). A comparison of the two techniques (Thornton and Guza, 1986) showed that simple monochromatic wave refraction was accurate enough on the days of very narrow-banded incident waves. The asterisks in Figure 15-2 correspond to the refracted S_{xy}^{T} values and compare well with the estimates from EMD. The refracted wave angle and wave height calculated using the pressure sensor at 4 m are used as the outer boundary wave input for the one-dimensional longshore current models (Table 15-1). A list of all the data for mean (68 minute) longshore currents and H_{rms} at respective measured depths for 3-6 February are given in §15.6 Appendix.

15.3.2 Two-Dimensional Nearshore Current Model

Analytic solutions of waves and nearshore currents are limited to beaches of simple plan form. For complex beach topography, a numerical model which

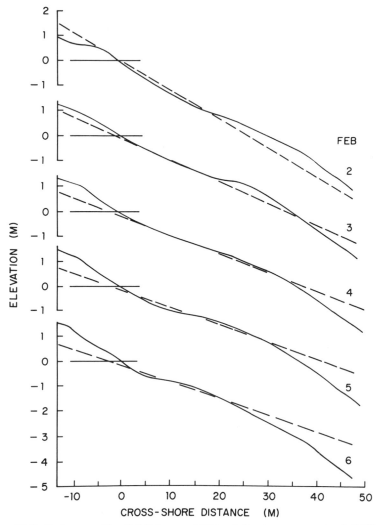

Figure 15-9. Bottom profiles 2-6 February 1980, relative to mean sea level (MSL). The mean bottom slope within the surf zone is indicated by dashed line. The shoreline is at the intersection on the mean water line and the beach profile. (Thornton and Guza, 1986)

takes into account the two-dimensional wave transformation and nearshore current circulation is needed. Several numerical models, which include realistic topography but ignore the effect of turbulent mixing, have been developed (Noda, 1974; Sasaki and Horikawa, 1975; Ebersole and Dalrymple, 1980). In general, these models predict unrealistic current intensity when compared with the results of field experiments. With the inclusion of a turbulent mixing term and convective accelerations, Kirby and Dalrymple (1982) obtained longshore current distribution comparable to field observations.

The nearshore current model developed by Wu and Liu (1985), using a finite element approach, and applied by Wu *et al.* (1985) to NSTS data is discussed here. This model includes linear bottom friction, lateral mixing, and nonlinear convective accelerations. The finite element model calculates steady currents using a flexible triangular grid to resolve the highly variable wave field and mean current distribution across the breaker zone. The finite element model was chosen because of its superior accuracy and efficiency compared with a finite difference model. This is particularly true for waves over bathymetry with local irregularities.

The conservation of mass and momentum equations (10A-10 and 10A-11) are used to describe the mean current motion. The frictional forces in the interior of the water column (i.e., horizontal turbulent mixing) are modeled by (10A-6) and (10A-18), and the linear bed shear stress formulation (10A-21) is used. The refraction of the two-dimensional wave field is described by the irrotational wave number equation (10A-7). The wave height is obtained by solving the energy conservation equation (10A-28) and the breaker height description (8.4-1), whichever is smaller.

The numerical approximations for the current model are based on the Galerkin Weighted Residual method and is described in detail in Wu and Liu (1985). A computation domain was chosen 120 meters alongshore to include the main range line of instruments, and 300 meters offshore covering the wave slope array. The model uses a 3-node linear triangular element for the wave set-up and a 6-node quadratic triangular element for current velocities U and V. To save storage and computational efforts, a coarse mesh is used in the offshore region where the currents vary slowly, while a finer mesh is used in the surf zone region.

Numerical results are obtained for the four days 3-6 February. Both waves and currents are simulated and compared with data. The wave conditions (height and incident wave angle) in 9 m depth and measured bathymetry on each day are used as model input. Periodic conditions at the lateral boundaries and a no flow condition at the beach are imposed. Typical H_{rms} contours are shown in Figure 15-10. The corresponding two-dimensional current velocity vectors are shown in Figure 15-11, where the currents are mostly confined within the surf zone region (in this case, X < 40 m) and smoothly distributed in the seaward direction. The meandering current pattern was also predicted for the other days during the experiment.

The measured rms wave heights and one-hour averaged longshore currents on the cross section $y = 0$ are compared with the model calculations in Figures 15-12 and 15-13 for 5 and 6 February. The maximum H_{rms} decreases about 40

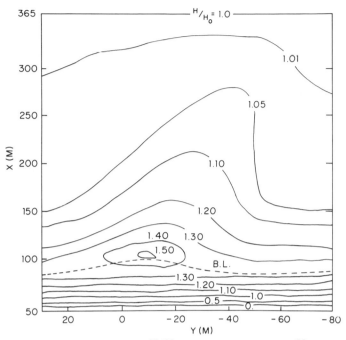

Figure 15-10. Wave height contours, H/H_o, on 4 February, where H_o = 0.52 m is the average rms wave height measured at the deep water slope array. The breaker line (BL) is noted by the dashed line. (Wu *et al.*, 1985)

percent from 5 February to 6 February; the corresponding longshore currents and surf zone width also show significant reduction. The predicted rms wave height transformation is in agreement with the data for 5 February, but there is a consistent overprediction on 6 February. The longshore current models have two free parameters (c_f and N), which are determined by least square fit of the model with the field data (see Table 15-2). The maximum currents usually occurred in the mid-surf zone, and decay mildly offshore and rather quickly toward the shore. The model comparisons are generally favorable seaward of the mean breaker line, but do not agree as well near the shoreline where the wave set-up, surf beat and edge waves may contribute. The effects of set-up and long waves on the incident waves and induced currents are ignored in the present model.

The importance of the convective inertial terms in the momentum balance equation is investigated by comparing linear and nonlinear model results using the same values of c_f and N (Figure 15-13). The nonlinear terms reduce the magnitude of the maximum longshore current and shift the location of the peak velocity slightly seaward. The effect of including the convective inertial terms

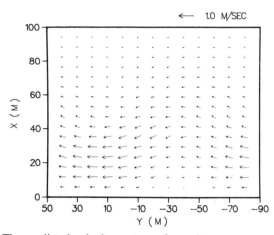

Figure 15-11. The predicted velocity vectors of nearshore current at Leadbetter Beach on 4 February. (Wu *et al.*, 1985)

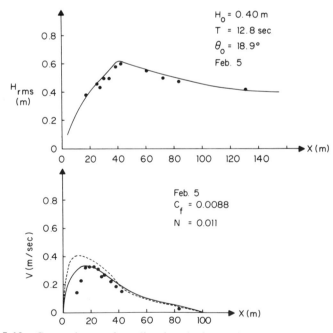

Figure 15-12. Comparisons of predicted and observed wave heights and longshore current velocities on 5 February. (Wu *et al.*, 1985)

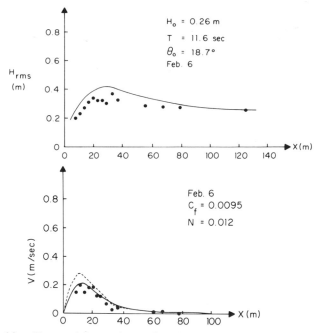

Figure 15-13. Comparisons of predicted and observed wave heights and longshore current velocities on 3 February; linear current (dotted line), non-linear current (solid line). (Wu *et al.*, 1985)

is most evident for the case of strong currents generated by the relatively large waves on 3 February.

In the next two sections, we consider depths shallower than 4 m, where the bottom contours are approximately straight and parallel. Due to the simplicity in bathymetry, one-dimensional longshore current theory can be used. The two-dimensional model described above shows (Figure 15-10) that the wave heights have some alongshore dependence due to refraction, even inside 4 m depth where the contours may be considered straight and parallel. Therefore, the accuracy of the one-dimensional models is probably limited by the alongshore variability of the wave heights.

15.3.3 Longshore Current Monochromatic Wave Model

Longshore current models derived assuming stationary, narrow-band waves (in frequency and direction) obliquely incident over straight and parallel bottom contours are discussed in this and the next section. The alongshore momentum balance (10A-11) simplifies to

$$\frac{\partial \tilde{S}_{xy}}{\partial x} + \frac{\partial S'_{xy}}{\partial x} = \tau_y^b = -c_f \, \overline{|\tilde{u}|} V \tag{15.3-3}$$

Table 15-2. Longshore Current Model Coefficients,
c_f and N, and Percent Errors

Date		Feb	3	4	5	6	Ave
		Two-Dimensional Monochromatic Wave Model					
	c_f		.009	.010	.008	.009	.009
	N		.011	.012	.018	.013	.014
		One-Dimensional Monochromatic Wave Model					
Plane Beach							
	c_f		.007	.0075	.0065	.0085	.007
	N		.003	.002	.004	.002	.004
	M		.11	.34	.55	.22	.35
	% error		32	9	14	14	17
Actual Bottom							
	c_f		.006	.0085	.008	.008	.008
	N		.002	.002	.001	.003	.003
	% error		35	13	27	23	25
		One-Dimensional Random Wave Model					
Plane Bottom, $N = 0$							
	c_f		.009	.009	.009	.010	.009
	% error		48	15	19	18	25
Actual Bottom, $N = 0$							
	c_f		.009	.009	.008	.009	.009
	% error		54	18	24	33	32
Actual Bottom							
	c_f		.007	.009	.007	.008	.008
	N		.004	.0001	.0004	.002	.002
	% error		51	18	24	28	30
Actual Bottom, $N = 0$, Nonlinear τ_y^b							
	c_f		.0065	.0060	.0055	.0075	.006
	% error		49	14	17	34	28

where the quadratic shear stress law (10A-20) has been assumed. It has been assumed in (15.3-3) that the wave-induced and turbulent velocity components are statistically independent, which appears to be a reasonable assumption within the surf zone (Thornton, 1979).

Longuet-Higgins (1970) obtained an analytical solution of (15.3-1) for the mean longshore current distribution assuming monochromatic waves. He modeled the integrated Reynolds stress, S'_{xy}, with an eddy viscosity parameterization (10A-18) and simplified the bed shear stress with the weak current assumption (10A-21), where

$$\overline{|\tilde{u}|} = \frac{H}{\pi}\sqrt{\frac{g}{h}} \tag{15.3-4}$$

In terms of the nondimensional longshore velocity, $V* = V/V_b$, and distance offshore, $X = x/x_b$, where subscript b refers to conditions at breaking

$$V* = B_1 X^{p_1} + AX \qquad 0 < X < 1$$
$$= B_2 X^{p_2} \qquad 1 < X < \infty \tag{15.3-5}$$

where A, B_1, B_2, p_1 and p_2 are all complicated functions of a nondimensional parameter

$$P = N\tan\frac{\beta}{(\gamma c_f)} \tag{15.3-6}$$

representing the relative importance of horizontal mixing and bottom friction. Values of P ranged between 0.1 and 0.4 for model fits of the laboratory data of Galvin and Eagleson (1965).

It was originally intended to directly compare the Longuet-Higgins (1970) analytical solution to field data. This approach was slightly modified, however, because the analytic solution (15.3-6) is cast in terms of breakpoint conditions, and the observed γ_b values at the breakpoint are not consistent with $\hat{\gamma}$ averaged over the inner surf zone. The γ_b values, defined as the ratio of the wave height to depth at the mean breakerline (maximum H_{rms}) are significantly less than the γ values defined over the saturation region only (see Figures 8-10 and 8-11 and Table 15-1). It will be seen this model is sensitive to $\hat{\gamma}$ values. Using the observed values of X_b (or H_b, h_b) as fundamental model parameters leads to unsatisfactory predictions of H_{rms} distributions; a scatter of c_f and N values resulted from fitting the longshore current distributions to compensate for the poor H_{rms} predictions.

Therefore, the incident wave height, frequency, and angle are prescribed at the outer boundary (4 m depth), and the longshore current equations solved numerically with the following boundary conditions

$$V \rightarrow 0 \quad \text{as} \quad h \rightarrow 0$$
$$V \rightarrow 0 \quad \text{at} \quad h = h_o \sim 4\text{ m}$$

Wave heights are obtained via conservation of energy or $H = \hat{\gamma}h$, whichever is smaller. Thus, breaker conditions in the model are prescribed by the offshore measurements and the shoaling model rather than direct measurements. Wave angles are obtained from Snell's law. A finite, central difference approximation of the second order differential equation leads to a set of linear algebraic equations, which are solved by Gaussian elimination (Gerald, 1978). This approach does not alter the physics of the Longuet-Higgins solution, only the location of boundary conditions is changed.

Longshore current distributions are shown in Figure 15-14. Solutions were calculated for both the plane beach approximation (dotted line) and the actual beach bathymetry (solid line). The distributions are plotted relative to distance offshore from the mean water level intersection on the actual beach. The velocities for the plane beach solutions do not go to zero at $x = 0$ because the selected plane topography beach usually intersects the mean water level at a location shoreward of the actual intersection point (see Figure 15-9).

The optimal c_f and N values were determined by iteratively solving for the minimum least square error between measured and calculated longshore current values shoreward of the mean breakerline (Table 15-2). The coefficients are similar for both plane and actual beach solutions with an average c_f of 0.008 and an average N of 0.003 for the actual beach. The c_f are relatively constant for all days, but N varies considerably. These values of N, and the measured values of $\hat{\gamma}$ and bottom slope $\tan\beta$ (Table 15-1) were used to solve for M (10A-19). The resulting values of M (Table 15-2) are roughly 0(1), consistent with Battjes (1974) derivation. However, M has slightly greater percentage variation than N indicating the Battjes formalism (10A-19) does not appear to improve over (10A-18) with N assumed constant.

The sensitivity of the solutions to variations in N and c_f are illustrated in Figures 15-15 (N constant, c_f varies) and 15-16 (N varies, c_f constant). Figure 15-15 shows the expected inverse relationship between c_f and V. Figure 15-16 shows that V is a nonlinear function of N, and is insensitive to large percentage changes in small N (i.e., 0.001-0.0001), but sensitive to changes in larger values of N ($N > 0.001$). The model is most sensitive to the value of $\hat{\gamma}$ which determines the breaker location and rate of energy dissipation inside the surf zone. For fixed c_f and N, decreasing $\hat{\gamma}$ moves the breaker line offshore, thus increasing the currents offshore, and reducing them inshore (Figure 15-17). Recall that $\hat{\gamma}$ is not a free parameter in the present model but determined from the measurements. Selecting the appropriate $\hat{\gamma}$ value is essential for the monochromatic models.

15.3.4 Longshore Current Random Wave Models

In the monochromatic models, the waves outside the surf zone are assumed nondissipative, $\tilde{S}_{xy}$ is conserved, and the driving force for longshore currents is zero. At breaking, the waves lose energy and there results an abrupt change in the $\tilde{S}_{xy}$ gradient. The resulting longshore current distribution is no current outside the surf zone, a maximum current at breaking and then gradually decreasing to zero at the beach. It is then necessary to introduce an eddy viscosity term to smooth the physically unrealistic shear currents at the breakerline.

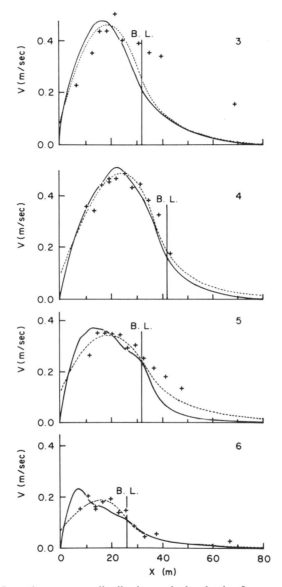

Figure 15-14. Longshore current distributions calculated using Longuet-Higgins (1970) formulation for both plane beach (dotted line) and actual beach (solid line). Measured longshore currents (68 min average) are denoted by (+) and the mean breaker line by BL.

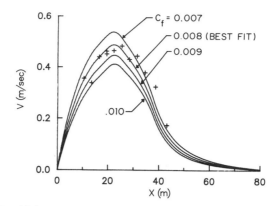

Figure 15-15. Sensitivity of model results to variations in c_f ($N = 0.0015$, $\hat{\gamma} = 0.45$) on 4 February, actual bathymetry.

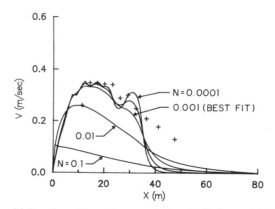

Figure 15-16. Sensitivity of model results to variations in N ($c_f = 0.008$, $\hat{\gamma} = 0.43$) on 5 February, actual bathymetry.

An improved formulation is to start out by describing the wave heights in terms of a probability distribution as described in Chapter 8, resulting in a distributed breaker region and gradual changes in eneregy dissipation and $\tilde{S}_{xy}$. The resulting longshore currents can have a smoothed profile. The need for including eddy viscosity to smooth out the velocity profile is reduced, or eliminated (Collins, 1972; Battjes, 1972).

In this section, the narrow-band, random wave transformation model described in Chapter 8 is used to describe the breaking waves and $\tilde{S}_{xy}$ in the longshore current formulation. For convenience, the radiation stress is rewritten

$$\tilde{S}_{xy} = \left[E \ C_g \ \cos \hat{\alpha} \right] \left[\frac{\sin \hat{\alpha}}{C} \right] \tag{15.3-7}$$

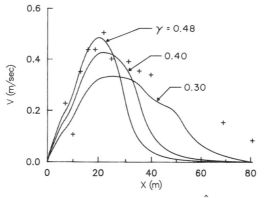

Figure 15-17. Sensitivity of model results to variations in $\hat{\gamma}$ ($c_f = 0.007$, $N = 0.004$) on 3 February, actual bathymetry.

where

$$\frac{\sin\hat{\alpha}}{C} = \text{const} = \frac{\sin\hat{\alpha}_o}{C_o} \tag{15.3-8}$$

by Snell's law of linear wave refraction. Therefore,

$$\frac{d}{dx}\tilde{S}_{xy} = \frac{\sin\hat{\alpha}_o}{C_o}\frac{d}{dx}(E\ C_g\ \cos\hat{\alpha}) = \frac{\sin\hat{\alpha}_o}{C_o}\langle\varepsilon_b\rangle \tag{15.3-9}$$

with the change in energy flux equal to the breaking wave dissipation $\langle\varepsilon_b\rangle$ for straight and parallel contours (8.8.1). Several longshore current solutions are considered, including linear and nonlinear bottom shear stress formulations, and plane sloping and actual bathymetry.

Longshore current formulations using linear bottom shear stress (10A-21) are described first. For the longshore current model which uses the linear wave height transformation scheme, the $|\tilde{u}|$ term is calculated using linear, shallow water, wave theory relationships between surface elevation and horizontal velocity, and the Rayleigh wave height distribution

$$|\tilde{u}| = \frac{1}{2}\left[\frac{g}{h}\right]^{1/2}\left[\int_0^\infty H\ p(H)\ dH\right]\left[\frac{1}{T}\int_T|\cos(kx-\omega t)|\,dt\right]$$

$$= \frac{1}{2\sqrt{\pi}}\ H_{\text{rms}}\ \sqrt{\frac{g}{h}} \tag{15.3-10}$$

Note that the constants multiplying H_{rms} (15.3-10) are greater than the analogous constants for monochromatic waves (15.3-5) by the factor $\sqrt{\pi}/2$, because a distribution of waves is considered here.

Solving (15.3-3) for the longshore current, neglecting turbulent lateral momentum flux, S'_{xy}, and substituting (15.3-9) yields

$$V = \frac{1}{\rho c_f \, |\tilde{u}|} \frac{\sin\hat{\alpha}_o}{C_o} \langle \varepsilon_b \rangle \tag{15.3-11}$$

For the linear wave model, $\langle \varepsilon_b \rangle$ and $\overline{|\tilde{u}|}$ are given by (8.8.1) and (15.3-10), and (15.3-11) reduces to

$$V = \frac{3}{8}\pi \frac{B^3 \hat{f}_p^{1/2}}{c_f \gamma^4} \frac{\sin\hat{\alpha}_o}{C_o} \frac{H_{rms}^6}{h^{9/2}} \tag{15.3-12}$$

An analytical solution is obtained for a plane sloping beach limited to shallow water by simply substituting H_{rms} (8.8.11) into (15.3-12)

$$V = \frac{23}{40}\pi^{1/2} g \frac{a^{1/5}}{c_f} \tan\beta \frac{\sin\hat{\alpha}_o}{C_o} h^{9/10} \cdot \tag{15.3-13}$$

$$\left[1 - h^{23/4}\left[\frac{1}{h_o^{23/4}} - \frac{1}{y_o^{5/2}}\right]\right]^{-6/15}$$

$$0 < h < h_o = \frac{L}{20}$$

with y_o defined in (8.8.11). A maximum velocity occurs at about mid-surf zone, $V \rightarrow 0$ as $h \rightarrow 0$, and V is small for $h \sim h_o$. For general bottom profiles, V can be solved for by first numerically integrating for H_{rms} (8.8.10) and then substituting into (15.3-12).

Observed and model longshore current and H_{rms} distributions for 4 February, assuming a planar beach and no eddy viscosity ($N = 0$), are compared in Figure 15-18. The analytical solutions for wave height and longshore current distributions can give quite reasonable comparisons with field measurements provided the beach is approximately planar and the appropriate $\hat{\gamma}$, B and c_f values are used.

Longshore current distributions over actual bathymetry, with and without eddy viscosity, are compared separately in Figure 15-19. Again, optimal c_f (and N) coefficients are determined by iteratively solving for the least square error between calculated and measured longshore current values shoreward of the mean breaker line. The inclusion of eddy viscosity smooths out sharp velocity gradients associated with rough bathymetry and shifts the location of the maximum longshore current, V_{max}, shoreward. Since the V_{max} predicted by linear theory without eddy viscosity is already usually slightly too far shoreward, eddy viscosity does not significantly improve the predictions of longshore current (Figure 15-19, Table 15-2).

The strength of the longshore current relative to the magnitude of the wave induced bottom velocity, $|\tilde{u}|$, calculated at V_{max} is $0(1)$ (Table 15-1),

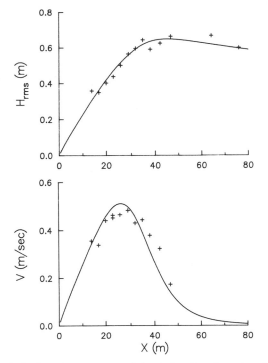

Figure 15-18. Analytical random wave model solution for planar beach of V and H_{rms} versus distance offshore compared with measurements, $N = 0.4$, 4 February. (Thornton and Guza, 1986)

suggesting that linearization of the bottom friction may be questionable. The general longshore bed shear stress formulation requires substituting (10A-23, 24) into (10A-20) and averaging over the wave period for a particular wave height to obtain $\tau_y^b(H)$. The ensemble average is then calculated

$$\tau_y^b = \int_0^\infty \tau_y^b(H) \, p(H) \, dH \qquad (15.3\text{-}14)$$

This results in a difficult nonlinear equation since V and H cannot be brought outside the averaging integrals. The linear solution is used as the starting values for a numerical iterative solution, which was verified against the analytical solution for small angles of wave incidence.

Longshore current solutions for linear and nonlinear τ_y^b are compared in Figure 15-20. Eddy viscosity is not included ($N = 0$). The nonlinear, and presumably more physically correct τ_y^b, increases the bottom stress compared with the linear τ_y^b, i.e., for the same c_f value, the nonlinear τ_y^b decreases the predicted longshore currents. The effect of the nonlinear τ_y^b increases with

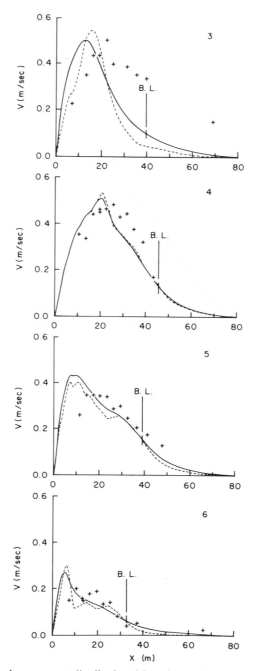

Figure 15-19. Longshore current distribution driven by random waves with (solid line) and without (dashed line) eddy diffusivity. (Thornton and Guza, 1986)

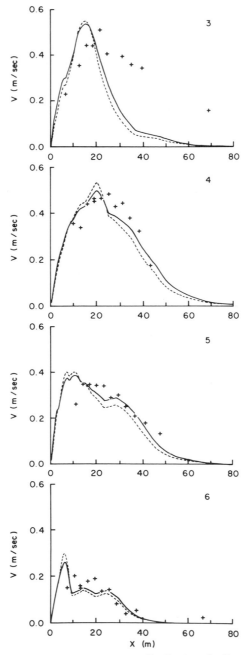

Figure 15-20. Comparison of longshore current distributions for linear (dashed line) and nonlinear (solid line) τ_y^b. Eddy viscosity is not included ($N = 0$). (Thornton and Guza, 1986)

increasing V_{max}. The resulting best fit c_f values for nonlinear τ_y^b are decreased on the average by one-third compared with the corresponding linear τ_y^b (Table 15-2). Using the best fit c_f values results in similar velocity distributions, but with the nonlinear τ_y^b, the velocity distribution is broadened with slightly increased velocities both shoreward and seaward of the mean breakerline (Figure 15-20). The nonlinear τ_y^b formulation gives a slightly improved fit.

15.4 Wide Band, Empirical Model

A wide variety of incident wave and longshore current conditions were observed. As was shown in §15.2, in many cases the total radiation stress contains important contributions from a wide range of frequencies. In a few instances, sea and swell approach the beach from different directional quadrants, resulting in a near zero S_{xy}^T. The strong velocity shears and current reversals which might be expected to occur in these circumstances were not observed. Empirical orthogonal eigenfunction (EOF) analysis is applied here to obtain a compact representation of the modes of variability of V. The temporal variability of V and S_{xy}^T will be shown (Figure 15-22) to be highly correlated. Several EOF representations were constructed. In the first case, the measured mean velocities for each run were divided into 10 equal 12 m bins starting from the beach. All sensors in a given bin, for each run, were averaged together. The 10 bins covered 120 m which is the approximate width of the cross-shore current meter array. The resulting 640 data points (64 runs × 10 spatial bins) formed the data matrix for EOF analysis (e.g., Kundu et al., 1975). The first eigenfunction contained 89 percent of the variance.

A second EOF decomposition used 10 spatial bins of variable length, which widen with offshore distance, giving better spatial resolution in shallow water where there are more sensors and greater velocity shears are anticipated. In addition, the offshore length scale was varied from run to run, depending on the wave height. That is, with x; the seaward boundary of the jth spatial bin

$$
\begin{aligned}
x; &= js, & 1 \le j \le 7 \\
&= 7S + (j-7)2s & 8 \le j \le 9 \\
&= 20s, & j = 10
\end{aligned}
\tag{15.4-1}
$$

with $s = 25\sigma$ ($\sigma^2 =$ variance of offshore sea surface elevation). The first spatial mode of this second EOF described 93 percent of the variance. Numerous other schemes for scaling the offshore coordinate were tried, including separate dependences on high, mid- and low-frequency incident wave components. In no case did the first EOF explain more than 95 percent of the variance.

The second EOF description is selected for detailed discussion because of its simplicity and correspondence with the intuition that the surf zone width (and offshore extent of the longshore current) increases with increasing wave height. The spatial variation of the first mode EOF is shown in Figure 15-21. The

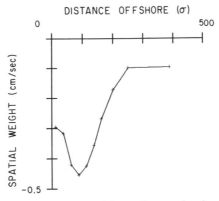

Figure 15-21. Spatial weights versus offshore distance for the 1st EOF (93% of variance). The offshore coordinate is in units of σ, where σ^2 is the offshore sea surface variance. If $H_s = 1$ m, then $\sigma = 0.25$ m and the maximum offshore distance shown is 125 m. (Guza *et al.*, 1986)

cross-shore dependence is basically parabolic, although the most shoreward position has a relatively stronger velocity than the narrow-band models of §15.3. The maximum longshore current occurs in spatial bin 4, which is centered at 87.5 σ. Using $H_s = 4\sigma$, and a nominal beach slope of 0.046, the bin is centered at a depth $h_4 = H_s$. With a nominal surf zone width based on $H_s = 0.6h_b$, the maximum current occurs in the seaward half of the surf zone.

The temporal variation of the longshore current spatial pattern is given by the temporal expansion coefficients (Figure 15-22). The temporal expansion coefficients have been multiplied by the weight of spatial bin 4 to scale the results. That is, Figure 15-22 shows the temporal variation of the maximum longshore current of the first EOF, $V(x_4, t_j)$, $j = 1, 2....64$. The scatter plot of $V(x_4, t_j)$ and $S_{xy}^T(t_j)$ (Figure 15-23) shows these variables to be highly correlated (correlation $= 0.97$). The slope of the best fit regression line constrained to pass through the origin is 0.641. Therefore, the spatial and temporal variation of mean longshore currents of this data set can be efficiently expressed as

$$V(x_i, t_j) = BS_{xy}(t_j)\frac{W(x_i)}{W(x_4)} \qquad (15.4\text{-}2)$$

where $B = 0.641$ (cm sec)$^{-1}$, $W(x_i)$ are the weights of the ith spatial bin (Figure 15-21), the bin boundaries depend linearly on H_s (15.4-1) and S_{xy}^T has units of cm^2 (no ρg factor). The temporal variability very strongly correlated with S_{xy}^T.

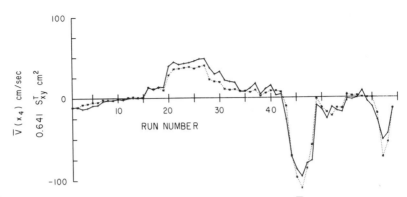

Figure 15-22. Maximum longshore velocity (spatial bin 4, $\overline{V}(x_4)$, solid line) and the best linear fit total radiation stress (0.641 S_{xy}^T, dotted line) versus run number. (Guza *et al.*, 1986)

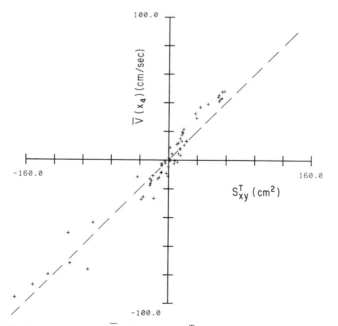

Figure 15-23. Scatter plot of $\overline{V}(x_4)$ versus S_{xy}^T for all 64 runs. The dashed line is $\overline{V}(x_4) = 0.641 \, S_{xy}^T$. (Guza *et al.*, 1986)

That is, the total radiation stress, and not the structure of $S_{xy}(f)$, is of primary importance to the magnitude of longshore currents.

15.5 Summary and Conclusions

A wide variety of wave conditions occurred during NSTS at Santa Barbara, California. All theoretical models assume the waves are narrow-banded, and for comparisons, four data days were selected when the waves were very narrow-banded in both frequency and direction in keeping with the model assumption. The wave angles at breaking were moderately large (α_b, 5°) resulting in sometimes strong longshore currents (up to 0.5 m/sec), depending on the wave height.

The theoretical models included a two-dimensional, finite element model for monochromatic waves, a classical one-dimensional monochromatic wave driven model and a one-dimensional random wave model. The models incorporated various simplifying assumptions. The two-dimensional model included nonlinear convective terms but assumes linear bottom friction. The one-dimensional random wave model compares linear and nonlinear bottom stress formulations. In the two-dimensional model, the boundary conditions were specified in 9 m depth where the bottom contours are nonparallel. The one-dimensional models were started in 4 m depth, inside which the bottom contours could be reasonably considered straight and parallel. Good comparisons with data were obtained for all models by adjusting the free parameters of c_f and N used in the bottom shear stress and eddy viscosity formulations. The capabilities of the models and the average c_f and N values for the four data days solved using a least square fit with the data are summarized in Table 15-3.

For the two-dimensional finite element model, the inclusion of nonlinear inertial terms for a nonuniform longshore current pattern is shown to reduce the magnitude of the calculated longshore current velocity. Numerical experiments indicate that the N value for the nonlinear model to fit the data is larger than that for the linear model, and the c_f value, on the contrary, is smaller for the nonlinear convective model. The reason for these differences are not clear. It is possible that the inertial terms tend to locally focus the momentum; larger N values are then required in the nonlinear model to suppress the stronger velocity gradients. The linear model compensates for the lack of inertial terms by increasing c_f values to fit the observations.

In general applications, the nonlinear convective acceleration terms can be significant as was demonstrated by the two-dimensional model solutions. However, the one-dimensional models assume straight and parallel contours, and, if the coordinates are parallel to the contours, the nonlinear convective terms are zero. Therefore, care was taken to find the best coordinate alignment to parallel contours to minimize the nonlinear term.

The one-dimensional monochromatic wave driven model equations (Longuet-Higgins, 1970) incorporating a linearized bed shear stress and eddy

Table 15-3. Model Capabilities and Average c_f and N Values

Model	Wave Description	Profile Description	Nonlinear Convection	Bottom Shear Stress	c_f	N	% error
2-D, finite element	Monochromatic	Actual	Yes	Linear	.009	.014	
1-D	Monochromatic	Plane	No	Linear	.007	.003	17
1-D	Monochromatic	Actual	No	Linear	.008	.002	25
1-D	Random	Plane	No	Linear	.009	No	25
1-D	Random	Actual	No	Linear	.009	No	32
1-D	Random	Actual	No	Lineaer	.008	.002	30
1-D	Random	Actual	No	Nonlinear	.006	No	28

viscosity varying approximately as $x^{3/2}$ were numerically integrated. By adjusting the bed shear stress coefficient c_f and the coefficient N associated with eddy viscosity, reasonable longshore current model-data comparisons were obtained. Using the actual bottom in the model, the c_f coefficients were almost constant, with an average of 0.008. The N coefficient varied considerably, having an average of 0.003 for the four days. The sensitivities of the model to variations in c_f, N and $\hat{\gamma}$ were examined. The calculated velocity field is inversely proportional to c_f, not sensitive to changes in small N but increasingly sensitive to large N ($N > 0.001$), and most sensitive to proper selection of $\hat{\gamma}$.

In the one-dimensional random wave model (and in nature) the location of initial wave breaking is spatially distributed, resulting in a smoothly varying wave induced momentum flux and longshore current distribution. Good agreement with measurements was obtained using an analytic solution incorporating linear wave theory and a plane beach assumption. In fact, the plane beach solution for no eddy viscosity gave the least error when compared to the four data days, suggesting the model is too sensitive to bottom changes. A numerical solution is required when applying to actual bottom profiles. A least square fit of the longshore current solution inside the mean breakerline resulted in an optimal mean bed shear stress coefficient, $c_f = 0.009$, for the actual beach, $N = 0$ case (Table 15-3) averaged over four days. Including eddy viscosity acted to smooth the longshore current distribution, but does not significantly improve the linear model fits. The eddy viscosity is described using the Longuet-Higgins formulation; the optimal N values were small with an average of 0.002.

The effect of including nonlinear τ_y^b in the formulation is significant, with increasing importance as wave height and V increase. The nonlinear τ_y^b decreased V_{max} (compared with using linear τ_y^b) by 11 to a maximum of 22 percent. The best fit c_f values for nonlinear τ_y^b are 0.006, significantly less than for linear τ_y^b, while the average N values were the same as for the linear case. The shape of longshore current distributions are virtually the same for linear and nonlinear τ_y^b after adjusting c_f. Because of its greater internal consistency, and fewer free parameters, the random wave driven model with nonlinear τ_y^b and N = 0 is our choice for narrow-band applications. For small angle of wave incidence and weak currents, the plane beach analytic solutions can be slightly more accurate than numerical solutions on real topography with or without eddy viscosity (Table 15-3). Thus, the analytic solutions may be useful for many situations.

Because of the lack of a theoretical formulation to describe the more general case of wideband wave days, an EOF analysis was performed on 17 data sections between 30 January and 15 February. The EOF decomposition of the mean longshore currents pattern showed that most of the current variation (93%) was contained in the first mode with a classical parabolic shape. The temporal expansion coefficients of the first mode were highly correlated (0.97) with fluctuations in S_{xy}^T. Qualitative comparison of $S_{xy}(f)$ and the associated longshore currents (e.g., Figures 15-14, 15-16, 15-17) suggest only a weak longshore current response to directional spectra which have small S_{xy}^T because of cancelling contributions from different quadrants. Therefore, it is concluded that the total radiation stress, and not the structure of $S_{xy}(f)$, is of primary importance in predicting the magnitude of longshore currents. The importance of S_{xy}^T in predicting longshore currents has implications about its importance to longshore sediment transport, as is pointed out in the next chapters.

15.6 Appendix

Average (68 minute) Wave Height and Current Data for 3-6 February

[1]SENSOR	FEB	START TIME	V (m/s)	H_{rms} (m)	DEPTH (m)	[2]OFFSHORE DISTANCE (m)
SXYW	3	849	--	0.50	9.27	308.40
COD	3	849	−0.04	0.49	6.51	128.60
CO1	3	849	0.08	0.49	3.80	80.70
PO1	3	849	0.08	0.60	3.80	80.70
CO3P	3	849	0.15	0.61	3.02	68.96
P12	3	849	--	0.68	2.28	57.16
CO7P	3	849	0.34	0.71	1.77	39.70
C11	3	849	0.35	0.68	1.65	35.12
C12	3	849	0.39	0.67	1.48	31.00
C14P	3	849	0.40	0.59	1.24	24.80
C15	3	849	0.50	0.60	1.11	21.88
C16P	3	849	0.44	0.54	0.94	18.75
C17	3	849	0.44	0.48	0.80	15.95
C18	3	849	0.35	0.50	0.65	12.99
C20	3	849	0.11	0.31	0.50	9.93
C21	3	849	0.23	0.40	0.37	6.93
SXYW	4	1042	--	0.52	9.26	311.86
COD	4	1042	−0.03	0.49	6.50	132.06
CO1	4	1042	0.06	0.52	3.79	84.16
PO1	4	1042	0.06	0.60	3.79	84.16
PO3	4	1042	--	0.60	3.03	72.42
P12	4	1042	--	0.67	2.37	60.62
CO7P	4	1042	0.17	0.66	1.70	43.16
C11	4	1042	0.32	0.63	1.56	38.58
C12	4	1042	0.38	0.59	1.42	34.46
C13	4	1042	0.44	0.64	1.34	31.31
C14P	4	1042	0.43	0.60	1.24	28.26
C15	4	1042	0.48	0.57	1.16	25.34
C16P	4	1042	0.46	0.50	1.01	22.21
C17	4	1042	0.45	0.44	0.89	19.41
C22	4	1042	0.46	--	0.89	19.36
C18	4	1042	0.44	0.40	0.78	16.45
C20	4	1042	0.34	0.35	0.65	13.39
C21	4	1042	0.36	0.36	0.52	10.39
SXYW	5	1101	--	0.40	9.11	309.83
COD	5	1101	−0.03	0.40	6.35	130.03
CO1	5	1101	0.04	0.43	3.64	82.13
PO1	5	1101	0.04	0.47	3.64	82.13
PO3	5	1101	--	0.49	2.90	70.39
P12	5	1101	--	0.55	2.21	58.59
CO4	5	1101	0.13	--	1.78	47.60
CO7P	5	1101	0.18	0.59	1.52	41.13
C11	5	1101	0.21	0.59	1.37	36.55
C12	5	1101	0.25	0.50	1.26	32.43
C13	5	1101	0.30	0.51	1.18	29.28
C14P	5	1101	0.29	0.48	1.12	26.23
C15	5	1101	0.34	0.47	1.07	23.31
C16P	5	1101	0.34	0.44	0.97	20.18
C17	5	1101	0.35	0.43	0.88	17.38
C18	5	1101	0.35	0.38	0.78	14.42
C20	5	1101	0.26	0.34	0.65	11.36
SXYW	6	1152	--	0.27	8.93	306.03
COD	6	1152	−0.03	0.26	6.16	126.23
CO1	6	1152	−0.00	0.25	3.45	78.33
PO1	6	1152	−0.00	0.28	3.45	78.33
CO3P	6	1152	0.02	0.28	2.79	66.59
P12	6	1152	--	0.30	2.22	54.79
CO7P	6	1152	0.05	0.34	1.38	37.33
C11	6	1152	0.04	0.37	1.18	32.75
C12	6	1152	0.08	0.31	1.03	28.63
C13	6	1152	0.14	0.32	0.95	25.48
C14P	6	1152	0.13	0.31	0.90	22.43
C15	6	1152	0.19	0.34	0.85	19.51
C16P	6	1152	0.18	0.27	0.77	16.38
C17	6	1152	0.15	0.27	0.71	13.58
C22	6	1152	0.16	0.24	0.70	13.23
C18	6	1152	0.20	0.23	0.65	10.58
C20	6	1152	0.15	0.20	0.50	7.53

[1]Sensor SXYW refers to measurements made at the S_{xy} west array. Prefix C refers to measurements using current meters, prefix P refers to measurements using pressure sensors, and prefix C with suffix P refers to H_{rms} averages using both the current meter and pressure sensor measurements.

[2]The distance offshore is measured relative to the mean shoreline during the measurement interval.

15.7 References

Battjes, J. A., 1972, Set-up due to irregular waves, *Proceedings*, Thirteenth Coastal Engineering Conference, July 10-14, 1972, Vancouver, B.C., Canada, American Society of Civil Engineers, New York: 1993-2004.

_____. 1974, Surf similarity, *Proceedings*, Fourteenth Coastal Engineering Conference, June 24-28, 1974, Copenhagen, Denmark, American Society of Civil Engineers, New York: 466-480.

Collins, J. I., 1972, Probabilities of breaking wave characteristics, *Proceedings*, Thirteenth Coastal Engineering Conference, July 10-14, 1972, Vancouver, B.C., Canada, American Society of Civil Engineers, New York: 399-412.

Ebersole, B. A. and R. A. Dalrymple, 1980, Numerical modeling of nearshore circulation, *Proceedings*, Seventeenth Coastal Engineering Conference, March 23-28, 1980, Sydney, Australia, American Society of Civil Engineers, New York: 2710-2725.

Galvin, C. J. and P. S. Eagleson, 1965, Experimental study of longshore currents on a plane beach, U. S. Army Coastal Engineering Research Center, Technical Memo 10.

Gerald, C. F., 1978, *Applied Numerical Analysis*, Addison-Wesley.

Guza, R. T., E. B. Thornton and Niels Christensen, Jr., 1986, Observations of steady longshore currents in the surf zone, *Journal of Physical Oceanography*, 16(11): 1959-1969.

Higgins, A. L., R. J. Seymour and S. S. Pawka, 1981, A compact representation of ocean wave directionality, *Applied Ocean Research*, 3: 105-111.

Kirby, J. T., Jr. and R. A. Dalrymple, 1982, Numerical modeling of the nearshore region, ONR Technical Report No. 11, Ocean Engineering Report 26, Department of Civil Engineering, University of Delaware, Newark, DE.

Kundu, P. K., J. S. Allen and R. L. Smith, 1975, Model decomposition of the velocity field near the Oregon Coast, *Journal of Physical Oceanography*, 5: 683-704.

Longuet-Higgins, M. S., 1970, Longshore currents generated by obliquely incident sea waves, *Journal of Geophysical Research*, 75: 6778-6801.

Noda, E. K., 1974, Wave-induced nearshore circulation, *Journal of Geophysical Research*, 79: 4097-4106.

Oltman-Shay, Joan and R. T. Guza, 1984, A data-adaptive ocean wave directional-spectrum estimator for pitch and roll type measurements, *Journal of Physical Oceanography*, 14(11): 1800-1810.

Sasaki, T. and K. Horikawa, 1975, Nearshore current system on a gently sloping bottom, *Coastal Engineering in Japan*, 18: 123-142.

Thornton, E. B., 1979, Energetics of breaking waves within the surf zone, *Journal of Geophysical Research*, 84: 4931-4938.

Thornton, E. B. and R. T. Guza, 1986, Surf zone longshore currents and random waves: field data and models, *Journal of Physical Oceanography*, 16(7): 1165-1178.

Wu, C.-S. and P. L.-F. Liu, 1985, Finite element analysis of nonlinear nearshore currents, *Journal of Waterway, Port, Coastal and Ocean Division*, American Society of Civil Engineers, 111: 417-432.

Wu, C.-S., E. B. Thornton and R. T. Guza, 1985, Waves and longshore currents: comparison of a numerical model with field data, *Journal of Geophysical Research*, 90: 4951-4958.

Chapter 16

STATE OF THE ART
IN OSCILLATORY
SEDIMENT TRANSPORT MODELS

David B. King Jr. and Richard J. Seymour
Scripps Institution of Oceanography

The chapters following will present the models for longshore and cross-shore surf zone transport that have been developed from the analyses of NSTS data. These are global models in which net transport is integrated in space and time across the surf zone. This chapter is intended to form an introduction to them by reviewing existing models for gross sediment transport, i.e., transport at a point over a half cycle of oscillation. None of these point models were developed within the NSTS program, but all of the integrated NSTS net transport models have been developed from them.

This review has been restricted to oscillatory sediment motion on a horizontal bed in a single plane. All velocities are assumed to be sinusoidal. Only models which relate the bedload or total load transport rate to local characteristics of the flow field, such as the local bottom shear stress, have been included. Specifically excluded are models which predict only the net direction of transport and those which relate transport rate to such parameters as wave height, wave length, etc. Models which only deal with suspended load transport have also been excluded.

16.1 Notation

Traditionally, it has been found necessary to use seven independent dimensional variables to define sediment transport rates in oscillatory flow. The ones used in this chapter will be:

a -- the horizontal fluid excursion amplitude just outside the boundary layer (L),

D -- the geometric mean grain diameter (L),

g -- the acceleration of gravity (L/T^2),

U -- the maximum oscillatory velocity just outside the boundary layer (L/T),

v -- the kinematic viscosity (L^2/T),

ρ -- the density of the fluid (M/L^3), and

ρ_s -- the density of the sediment (M/L^3).

These are used to define other dimensional variables used in this chapter:

τ, the bottom shear stress (M/LT^2)

$$\tau = (1/2)f_w \rho U^2$$

where f_w is the dimensionless Jonsson (1966) friction factor,

w, the grain fall velocity (L/T)

$$w = \left[(4/3C_D)(s-1)gD\right]^{1/2},$$

where C_D is the drag coefficient on a grain,

W, the rate at which energy is dissipated by bottom friction (M/T^3)

$$W = \tau U,$$

and ω, the wave frequency ($1/T$).

$$\omega = U/a.$$

The seven primary dimensional variables can be combined to form four independent dimensionless parameters in several ways. The four primary dimensionless parameters used in this chapter are:

a/D, an inverse Strouhal number, sometimes called the Keulegan-Carpenter number, the ratio of drag to inertia forces on a grain,

s, a density ratio $= \rho_s/\rho$,

D^*, a dimensionless grain number

$$D^* = \left[\frac{(s-1)g}{v^2}\right]^{1/3} D$$

the ratio of buoyancy to viscous forces on a grain, and

θ, the Shield's parameter

$$\theta = \frac{\tau}{\left[(\rho_s - \rho)gD\right]}$$

the ratio of drag to buoyancy forces on a grain. Several modified Shield's parameters are discussed in the text. θ' is used to designate those based upon velocity rather than shear stress.

$$\theta' = \frac{U^2}{(s-1)gD}$$

Since values of f_w are typically in the range of 0.01 to 0.07, values of θ are a few percent of the values of θ'.

Authors have usually used one of two dimensional variables to express the sediment transport rate: i, the immersed weight sediment transport rate per unit width (M/T^3), or Q, the grain volume transport rate per unit width (L^2/T). These are related as

$$i = (s-1)\rho g Q,$$

or equivalently

$$i = (\frac{\tau}{\theta D}) Q.$$

16.2 Sediment Transport Regimes

In the nearshore environment, there are two major sediment transport regimes; the ripple regime and the sheetflow regime. [For a discussion, see Dingler and Inman (1976)]. The division between these two zones is more properly considered as a transition zone than a sharp boundary. It occurs somewhat seaward of the breaker line. Several authors have proposed criteria for the division between these two regimes. Most of the formulas are similar. In the following discussion, it will be easiest to use the criterion of Nielsen (1979), $\theta \approx 1$. The ripple regime occurs between $\theta = \theta_c$ and $\theta = 1$, where θ_c is the Shield's parameter for the threshold of motion. (While θ_c is not a constant, a typical value of 0.04 will be used in this chapter). The sheetflow regime occurs for values of $\theta > 1$. For extreme surf zone conditions, θ may reach maximum values of 5 to 10. Other transition criteria have been given by Manohar (1955), Kennedy and Falcon (1965), Carstens (1966), Carstens et al. (1969), Dingler (1974), Komar and Miller (1974), and Lofquist (1978). These other criteria are typically mostly functions of a/D and θ, since these two dimensionless parameters change substantially across the surf zone, while s and D^* remain relatively constant.

Equilibrium ripples take considerable time to become established. Therefore, transport experiments can be conducted in the laboratory on flat beds under flow conditions within the ripple regime before the ripples are formed. There is disagreement as to whether the transport rates on rippled and flat beds are equal for the same flow conditions. Some investigators have found that their model predictions did not disagree with both flat bed and rippled bed transport data. Thus the implication is that the transport rate is independent of bed condition. However, the mechanisms of transport are quite different. [For a discussion of the complicated transport mechanism associated with ripples, see Bagnold (1946), Inman and Bowen (1962), or Sleath (1982)]. Therefore, it would appear to be serendipitous if transport rates were unaffected by bed condition. The critical experiments to resolve this question have not been performed.

16.3 Transport Measurements

In a comprehensive survey, Hallermeier (1982) discussed 20 oscillatory flow bedload sediment transport data sets. These sets, by a number of authors using various techniques, contain over 700 data points. Essentially all of these studies were performed in the laboratory. The sediment transport rates in these experiments were variously measured with traps, with photographs, or with sediment tracer. The flow fields in these experiments were produced either in wave flumes or by using an oscillating bed in a tank of otherwise still water. Hallermeier (1982) formulated a model using all 20 data sets but with particular emphasis on only a few. His model, and the others discussed in this chapter, emphasised or used only the data of Manohar (1955), Kalkanis (1964), Abou-Seida (1965), and/or Sleath (1978) in their calibration.

The experiments of Manohar (1955), Kalkanis (1964) and Abou-Seida (1965) were all performed using the same facility. These experiments used an oscillatory bed and measured the sand transport with traps. In Manohar's transport study the motion of the plate was periodic, but not sinusoidal. He measured both sediment transport and ripple velocity, so it is possible to determine from his data when bed forms were present (about half the time). The rest ranged throughout the ripple regime. Kalkanis (1964) and Abou-Seida (1965) measured transport rates under sinusoidal flow conditions. However, they did not report whether bedforms were present.

Sleath (1978) also conducted a sediment transport experiment using an oscillating plate. To measure transport rate, he photographed and counted individual grains as they passed the end of the bed. Thus, he was able to measure instantaneous transport rate. However, this technique limited him to a single layer of moving sediment, and thus to very weak flow conditions (only slightly above threshold). All of his experiments were performed on flat beds.

The ranges of their experimental data for typical sand density grains are shown in Table 16-1. It should be noted that the range for each parameter is shown but that not all combinations of parameter space were examined. This table shows that essentially all of the experiments of these investigators were done in the ripple regime (i.e., $\theta < 1$). Kalkanis and Sleath in particular performed their experiments with very gentle flows, i.e., velocities just above the threshold of motion. Essentially none of the other data points mentioned by Hallermeier (1982) were in the sheetflow regime either. This is unfortunate, but understandable. Sheetflow conditions are difficult to produce in the laboratory

Table 16-1. Range of Experimental Data

	Number of Exp.	a/D	D^*	θ	Bedforms
Manohar	142	300-2000	7.6-27	0.12-1.2	Rippled and Flat
Kalkanis	27	81-230	44-74	0.021-0.15	?
Abou-Seida	46	94-2600	3.8-69	0.76-0.81	?
Sleath	22	6.1-53	44-100	0.041-0.11	Flat

using quartz density grains. Recently, Horikawa *et al.* (1982) have reported a few data points for θ values in the range of 1 to 3.

16.4 Bagnold (1963) Model

Bagnold derived a theoretical steady flow model in which the sediment transport rate is proportional to the energy dissipated per unit of bottom area (Bagnold 1956, 1966.) His oscillatory flow model (Bagnold, 1963) was derived from this steady flow model. His oscillatory flow total load transport model for zero bed slope is

$$i = W(K_b + K_s) \tag{16-1}$$

$K_b = \varepsilon_b/\tan\phi$ where ε_b is an efficiency factor (a constant less than one), and $\tan\phi$ is the ratio of the tangential and the normal stresses along a shear plane in the sediment. Bailard and Inman (1979) equate $\tan\phi$ with the angle of repose of the sediment. $K_s = \varepsilon_s(1-\varepsilon_b)\,U/w$ where ε_s is another efficiency factor, again less than 1. The subscripts b and s refer to bedload and suspended load, respectively.

The only investigators who have attempted to calibrate Bagnold's model in the context of cross-shore transport were Inman and Bowen (1962). Their experiments were carried out on a rippled bed and were inconclusive. They concluded that Bagnold's model is only applicable to flat bed conditions. Their values of K_b ranged from 0.01 to 0.3.

A number of investigators have started with Bagnold's formulation (frequently only his bedload model) for transport at a point and developed global net transport models. Some of these are longshore transport models which have been calibrated using field data. [e.g., Komar and Inman (1970), CERC (1973) and Inman *et al.* (1980)]. Bowen (1980), Bailard (1981), and Bailard and Inman (1981) present models for cross-shore transport in generally unsymmetrical flows which reduce to Bagnold's formulation under sinusoidal flow conditions.

16.5 Einstein (1972) Model

The models developed by Kalkanis (1964) and Abou-Seida (1965) were summarized by Einstein (1972). They were based upon the steady transport model of Einstein (1950). The model that Einstein (1972) presented was in the form of a double integral with three calibration constants. It was solved numerically using the data of Kalkanis and Abou-Sieda to determine appropriate values for the constants. The result of the integration was presented in graphical form (see his Figure 6) and showed a relationship between the dimensionless transport rate

$$\frac{Q}{[(s-1)gD^3]^{.5}} \tag{16-2}$$

($= \Phi$ in Einstein's notation) and a modified Shield's parameter

$$\theta_a' = \frac{U_a^2}{(s-1)gD} \tag{16-3}$$

($= 1/\psi$ in Einstein's notation). U_a is the maximum velocity at a distance of 0.35 D above the bed. This velocity was chosen based upon experimental work done by Einstein and El Samni (1949). Einstein did not discuss the applicability of this model to a rippled or flat bed.

16.6 Madsen and Grant (1976) Model

Madsen and Grant (1976) proposed a model based upon the empirical work of Brown (1949). The Brown (1949) sediment transport relationship for steady flow conditions is

$$\frac{Q}{wD} = 40 \,\overline{\theta}^3 \tag{16-4}$$

Here, $\overline{\theta}$ is the Shield's parameter for steady flow. Madsen and Grant assumed a time varying shear stress of the form

$$\tau(t) = .5f_w \rho \,|u(t)|u(t) \tag{16-5}$$

where $u(t)$ is the time varying horizontal velocity at the top of the bottom boundary layer. Using an instantaneous Shield's parameter, Madsen and Grant integrated Brown's formula over the portion of a half wave cycle when the velocity was above critical. Thus their model is

$$\frac{Q}{wD} = C\theta^3 \tag{16-6}$$

For θ less than or equal to θ_c, $C = 0$. For θ greater than about four times θ_c, they found that the sediment was essentially always in motion and $C = 12.5$. They presented a table (their Table 2) showing how their coefficient increased as θ grew above θ_c. They compared their model with the data of Kalkanis (1964) and Abou-Seida (1965) (their Figure 7). More recently, Horikawa et al. (1982) have shown data that are in agreement with Madsen and Grant's model. By comparing their model with some of the data of Manohar (1955), Madsen and Grant claimed that their model was valid for ripple and non-ripple conditions.

16.7 Sleath (1978) Model

Sleath (1978) presented a model without theoretical development which was justified by a graphical comparison of the model with his data, plus some data of Kalkanis and Abou-Seida. His model is

$$\frac{Q}{\omega D^2} = 47 \ (\theta^* - \theta_c^*)^{3/2} \tag{16-7}$$

where θ^* is a Shield's parameter, $\theta^* = .5f_1 U^2/ (s-1)gD = \theta f_1 / f_w$, f_1 is a coefficient of friction graphically defined in his paper (similar to f_w), and θ_c^* is his Shield's parameter for threshold of motion. He assumed that most of the data of Abou-Seida were for ripple beds. His model showed better agreement with his own data and that of Kalkanis, and Sleath thus concluded that his model was a flat bed transport model. Sleath (1982) made detailed measurements of bedload transport rate over ripple crests. Comparing these data to his model, he found poor agreement.

16.8 Shibayama and Horikawa (1980) Model

Shibayama and Horikawa (1980) started with the Brown (1949) steady flow model (16-4), following Madsen and Grant (1976). However, they integrated the time varying form of Brown's equation in a different way. They assumed that once a grain started to move, it would not stop until the flow had reversed direction. Therefore, their equation has a different coefficient than the equation of Madsen and Grant (1976). It is

$$\frac{Q}{wD} = 19 \ \theta^3 \tag{16-8}$$

They compared their model with a number of data points from their own experiments on flat beds. Also Shibayama and Horikawa (1982) compared this model with the data of Horikawa *et al.* (1982).

16.9 Hallermeier (1982) Model

Hallermeier (1982) presented an empirical model that was similar in form to Sleath's. The major difference was the absence of a threshold of motion term in Hallermeier's model. His model is

$$\frac{Q}{\omega D^2} = .03 \ \theta'^{3/2} \tag{16-9}$$

This model was calibrated graphically by selectively using the 20 data sets discussed above. Since these data sets included both rippled and flat beds, he concluded that his model was valid in both regions.

16.10 Kobayashi (1982) Model

Kobayashi (1982) proposed a model in which he equated the instantaneous transport rate with the number of grains moving times the mean particle velocity. He based his development upon the work of Kalinske (1947) and Bagnold (1956). His equations predict the instantaneous two dimensional transport rate over a gently sloping bed. This model was integrated over a half sine wave on a horizontal bed to obtain an average transport rate equation. This equation is

$$
\frac{Q}{wD} = 3.20 \, (\theta - \theta_c)^{.5} \left[1 - (1.5) \frac{\theta_c}{\theta} \right.
$$

$$
\left. + 0.53 \left[\frac{\theta}{\theta_c} - 1 \right]^{-.5} \left[\frac{2\theta_c}{\theta} - 1 \right] \cos^{-1} \left[\frac{\theta_c}{\theta} \right]^{.5} \right]
\qquad (16\text{-}10)
$$

Kobayashi compared his instantaneous model with the instantaneous data of Sleath. He compared the model above with the data of Kalkanis and Abou-Seida. He stated that his model was valid for flat beds and suggested how it could be extended to include rippled beds.

16.11 Comparison of Models

Figure 16-1 shows a comparison of the seven bedload models for a condition typical of the ones for which these models have been calibrated (i.e., gentle, laboratory type flows). To compare the model of Bagnold (1963) with the others in Figures 16-1 and 16-2, values for his coefficients were needed. For simplicity, $\varepsilon_b = 0.1$ and $\varepsilon_s = 0.0$ were used. Predictably, all of the models agree reasonably well under conditions very similar to those used for calibration. Figure 16-2 shows the model predictions for typical surf zone conditions. For example, at a value of $\theta = 1$ and D = 0.2 mm, Figure 16-2 implies a wave period of 12 seconds and a maximum horizontal water particle full excursion of 4 m. This would result from a wave of 1 m height in a depth of 2 m. At this point, the model predictions cover a range of two orders of magnitude. At larger values of θ, the discrepancies between the models expand to almost four orders of magnitude.

In all of the bedload models mentioned above, there is an analytic relationship between sediment transport rate and grain, fluid, and flow parameters except for the model of Einstein (1972). The rest can be compared when put in the same form. For this purpose, $\theta >> \theta_c$ is assumed. The six remaining models then become

$$
\frac{i}{W} = 10 \left[\frac{f_w}{C_D} \right]^{1/2} \theta^{3/2} \qquad [\textit{Madsen and Grant} \ (1976)] \qquad (16\text{-}12a)
$$

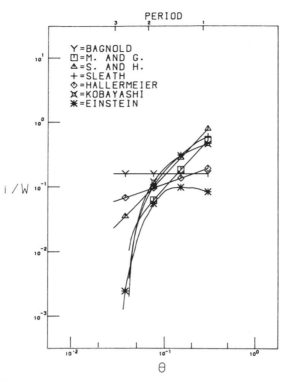

Figure 16-1. Comparison of oscillatory sediment transport models for $D^* = 70$, $a/D = 50$.

$$\frac{i}{W} = 16\left[\frac{f_w}{C_D}\right]^{1/2} \theta^{3/2} \quad [\textit{Shibayama and Horikawa (1980)}] \quad (16\text{-}12\text{b})$$

$$\frac{i}{W} = 47\left[\frac{D}{a}\right]\left[\frac{f_1}{f_w}\right]^{3/2} \theta^{1/2} \quad [\textit{Sleath (1978)}] \quad (16\text{-}12\text{c})$$

$$\frac{i}{W} = .09\left[\frac{D}{a}\right]\left[\frac{1}{f_w}\right]^{3/2} \theta^{1/2} \quad [\textit{Hallermeier (1982)}] \quad (16\text{-}12\text{d})$$

$$\frac{i}{W} = K_b \quad [\textit{Bagnold (1963)}] \quad (16\text{-}12\text{e})$$

$$\frac{i}{W} = 2.6\left[\frac{f_w}{C_D}\right]^{1/2} \quad [\textit{Kobayashi (1982)}] \quad (16\text{-}12\text{f})$$

In these models C_D is dependent upon D^*, f_w is dependent upon a/D, f_1 is dependent upon D^*, a/D, and θ, and K_b is dependent upon $\tan\phi$. However, the models are not strongly dependent upon D^* or a/D. [If f_w is replaced by a/D using either the approximation of Kamphuis (1975) ($f_w \propto (a/D)^{-3/4}$), or Nielsen

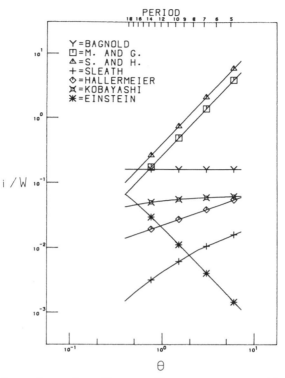

Figure 16-2. Comparison of oscillatory sediment transport models for $D^* = 5$, $a/D = 10,000$.

(1979) $(f_w \propto (a/D)^{-1/3})$, the models are seen to depend upon $(a/D)^n$, where n ranges from $-1/2$ to $1/8$]. The major difference in the formulas is seen to be in the power of the Shield's parameter, θ. Since θ is proportional to $(velocity)^2$ and W is proportional to $(velocity)^3$, the models of Madsen and Grant (1976) and Shibayama and Horikawa (1980) predict that:

$$\text{transport rate} \propto (velocity)^6,$$

the models of Sleath (1978) and Hallermeier (1982) predict that:

$$\text{transport rate} \propto (velocity)^4,$$

and the models of Bagnold (bedload) (1963) and Kobayashi (1982) predict that:

$$\text{transport rate} \propto (velocity)^3.$$

Since Bagnold's suspended load model has another velocity term in it, his total load transport model predicts that transport is proportional to a power of velocity between three and four.

In Einstein's (1972) model, there is not a direct relationship between transport rate and $(velocity)^n$. However, for the conditions given in Figure 16-2 it is seen that his model predicts that transport rate is approximately independent of velocity. This large variation in the exponent of velocity between models is unsatisfactory, but it is also characteristic of steady flow model comparisons.

It should be pointed out that there is an ambiguity in the relationship between transport rate and velocity, depending upon the parameters one chooses for non-dimensionalizing. Sleath and Hallermeier (1978) used ω rather than a as a primary variable and thus the transport relationship in their notation showed a $(velocity)^3$ dependence. One of the reasons for the choice of parameters in this chapter was because this choice lumped much disagreement in the models into one parameter, (θ) and therefore the differences could be more easily visualized. The important point to note is that a change of parameters will not reduce the disagreements in the models.

16.12 Discussion of Models

The models discussed above are generally of two types. The models of Bagnold (1963), Einstein (1972) and Kobayashi (1982) are theoretical, while the models of Madsen and Grant (1976), Sleath (1978), Shibayama and Horikawa (1980) and Hallermeier (1982) are empirical or based upon other empirical models. The empirical models would not be expected to have great predictive power outside the range for which they were calibrated.

The theoretical models of Einstein (1972) and Kobayashi (1982) are based upon a model of the fluid boundary layer. However, Bagnold (1956) suggested that at high sediment concentrations a granular-fluid boundary layer behaves very differently than a fluid boundary layer. For instance, he argued that the shear stresses in the boundary layer were essentially completely transmitted by grain to grain contact. Therefore, the models of Einstein (1972) and Kobayashi (1982) would be expected to have the greatest predictive value at very low sediment transport rates, where the grains would not be in high enough concentrations to affect the fluid boundary layer. The model of Bagnold (1963) attempts to incorporate much of the underlying physics, however, it is incomplete. The efficiency factor is assumed to be a constant, but there is little justification for this assumption.

A few other comments about the models deserve mention. Three of the models, Madsen and Grant (1976), Sleath (1978) and Kobayashi (1982), incorporate the effects of the threshold of motion. That is, they all predict zero transport at finite fluid velocities in agreement with observation. However, they do this in different ways. Figure 16-3 shows how the threshold term becomes less important as θ increases above θ_c for these three models. The vertical axis is the transport rate divided by what the transport rate would be if the threshold term were ignored. The horizontal axis is θ, or for Sleath's model, θ^*.

Sleath's model incorporates the threshold term in the way that it is done in many steady flow transport rate models, i.e., transport $\propto (\theta - \theta_c)^n$. If one believes that, above threshold, transport rate is proportional to θ^n, then this

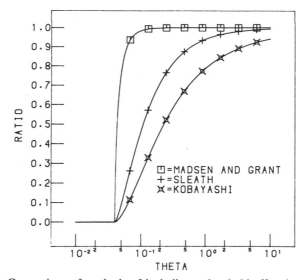

Figure 16-3. Comparison of methods of including a threshold effect in sediment transport equations. The vertical axis is a ratio of the model prediction to what the same model would predict without its threshold term. The horizontal axis is the Shield's parameter, θ; or for Sleath's model, a modified Shield's parameter, $\theta*$. $\theta_c = \theta_c^* = 0.04$ is assumed.

method of incorporation of the threshold term is clearly inappropriate. Kobayashi's threshold term dies away even more slowly.

In contrast, Madsen and Grant's threshold effect is incorporated in a coefficient whose value is derived mathematically from a physical understanding of the problem. Their threshold effect dies away very rapidly as one would expect. In fact Sleath (1978) has shown that the effects of threshold die away even more rapidly than is modeled by Madsen and Grant. Using his instantaneous measurements of transport rate, he has shown that for values of θ just above θ_c, transport occurs over more of the cycle than just during the time when $u(t) > U_c$ (where U_c is the threshold velocity). He attributes this to the fact that if grains are transported, some will be deposited in very unstable positions at the end of a cycle and will be available for transport in the opposite direction at flow rates below nominal U_c.

The effects of these threshold terms are evident in Figures 16-1 and 16-2. Madsen and Grant's model becomes parallel to Shibayama and Horikawa's model at low values of θ. However, only at very high values of θ does Sleath's model become parallel to Hallermeier's model or does Kobayashi's model become parallel to Bagnold's (horizontal).

There are very few experiments that have been performed to help resolve the uncertainty in the exponent of the velocity. The most important of these are the measurements of Sleath (1978) on instantaneous transport rates. For a velocity field of $u(t) = U \cos(\omega t)$ he found that sand and gravel transport rates fit a $\cos^4$ curve very closely.

$$q(t) = \left[\frac{8}{3}\right] Q \ \cos^3(\omega t + \alpha) |\cos(\omega t + \alpha)| \qquad (16\text{-}15)$$

where, $q(t)$ is the instantaneous transport rate per unit width and α is a slight phase lead which was always less than $\pi/8$. Thus, at very low transport rates on flat beds, there is strong empirical evidence that a fourth power law is most appropriate.

16.13 Conclusions

It is evident that there is still much work to be done in establishing oscillatory sediment transport models that are useful in the surf zone. A much more complete understanding of the granular mechanics involved in sediment transport must be combined with more high quality data sets under a broader range of conditions. Until one of these models is shown to be broadly applicable, researchers must establish criteria for determining which model best meets their needs.

16.14 References

Abou-Seida, M. M., 1965, Bedload function due to wave action, University of California, Berkeley, HEL 2-11, 78 pp.

Bagnold, R. A., 1946, Motion of waves in shallow water, interaction between waves and sand bottoms, *Proceedings*, Royal Society of London, Series A (187): 1-18.

_____. 1956, The flow of cohesionless grains in fluids, Philosophic Transcript, Royal Society of London, Series A (249): 239-297.

_____. 1963, Mechanics of marine sedimentation, *The Sea*, Volume 3, Wiley-Interscience, New York: 507-528.

_____. 1966, An approach to the sediment transport problem from general physics, Geological Survey Professional Paper 422-I, U. S. Department of the Interior.

Bailard, J. A., 1981, An energetics total load sediment transport model for a plane sloping beach, *Journal of Geophysical Research*, 86: 10938-10954.

Bailard, J. A. and D. L. Inman, 1979, A reexamination of Bagnold's granular fluid model and bedload transport, *Journal of Geophysical Research*, 84(C12) Paper 8C1218: 7827-7833.

_____. 1981, An energetics bedload model for a plane sloping beach: local transport, *Journal of Geophysical Research*, 86(C3) Paper 80C1291, March 20: 2035-2043.

Bowen, A. J., 1980, Simple models of nearshore sedimentation; beach profiles and longshore bars, *The Coastline of Canada*, S. B. McCann, ed, Geological Survey of Canada, Paper 80-10: 1-11.

Brown, C. B., 1949, Sediment transportation, *Engineering Hydraulics*, Chapter XII, Sections A-C, Proceedings of Fourth Hydraulics Conference, Iowa Institute of Hydraulic Research, H. Rouse, ed., John Wiley & Sons, Inc., New York: 769-804.

Carstens, M. R., 1966, Similarity laws for localized scour, *Journal Hydraulics Division*, American Society of Civil Engineers, 92(HY3) May: 13-36.

Carstens, M. R., F. M. Neilson and H. D. Altinbilek, 1969, Bed forms generated in the laboratory under an oscillatory flow: analytical and experimental study, TM-28, Coastal Engineering Research Center, U. S. Army Corps of Engineers, June, 83 pp.

Coastal Engineering Research Center, 1973, *Shore Protection Manual*, U. S. Army Corps of Engineers, Washington, D. C., 3 volumes.

Dingler, J. R., 1974, Wave-formed ripples in nearshore sands, Ph.D. dissertation, Department of Oceanography, University of California, San Diego, California, 136 pp.

Dingler, J. R. and D. L. Inman, 1976, Wave-formed ripples in nearshore sands, *Proceedings*, Fifteenth Coastal Engineering Conference, July 11-17, 1976, Honolulu, Hawaii, American Society of Civil Engineers, New York: 2109-2126.

Einstein, H. A., 1950, Bedload function for sediment transportation in open channel flows, U. S. Department of Agriculture, S.C.S. Technical Bulletin No. 1026.

_____. 1972, A basic description of sediment transport on beaches, *Waves on beaches and resulting sediment transport*, R. E. Meyer, ed., Academic Press, New York: 53-92.

Einstein, H. A. and A. El Samni, 1949, Hydrodynamics forces on a rough wall, *Review of Modern Physics*, 21(3): 520-524.

Hallermeier, R. J., 1982, Oscillatory bedload transport: data review and simple formulatoin, *Continental Shelf Research*, 1(2): 159-190.

Horikawa, K., A. Watanabe and S. Katori, 1982, Sediment transport under sheet flow conditions, *Proceedings*, Eighteenth Coastal Engineering Conference, November 14-19, 1982, Cape Town, Republic of South Africa, American Society of Civil Engineers, New York: 1335-1352.

Inman, D. L. and A. J. Bowen, 1962, Flume experiments on sand transport by waves and currents, *Proceedings*, Eighth Conference on Coastal Engineering, Council on Wave Research, J. W. Johnson, ed., Berkeley, California, November, 1962, Mexico City: 137-150.

Inman, D. L., J. A. Zampol, T. E. White, B. W. Waldorf, D. M. Hanes and K. A. Kastens, 1980, Field measurements of sand motion in the surf zone, *Proceedings*, Seventeenth Coastal Engineering Conference, March 23-28, 1980, Sydney, Australia, American Society of Civil Engineers, New York: 1215-1234.

Jonsson, I. G., 1966, Wave boundary layers and friction factors, *Proceedings*, Tenth Conference on Coastal Engineering, September, 1966, Tokyo, Japan, American Society of Civil Engineers, New York: 127-148.

Kalinske, A. A., 1947, Movement of sediment as bed in rivers, *Transcript, American Geophysical Union*, 28: 615-620.

Kalkanis, G., 1964, Transport of bed material due to wave action, CERC TM-2.

Kamphuis, J. W., 1975, Friction factor under oscillatory waves, *Journal Waterways, Harbors and Coastal Engineering Division*, American Society of Civil Engineers, 101(WW2) May: 135-144.

Kennedy, J. F. and M. Falcon, 1965, Wave generated sediment ripples, MIT Hydraulics Lab, Report #86, August.

Kobayashi, N., 1982, Sediment transport on a gentle slope due to waves, *Journal Waterway, Port, Coastal and Ocean Division*, American Society of Civil Engineers, 108(WW3): 254-271.

Komar, P. D. and D. L. Inman, 1970, Longshore sand transport on beaches, *Journal of Geophysical Research*, 75(30): 5914-5927.

Komar, P. D. and M. C. Miller, 1974, Sediment threshold under oscillatory waves, *Proceedings*, Fourteenth Coastal Engineering Conference, June 24-28, 1974, Copenhagen, Denmark, American Society of Civil Engineers, New York: 756-775.

Lofquist, K. E. B., 1978, Sand ripple growth in an oscillatory-flow water tunnel, Coastal Engineering Research Center, TP 78-5, August, 103 pp.

Madsen, O. S. and W. D. Grant, 1976, Sediment transport in the coastal environment, MIT Report No. 209, January, 120 pp.

Manohar, M., 1955, Mechanics of bottom sediment motion due to wave action, BEB TM-75, June.

Nielsen, P., 1979, Some basic concepts of wave sediment transport, Series Paper 20, ISVA, Technical University, Denmark.

Shibayama, T. and K. Horikawa, 1980, Bedload measurement and prediction of two-dimensional beach transformation due to waves, *Coastal Engineering in Japan*, 23: 179-190.

_____. 1982, Sediment transport and beach transformation, *Proceedings*, Eighteenth Coastal Engineering Conference, November 14-19, 1982, Cape Town, Republic of South Africa, American Society of Civil Engineers, New York: 1439-1458.

Sleath, J. F. A., 1978, Measurements of bedload in oscillatory flow, *Journal Waterway, Port, Coastal and Ocean Division*, American Society of Civil Engineers, 104(WW4) Paper 13960, August: 291-307.

_____. 1982, The suspension of sand by waves, *Journal of Hydraulic Research*, 20(5): 439-452.

Chapter 17

MODELING CROSS-SHORE TRANSPORT

Richard J. Seymour and David Castel
Institute of Marine Resources
University of California

Seymour and King (1982) evaluated a number of existing models for cross-shore transport by testing their skill in predicting the observed beach excursions at Torrey Pines. The results were discouraging, in that none of the models showed a useful skill level. As discussed in Chapter 12, the Torrey Pines data set may not have provided a realistic test for these models. Therefore, the other three data sets described in Chapter 12 (Scripps Beach, Santa Barbara and Virginia Beach) were employed to evaluate models for cross-shore transport.

Seymour and King (1982) attempted to review the historical literature and to identify all proposed mechanisms. Some of these references were anecdotal in character and did not lend themselves to testing against field data. In this study, we have included only those models that will predict some measurable attribute of the shoreline change. In addition, we have included some additional models that were not treated in the earlier work. The models tested here fall into three categories. The first predicts threshold conditions, that is, the combination of beach and incident wave conditions that defines the transition between onshore and offshore transport. The second category is comprised of numerical models that predict the beach profile as a function of time, given the incident wave and sea level histories. Finally, a third category was considered of models that predict the slope or the general shape of the beach.

17.1 Threshold Models

Six models were tested for their ability to predict whether the wave-driven cross-shore transport would be towards or away from the beach. Four of these had been evaluated in Seymour and King (1982): Dean (1973), the two models contained in Short (1978), and Hattori and Kawamata (1980). The fifth, Quick and Har (1985), and the sixth, Sunamura and Horikawa (1974), were added for this study. Table 17-1 describes the characteristics of each model.

It can be noted from Table 17-1 that several of the models are quite similar. The Dean formulation, by simple substitutions, can be reduced to a factor $H/(wT)$ in which H is the significant wave height, w is the fall speed of the

Table 17-1. Models for Predicting Threshold of Cross-shore Transport

MODEL	FORMULATION		
Dean (1973)	$(2R)\dfrac{H}{wT}$	$<$	1 ONSHORE
		$>$	1 OFFSHORE
Short (1979) Height model	H	$<$	120 cm ONSHORE
		$>$	120 cm OFFSHORE
Short (1979) Power model	$\left[\dfrac{\rho g^2}{16\pi}\right]H^2 T$	$<$	30 Kw/m ONSHORE
		$>$	30 Kw/m OFFSHORE
Hattori and Kawamata (1980)	$2\,\dfrac{H\beta}{wT}$	$<$	0.5 ONSHORE
		$>$	0.5 OFFSHORE
Quick and Har (1985)	$\left[\dfrac{H}{wT}\right]_{INITIAL}$	$>$	$\left[\dfrac{H}{wT}\right]_{FINAL}$ ONSHORE
		$<$	$\left[\dfrac{H}{wT}\right]_{FINAL}$ OFFSHORE
Sunamura and Horikawa (1974)	$\dfrac{1.845}{g^{0.33}}\,H\dfrac{\beta^{0.27}}{(Td)^{0.67}}$	$<$	4 ONSHORE
		$>$	8 OFFSHORE

where: R = arbitrary constant
H = deep water significant wave height
w = sediment fall speed
T = period of spectral peak
g = gravitational constant
β = beach slope
ρ = fluid density
d = diameter of sand grains

median sediment size and T is a characteristic period of the incident waves. If this factor exceeds some threshold, erosion occurs. Although this is often referred to as a dimensionless fall speed in the literature, it will be called the wave-sediment parameter, or WSP, here. Hattori and Kawamata modify the Dean criterion by multiplying the WSP with the beach slope. Quick and Har compares the WSP on the previous day with the one being investigated. If the WSP today is bigger than the one yesterday, the beach is predicted to erode.

To evaluate these models, time histories of shoreline volume change were calculated from the data sets of Chapter 12. For each set, the profile change relative to the previous day was summed to yield a net volume change expressed in square meters (or cubic meters per meter). The rationale for interpreting these changes as cross-shore transport induced, rather than as caused by gradients in longshore transport, is described in Chapter 12. The sign convention gave negative values for erosion or decreases in volume. These time histories are shown in Figure 17-1. The figure shows that the Scripps data evidenced the least variability and the Santa Barbara data the greatest. The wave height and period, beach slope and sediment size were obtained as described in Seymour (1986).

Each model listed in Table 17-1 provides a threshold condition for reversal of the cross-shore transport from shoreward to offshore. In all of the models except that of Quick and Har, a value of the predictor greater than the threshold denotes offshore transport, or erosion. The Sunamura and Horikawa model brackets the threshold in recognition that near the conditions where transport direction changes it may be difficult to predict. None of the sources for these predictors claims an ability to predict the magnitude of the transport - only its direction. In Figure 17-2, Dean's predictions and the observed results of Figure 17-1 are compared. The threshold value of 1.0 for the predictor has been subtracted from each prediction and the signs have been reversed to agree with the sign convention for the observations. Therefore, when the sign of the predictor and the sign of the observation are the same, the prediction has been successful. This can be seen in the lower left quadrant for erosive events and the upper right quadrant for accretion. Figure 17-2 shows that Dean's formulation ($R = 0.6$) predicts erosion for 70 out of the 71 events observed in the three data sets, and that the single prediction of accretion is incorrect. Therefore, it would appear to have no value as a predictive tool - even though it displayed a skill factor (ratio of correct predictions to total observations) of 0.57. Because there are more erosionary than accretionary events, it requires no skill for a predictor to be right more than half the time if it consistently picks erosion. However, inspection of Figure 17-2 indicates that a higher value for the threshold would result in a number of correct predictions of accretion events - at the expense of failing to predict some occurrences of erosion.

The possible values of threshold were searched to determine that level that would most nearly produce the same skill factors in predicting erosion and accretion. For Dean's formulation, that value is a threshold of 2.62 (compared to 1.0 suggested by the model). This change amounts to an adjustment of the arbitrary constant to $R = 0.229$ in Dean's model. This results in an overall predictive skill of 0.62, about evenly divided between accretion and erosion. This would seem to be both a realistic and a useful assessment of the underlying predictive skill of Dean's model which utilizes wave height and period and sand size (indirectly, through fall speed).

One of Short's two formulations utilizes only wave height. With the suggested threshold of 120 cm for incident wave height, the skill factor for accretion is 1.0 while the skill for predicting erosion is 0.25. Following the

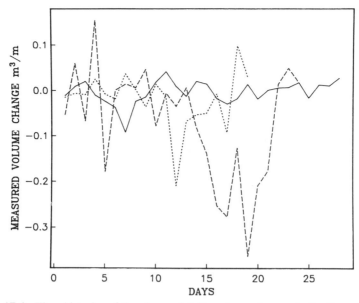

Figure 17-1. Time histories of the observed daily volume changes in the three data sets. Solid line is Scripps Beach, dashed line is Santa Barbara and dotted line is Virginia Beach.

previous procedure, threshold space was searched to find the point where equal skills were obtained in predicting direction. Reducing the threshold to 66 cm resulted in an overall skill index of 0.65, again about evenly divided between the two possible directions. Thus, a simple model containing only wave height very slightly outperforms the more complicated model of Dean. The adjusted threshold results for Short's height formulation are shown in Figure 17-3.

Short's second formulation is based upon the incident wave power at a depth of 10 m. The threshold of the predictor is 300 w/cm. Note that this is equivalent to the height-only predictor when the wave period is about 18 seconds. Short's height threshold of 120 cm produces a power density of only about 80 w/cm when the significant period is 5 seconds. This inconsistency suggests that the thresholds were established in a region where long period waves are responsible for erosion. The skill at predicting accretion, using the recommended threshold, is again 1.0 - but the skill for erosion drops to 0.00. The threshold which results in approximately equal skills is reduced to 43 w/cm. The skill associated with this value is about 0.63 - equal to the Dean performance. Again attempting to show consistency between the two formulations of Short, the revised threshold of 66 cm for the height-only approach corresponds to the revised power level of 43 w/cm when the period is about 10.5 seconds - not unreasonable for this data set.

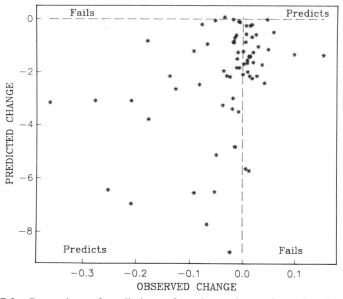

Figure 17-2. Comparison of predictions of erosion and accretion using the model of Dean (1973) with observations. All three data sets are included.

The model of Hattori and Kawamata, an extension of Dean's formulation by including beach slope, predicts a threshold of 0.5. At this value, it predicts accretion perfectly but only has a skill factor of 0.05 for erosion. Adjusting the threshold to achieve balanced skills yields a threshold value of 0.064 and a skill factor of 0.68 - showing a slight improvement over Dean because of the slope inclusion. However, it should be noted that both the sediment fall speed and the beach slope tend to increase with increasing sand size. Therefore, the addition of the beach slope in the numerator tends to counteract to some extent the fall speed in the denominator. This model more closely approaches a measure of wave steepness, H/T. The performance of the adjusted threshold in the Hattori and Kawamata model is shown in Figure 17-4.

The formulation of Quick and Har (1985) has a threshold of 1.0. At this value the skills for predicting accretion and erosion are nearly equal at 0.52 and 0.45. A very slight adjustment of the threshold to 1.01 yields equal skill factors of 0.49. As this performance is less than can be expected by chance alone, the Quick and Har formulation exhibits no useful predictive capability. This model depends completely upon the concept of equilibrium. It assumes that the beach acquires an equilibrium shape with each wave event and the new wave field drives it towards another appropriate position depending upon whether the WSP is larger or smaller than the previous event. Thus, a 200 cm incident wave

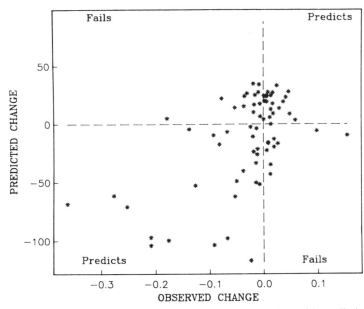

Figure 17-3. Comparison of observations of erosion and accretion with predictions using the height model from Short (1978) with the threshold adjusted to produce approximately equal skills in predicting the sign of the change. All three data sets are included.

following a 300 cm event (with no change in period) would be predicted to result in accretion. Because, in this model, the change in beach form is predominately a slope change, or rotation about a point, the predicted steepening of the beach under smaller waves would be expressed always as accretion. In the same sense, a small increase in wave height from 10 to 20 cm would predict erosion, which does not match with empirical observations. Although this model may have some success under laboratory conditions, it does not describe the behavior of natural beaches satisfactorily.

The final model evaluated, Sunamura and Horikawa, was tested using the suggested dual threshold limits of 18 and 9. When the model predicts a value between these values, no prediction is made. To allow a meaningful comparison with the other models, all of these non-predictions were treated as failures. The result was a skill factor for accretion of 0.23 and for erosion of 0.43, obviously less than by chance alone. However, if the two limits are collapsed on a value of 13.2 - that is, a single value threshold is used as in the other models - both acccretion and erosion events are predicted with a skill factor of about 0.6. Thus this model achieves a slightly better than chance performance with this alteration of threshold.

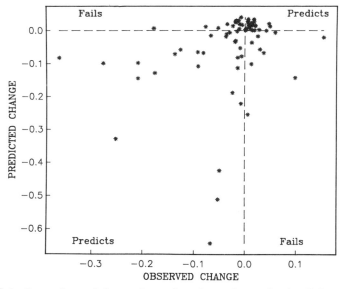

Figure 17-4. Comparison of observations of erosion and accretion in all three data sets with predicitons from the model of Hattori and Kawamata (1980). The threshold has been adjusted to equal skill in predicting the sign of the change.

17.2 Profile Prediction Models

Two models were found that attempt to predict the change in the shape and position of the beach profile in response to changing wave and tide forcing. The first of these is Quick and Har (1985), which extends the threshold prediction to a profile model. The second model was developed by Swart (1976).

The Quick and Har model contains the following assumptions:

(1) There is a certain depth, occurring just outside the breaker zone, at which there is no vertical change. The maximum sediment transport rate occurs at this depth, which functions as a pivot point for profile realignment.
(2) The slope at mean sea level is related to the wave-sediment parameter following Dalrymple and Thompson (1976).
(3) The profile follows the power law suggested by Dean (1977). That is, the depth is proportional to the offshore distance to the 2/3s power. Further, it approaches this equilibrium exponentially with an expected half-life on natural beaches of "a few hours."
(4) Allowing for tides, the numerical calculation would be made stepwise with adjustments within each time step for the effects of tide on the exponential approach to equilibrium.

In light of the lack of skill exhibited by the Quick and Har threshold predictor, it will be instructive to explore the assumptions of the profile prediction model. Referring to the results of the objective analyses in Chapter 12, it was shown that, in the Santa Barbara set, 97% of the variation about the mean beach position was caused by horizontal motion of the profile without change in slope or shape. At Virginia Beach 92% of the variation can be attributed to the same mechanism. Only at Scripps Beach was there a substantial contribution of slope change to the observed variation. About 85% of the variation here was caused by a change in concavity and slope. However, the null point around which the beach slope rotated was at an elevation of approximately 180 cm above mean sea level (MSL) rather than about 100 cm below as predicted by Quick and Har. Further, the Quick and Har model relies heavily upon a prediction of the slope at MSL that is based upon the value of WSP (the formulation of Dalrymple and Thompson). As shown in the next section, the model of Dalrymple and Thompson appears to have no validity for this data set. It would therefore appear that the underlying concepts for the Quick and Har model do not describe beach form change in these data sets. Considering both the low likelihood of success and the programming complexities associated with dealing with tides in this model, a decision was reached not to attempt further evaluation.

The second approach, following Swart (1976), is an empirical, numerical scheme for predicting beach form response with the following characteristics:

(1) The upper limit of wave action is calculated as a function of wave height and period and of the median sediment size. This upper extent is time-varying, relative to MSL, because of tides and storm surge. The backshore, above this upper limit, is assumed to remain unchanged.

(2) A lower limit for the developing profile (which Swart calls a D-profile) is defined by another empirical expression dependent upon the same parameters as the upper limit. It too is time-variant if sea level is changing. In the intervening zone between the two limits both bedload and suspended transport are expected to be significant.

(3) This developing profile is expected to be driven towards, but not necessarily achieve, an equilibrium position and shape. The shape is defined for each dimensionless depth on the profile solely on the basis of the median sediment size. Given no tides and a constant wave regime, the developing profile will approach its equilibrium exponentially.

(4) Below the lower limit of the developing profile, bedload is assumed to dominate. The model forces sediment across the lower limit boundary as necessary to satisfy continuity above the boundary. The profile shape below the lower limit is determined iteratively by calculating an equilibrium slope at each location following Eagleson et al. (1963).

The Swart model has been evaluated under laboratory and field conditions, including the NSTS Santa Barbara data, by Swain and Houston (1983, 1984) and Swain (1984). These evaluations have shown very skillful predictive capabilities, prompting further testing here with the complete NSTS data set.

A comparison of the Scripps Beach measured profiles and the predictions from Swart's model are shown in Figure 17-5. The predictions are not updated. That is, the model runs without correction through the complete data set. In certain cases, the predicted profiles extend beyond the limits of the measured profiles. In others they are shorter. The predicted profiles are always terminated at the lower limit for transport as defined by the model. For this data set, which contains the least variation of the group, Swart's model gives a reasonable prediction. A tendancy to predict a flatter than observed slope can be detected. In all cases, the model appears to successfully predict the position of the MSL intercept.

A similar comparison for Santa Barbara is shown in Figure 17-6. Here, the predicted response remains close to the measured profile until about 13 February. At this point, the predictor produces a rapid change in beach slope and an almost 30 m landward shift in the MSL intercept which occurs much more quickly than in nature. By 19-20 February, the beach has caught up to the predictor and the two curves remain reasonably close for the remainder of the data set.

Figure 17-7 shows the comparison for Virginia Beach. This data set departs markedly from the previous two in that it contains a very well-defined bar-terrace system that moves rapidly. The Swart model does not claim to predict barred beaches, but its predictive ability was tested on these data, nonetheless. As the figure shows, the model predicts the MSL intercept (which moved only slightly) rather well, but continues to flatten the beach with time such that there is a substantial departure between predicted and observed profiles at the end of the data set.

17.3 Models Predicting Beach Slope

Two models, each of which is embedded in at least one of the models discussed earlier, attempt to describe the general slope or shape of the beach. The first, by Dalrymple and Thompson (1976), gives a graphical relationship between the beach slope at MSL and the wave-sediment parameter. The second, from Dean (1977), describes a general exponential form for the nearshore profile. To test the validity of the Dalrymple and Thompson assumption, the slopes were calculated for all of the data sets and compared to the corresponding values of WSP. The results are plotted in Figure 17-8. Based upon these data, the wave-sediment parameter has no significant influence on beach slope.

The Dean model for nearshore slopes takes the form

$$Y = AX^m \qquad\qquad (17\text{-}1)$$

Beach profiles from all of the data sets were tested for fits to this model. The

Scripps Beach 10/08/80 − 11/04/80

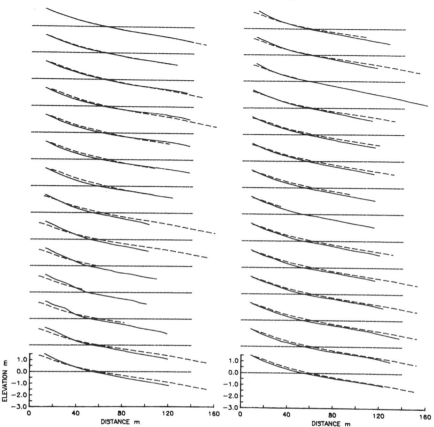

Figure 17-5. Comparison of observed beach profiles at Scripps Beach with the predictions using the model of Swart (1976). Dashed lines are predictions. Sequence starts at upper left and proceeds downward in each column. Horizontal lines are MSL.

upper limit was taken as MSL and the lower limit of the profile at an estimated depth of breaking. Best fit values of the coefficient A and the exponent m were determined for each profile. A smoothed histogram of the distribution of resulting values is shown in Figure 17-9. Dean (1977) gives a typical value of 0.67 for m, based upon Gulf and Atlantic Coast data. Figure 17-9 shows that this is a very reasonable value for these data sets, as well. The peaks of the histograms range from 0.63 to 0.67 for the three sites, with a mean value of about 0.65. The expected value of the coefficient A varies from about 0.065 for

Santa Barbara 2/1/80 — 2/25/80

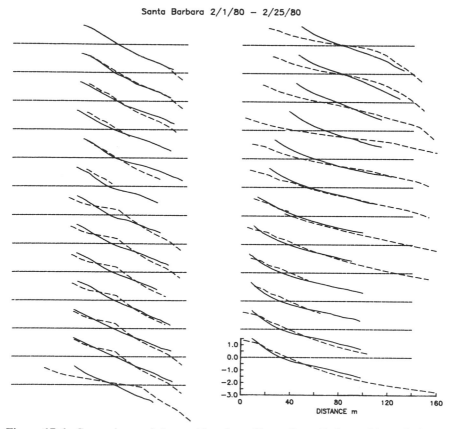

Figure 17-6. Comparisons of observed beach profiles at Santa Barbara with predictions from the model of Swart (1976).

Scripps Beach to about 0.13 at Virginia Beach. The value of the coefficient increases with increasing sediment size, as anticipated.

17.4 Discussion and Conclusions

The six models tested for their ability to predict the threshold condition for changing the sense of cross-shore transport were correct between about one-half and two-thirds of the time. Five of them, with revised thresholds, showed about equal skills in predicting both erosion and accretion - ranging only from 0.6 to 0.68 - hardly significant differences. Of the six, only two contain any history term. In the case of Hattori and Kawamata, the beach slope today can be

Virginia Beach 10/1/80 — 10/20/80

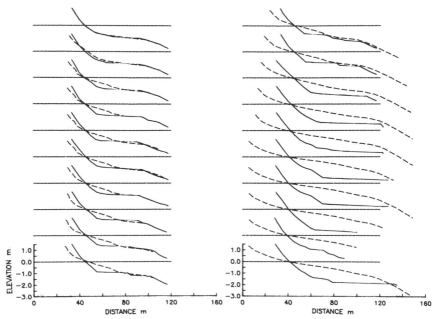

Figure 17-7. Comparisons of observed beach profiles at Virginia Beach with predictions from the model of Swart (1976).

assumed to have some memory of the previous wave climate. The model of Quick and Har depends upon the slope of the beach on the previous day to determine which direction cross-shore transport will proceed today.

Considering the results of these models when applied to this data set, the following general conclusions are made:

(1) Estimators with a success ratio of only two-thirds may be of some value, but must be used with appropriate caution.

(2) The estimators did not fail to predict the maximum erosion events in this series when adjusted to equal skill for each direction of transport. However, there were some failures to predict the two largest accretionary events.

(3) The differences between the five most successful models were too small to indicate whether the simplest model (wave height only) was any less successful than the most complicated (wave height, period, sand size and beach slope.) In particular, these tests shed no light on the importance of a memory term in the formulation.

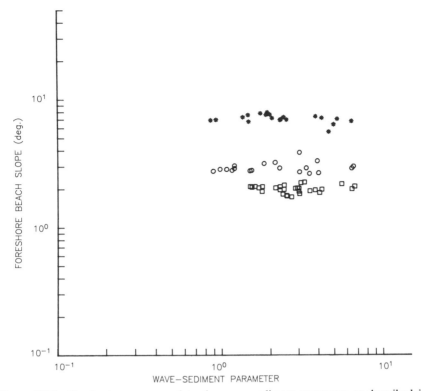

Figure 17-8. Beach slope compared to the wave sediment parameter as described in Dalrymple and Thompson (1976). Squares are Scripps Beach, circles Santa Barbara, and stars are Virginia Beach.

Both the threshold and the profile prediction models of Quick and Har (1985) appear to suffer from conceptual difficulties when applied to data from natural beaches.

Swart's model appears to have some interesting capabilities to predict shoreline evolution for beaches without bars (the condition for which it was formulated.) It is stable for periods of a few weeks and needs to be tested over much longer periods to verify its long-term stability. It appears to be the only model available today which provides reasonably reliable estimates of the shape and position of beaches subject to variation from cross-shore transport.

The findings of Dalrymple and Thompson (1976) on a relationship between beach slope and WSP is not verified by these field data. The exponential shape model of Dean (1977) describes these profiles satisfactorily to the depth typically achieved by wading surveys.

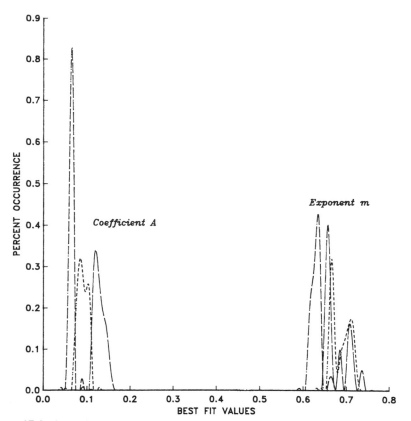

Figure 17-9. Smoothed histogram of occurrences of the coefficients and exponents in the power law for beach contours from Dean (1977) in the form $Y = AX^m$. [Data from Santa Barbara have the shortest dash length, Scripps Beach has intermediate dash length and Virginia Beach the largest dashes.]

As a general finding, models that are developed from laboratory data using monochromatic waves do not perform well on field data from natural beaches.

17.5 References

Dalrymple, R. A. and W. W. Thompson, 1976, Study of equilibrium beach profiles, *Proceedings*, Fifteenth Coastal Engineering Conference, July 11-17, 1976, Honolulu, Hawaii, American Society of Civil Engineers, New York: 1277-1296.

Dean, R. G., 1973, Heuristic models of sand transport in the surf zone, Proceedings, Conference on Engineering Dynamics in the Surf Zone, Sydney, Australia, 7 pp.

_____. 1977, Equilibrium beach profiles: U. S. Atlantic and Gulf Coasts, Ocean Engineering Report No. 12, Department of Civil Engineering, University of Delaware, Newark, Delaware.

Eagleson, P. S., B. Glenne and J. A. Dracup, 1963, Equilibrium characteristics of sand beaches, Proceedings, *Journal of Hydraulics Division*, American Society of Civil Engineers, New York, 89: 35-57.

Hattori, M. and R. Kawamata, 1980, Onshore-offshore transport and beach profile change, *Proceedings*, Seventeenth Coastal Engineering Conference, March 23-28, 1980, Sydney, Australia, American Society of Civil Engineers, New York: 1175-1194.

Quick, M. C. and B. C. Har, 1985, Criteria for onshore-offshore sediment movement on beaches, *Proceedings*, Canadian Coastal Conference, St. Jobus, Newfoundland: 257-269.

Seymour, R. J., 1986, Results of cross-shore transport experiments, *Journal Waterway, Port, Coastal, and Ocean Engineering*, American Society of Civil Engineers, 112(1): 168-173.

Seymour, R. J. and D. B. King Jr., 1982, Field comparisons of cross-shore transport models, *Journal of the Waterway, Port, Coastal and Ocean Division*, Proceedings of the American Society of Civil Engineers, 108(WW2), May, 1982: 163-179.

Short, A. D., 1978, Wave power and beach stages: a global model, *Proceedings*, Sixteenth Coastal Engineering Conference, August 27-September 3, 1978, Hamburg, Germany, American Society of Civil Engineers, New York, 2: 1145-1162.

Sunamura, T. and K. Horikawa, 1974, Two-dimensional beach transformation due to waves, *Proceedings*, Fourteenth Coastal Engineering Conference, June 24-28, 1974, Copenhagen, Denmark, American Society of Civil Engineers, New York: 920-938.

Swain, A., 1984, Additional results of a numerical model for beach profile development, Proceedings of the Annual Conference, CSCE, Halifax, Nova Scotia, Canada, May.

Swain, A. and J. R. Houston, 1983, A numerical model for beach profile development, Sixth Canadian Hydrotechnical Conference, CSCE, Ottawa, Canada, June, 2: 777.

_____. 1984, Discussion to the Proceeding Paper 17749 by Richard J. Seymour, The Nearshore Sediment Transport Study, *Journal Waterway, Port, Coastal and Ocean Engineering*, American Society of Civil Engineers, February, 110 (1): 130-133.

Swart, D. H., 1976, Predictive equations regarding coastal transports, *Proceedings*, Fifteenth Coastal Engineering Conference, July 11-17, 1976, Honolulu, Hawaii, American Society of Civil Engineers, New York: 1113-1132.

AUTHOR INDEX

SUBJECT INDEX